Information Systems Engineering and Management

Volume 64

The book series "Information Systems Engineering and Management" (ISEM) publishes innovative and original works in the various areas of planning, development, implementation, and management of information systems and technologies by enterprises, citizens, and society for the improvement of the socio-economic environment.

The series is multidisciplinary, focusing on technological, organizational, and social domains of information systems engineering and management. Manuscripts published in this book series focus on relevant problems and research in the planning, analysis, design, implementation, exploration, and management of all types of information systems and technologies. The series contains monographs, lecture notes, edited volumes, pedagogical and technical books as well as proceedings volumes.

Some topics/keywords to be considered in the ISEM book series are, but not limited to: Information Systems Planning; Information Systems Development; Exploration of Information Systems; Management of Information Systems; Blockchain Technology; Cloud Computing; Artificial Intelligence (AI) and Machine Learning; Big Data Analytics; Multimedia Systems; Computer Networks, Mobility and Pervasive Systems; IT Security, Ethics and Privacy; Cybersecurity; Digital Platforms and Services; Requirements Engineering; Software Engineering; Process and Knowledge Engineering; Security and Privacy Engineering, Autonomous Robotics; Human-Computer Interaction; Marketing and Information; Tourism and Information; Finance and Value; Decisions and Risk; Innovation and Projects; Strategy and People.

Indexed by Google Scholar. All books published in the series are submitted for consideration in the Web of Science.

For book or proceedings proposals please contact Alvaro Rocha (amrrocha@gmail.com).

Kirti Khanna · Md Shabbir Alam · Mohammad Irfan · Hasan Dincer
Editors

From Poverty to Prosperity

Leveraging Capital Markets for Sustainable Development

Springer

Editors
Kirti Khanna
Jaipuria Institute of Management
Ghaziabad, Uttar Pradesh, India

Mohammad Irfan
School of Business
Chandigarh University
Lucknow, Uttar Pradesh, India

Md Shabbir Alam
Department of Economics and Finance
College of Business Administration
University of Bahrain
Zallaq, Bahrain

Hasan Dincer
School of Business
İstanbul Medipol University
İstanbul, Türkiye

ISSN 3004-958X ISSN 3004-9598 (electronic)
Information Systems Engineering and Management
ISBN 978-3-031-99165-3 ISBN 978-3-031-99166-0 (eBook)
https://doi.org/10.1007/978-3-031-99166-0

This Springer imprint is published by the registered company Springer Nature Switzerland AG
The registered company address is: Gewerbestrasse 11, 6330 Cham, Switzerland

This book shines a spotlight on the visionaries behind sustainable development goals. Their tireless pursuit of innovation lays the groundwork for a greener, fairer tomorrow. We celebrate our families and loved ones; their unwavering support has been our driving force. Your encouragement lights our path.

To the future generations of thinkers, innovators, and leaders: embark on a bold journey. Challenge the status quo and push boundaries with fearless curiosity. Embrace technology's transformative power to shape a brighter, better world for all.

Preface

In a world eager for economic inclusivity and eco-friendly futures, capital markets shine. *From Poverty to Prosperity: Leveraging Capital Markets for Sustainable Development* reveals their power to transform. This captivating collection intertwines wisdom from scholars, policymakers, and industry sages. They show how capital markets drive sustainable growth worldwide.

From Poverty to Prosperity: Leveraging Capital Markets for Sustainable Development shows how capital markets can create positive change. This highlights how financial systems can drive sustainable growth and reduce poverty. Wealth gaps and environmental issues are growing, making this approach crucial. This book offers strategies to unlock inclusive economic opportunities. The book discusses many topics. It looks at how capital markets and financial implications can reduce poverty. It also explores green finance's role in fighting climate change. It covers new financial tools that help with social and economic inclusion. The contributors analyze case studies from developed and emerging economies.

The book explains how capital markets support public policy goals for sustainable development. It looks at how governments and international organizations can use capital markets. They can fund infrastructure projects, boost green technologies, and help social enterprises. The book shares insights on new financial tools. These include green bonds and social impact bonds. They aim to draw private investment for public benefits.

It lays out clear steps for nurturing sustainable finance. With insightful discussion, it reveals how smart government policies promote responsible investing. By guiding the growth of sustainable capital markets, it charts a brighter future.

This book highlights the vital collaboration between the public and private sectors. Together, they tackle pressing global challenges and champion sustainable development. It is a treasure trove for scholars, policymakers, and finance enthusiasts alike.

Rich with actionable ideas, it deftly navigates the capital markets. By doing so, it fuels economic growth, curbs inequality, and paves the way for a sustainable future for all.

Ghaziabad, India — Dr. Kirti Khanna
Zallaq, Bahrain — Dr. Md Shabbir Alam
Lucknow, India — Dr. Mohammad Irfan
Istanbul, Turkey — Dr. Hasan Dincer

Acknowledgements

We would like to thank everyone who helped create this book, *From Poverty to Prosperity: Leveraging Capital Markets for Sustainable Development*. Your support made it possible.

First and foremost, we send heartfelt thanks to our contributors. Your insights, skills, and experiences shine in this volume like a guiding star. Your dedication and hard work have woven a rich tapestry of knowledge. We are truly grateful for your unwavering commitment to expanding understanding. Together, we explore the intricate dance of sustainable and economic development.

We extend our deepest gratitude to our families and loved ones. Your steadfast support and encouragement have been our bedrock throughout this journey. With your patience and understanding, you became our guiding light. This strength allowed us to pour our hearts into crafting this book. Thank you for being a constant source of inspiration!

We thank our publisher for believing in this project. Their resources and guidance helped us finish it. Your professional support has been instrumental in bringing our vision to life.

Thanks to the peer reviewers for their careful and helpful feedback. It has really improved the content's quality and clarity. Your critical insights and suggestions have been invaluable in shaping the final manuscript.

At last, we celebrate the trailblazing minds of researchers across the globe. Their groundbreaking endeavors have forged the path for visualizing sustainable development goals. Your vision and fresh ideas ignite our inspiration. May this book stand as a tribute to your invaluable contributions.

To everyone who has shaped this book, thank you from the heart. Your combined efforts have illuminated the powerful impact of harnessing the capital market. Together, we have unveiled pathways for sustainable development in emerging economies. Your contributions have not only inspired us but also sparked hope for the future.

About This Book

This book, entitled *From Poverty to Prosperity: Leveraging Capital Markets for Sustainable Development*, discusses how capital markets can drive economic growth that benefits everyone and supports sustainability. The book sits at the crossroads of finance and development. It looks at how capital markets align with the sustainable development goals (SDGs). The author discusses new financial tools. These include green bonds, carbon credits, and ESG-focused ETFs. These tools are making a real difference in social and environmental impact. The book uses real-world examples from India and SAARC nations. These tools show how to lower poverty, strengthen communities, and adapt to climate change. It looks closely at corporate sustainability practices and ESG investing. It also examines the challenges of implementing these ideas in emerging markets.

The volume also looks at how technology changes finance. It helps include more people and makes markets work better. Chapters look at how blockchain, cryptocurrencies, and fintech change financial systems. Contributors talk about digital lending and payment platforms. These tools support MSMEs and rural communities. Studies analyze the barriers and opportunities for women. The book looks at behavioral finance topics. It covers mutual fund flow patterns and trends in digital use since the pandemic. This work combines entrepreneurship, infrastructure, and the circular economy. It offers a clear plan for using capital markets to create sustainable prosperity.

Contents

Editors and Contributors

About the Editors

Dr. Kirti Khanna is an Associate Professor at Jaipuria Institute of Management (JIM), Ghaziabad, Uttar Pradesh, India. She holds a Ph.D. and has over a decade of experience in academia, focusing on research and teaching. Dr. Khanna has been honored with the Best Paper Award at an international conference organized by the Research Development Association (RDA) and RDRF, Jaipur, in collaboration with the Rajasthan Chamber of Commerce and Industry. Her research interests include macroeconomic variables, stock markets, and corporate reporting practices, utilizing analytical software such as STATA, SPSS, and EVIEWS. She has published two patents and two copyrights. Her scholarly work focuses on macroeconomic trends, capital markets, green finance, sustainable investing, and the evolving landscape of digital finance. Her recent contributions explore sustainability in finance, the impact of AI on investment decisions, and digital financial innovations such as e-rupee and robo-advisors. Her research is published in reputed journals and edited volumes by publishers such as IGI Global and Emerald Publishing.

Dr. Md Shabbir Alam is a faculty member in the Department of Economics and Finance at the University of Bahrain. He holds both a master's and Ph.D. in economics from Aligarh Muslim University, India, and completed a post-doctoral fellowship at the Indian Council of Social Science Research (ICSSR). Recognized among the top 2% of global scientists in economics and business by Stanford University and Elsevier, Dr. Alam has over 14 years of teaching and research experience across Bahrain, Oman, Saudi Arabia, and India. His research focuses on development economics, energy and environmental economics, and international economics. With over 80 publications, including 60+ in high-impact journals, his work has received over 3500 citations. Dr. Alam is also an active member of editorial boards and conference committees.

Dr. Mohammad Irfan is presently working as an Professor at Chandigarh University, Uttar Pradesh, India. Dr. Irfan has done his Ph.D. from the Central University of Haryana. He is MBA (Finance), M.Com. (Account and Law), and MA (Economics). He has qualified UGC-JRF/SRF/NET in Management and Commerce. Dr. Irfan certified NSEs (NCFM) and BSEs certification. He has experience of more than sixteen years in the area of SAPM, artificial intelligence, machine learning, blockchain, cryptocurrency, financial engineering, Fintech, green finance, and alternative finance. He has to his credit more than 73+ Scopus-indexed articles, including *The Journal of Economic Cooperation and Development (Q2)*, *International Journal of Business Excellence (IJBEX)*, *International Journal of Economics and Management (IJEM)*, *Montenegrin Journal of Economics (Q2)*, *Cogent Business and Management (Taylor & Francis) (Q2)*, *Indian Journal of Finance (IJF)*, and *Journal of Islamic Monetary Economics and Finance (JIMF)*. His citations reached 546+ along with 13 H-index. Dr. Irfan has published 11 books in Springer, IGI Global Publication (Scopus-indexed).

Prof. Hasan Dincer is a distinguished academician in the interdisciplinary fields of business and finance, holding the position of Professor at Istanbul Medipol University in Turkey. He completed his B.A. and M.A. from Marmara University and later went on to receive his Ph.D. from the same university. With extensive international experience, Prof. Dincer has conducted research work at Hamburg University in Germany and served as a research professor at the University of Hertfordshire in the United Kingdom during the spring semester of 2019. He has also worked as a lecturer at Beykent University in Turkey from 2008 to 2015. Professor Dincer has a remarkable record of supervising theses at both the Master's and Doctoral levels. He has supervised successful theses on various topics of engineering, environmental, and social sciences. With over 400 scientific articles to his name, many of which are indexed in high-impact journals like SSCI, SCI-expanded, and Scopus, Prof. Dincer has contributed significantly to the academic literature. He has focused his research work on a range of topics such as strategic management, innovation, sustainability, circular economy, decision-making, as well as renewable energy investments, finance, and banking. Professor Dincer currently serves as an associate editor and editorial board membership of outstanding SSCI/SCI/Scopus Journals. Apart from being the executive editor of the *International Journal of Finance and Banking Studies (IJFBS)*, Prof. Dincer is also a founder member of the Society for the Study of Business and Finance (SSBF). He has also served as an editor of several books published by Springer, Emerald, and IGI Global. With exceptional teaching skills and academic achievements, Prof. Dincer is highly respected in the academic community. His valuable contributions to the interdisciplinary field of environmental and social sciences make him an asset to Istanbul Medipol University and the academic community at large.

Contributors

Fatima Mansour Alkhammash Yanbu Industrial College, Yanbu, Saudi Arabia

C. S. Amit Kumar School of Leadership and Management, Manav Rachna International Institute of Research and Studies, Faridabad, India

Richa Arora Sushant University, Gurugram, India

Shreya Arora Assistant Professor, Mangalayatan University, Jabalpur, MP, India

Neetha Mary Avanesh School of Business and Management, Christ University, Bangalore, India

Alireza Baghchesaraei Infrastructure Management (Engineering), Construction Industry, Sydney, Australia

Omid Reza Baghchesaraei Centre for Infrastructure Engineering, Sydney, Australia

Kulbir Kaur Bhatti Manav Rachna International Institute of Research and Studies, Faridabad, Haryana, India

Nikolay Bodrov Department of Forensic Science, Kutafin Moscow State Law University, Moscow, Russia

Priyanka Chadha Manav Rachna International Institute of Research and Studies, Faridabad, India

Apurva A. Chauhan School of Leadership and Management, Manav Rachna International Institute of Research and Studies, Faridabad, India

Meenakshi Chawla St. John's College, Agra, India

Priyanka Chawla Prestige Institute of Management and Research, Gwalior, India

Vinita Choudhary School of Leadership and Management, Manav Rachna International Institute of Research and Studies, Faridabad, Haryana, India

Salu Dsouza School of Law, Christ University, Bengaluru, India

Pallavi Dua Manav Rachna International Institute of Research and Studies, Faridabad, India

Mohd Farhan Mittal School of Business, Lovely Professional University, Phagwara, Punjab, India

Deepali M. Gala Bharati Vidyapeeth (Deemed to be University), Institute of Management, Kolhapur, India

Parul Gaur GLA University, Mathura, India

Akhil Goyal NIMS Institute of Business Studies, NIMS University, Rajasthan, Jaipur, India

Neha Kamboj School of Management, IILM University, Gurugram, Haryana, India

Amirali Khademi Department of Civil Engineering, Sharif University of Technology, Tehran, Iran

Khaliquzzaman Khan School of Business, American University in Dubai, Dubai, United Arab Emirates

Mohammad Aamir Khan Yanbu Industrial College, Yanbu, Saudi Arabia

Nabiha Khan Arab Open University, Riyadh, Saudi Arabia

Kirti Khanna Jaipuria Institute of Management, Ghaziabad, Uttar Pradesh, India

R. B. Lakshmi School of Business and Management, Christ University, Bangalore, India

Palanisamy Manigandan Department of Statistics, Periyar University, Salem, Tamil Nadu, India

Richa Nangia Sushant University, Gurugram, India

Anand Patil School of Business and Management, Christ University, Bangalore, India

Bhakti Ranjit Pawar School of Leadership and Management, Manav Rachna Educational Institutions, MRIIRS, Faridabad, Haryana, India

Ankit Saxena GLA University, Mathura, India

Pravin Saxena Dayalbagh Educational Institute, Deemed University, Agra, India

Hiten N. Shah Research Scholar, Bharathiyar University, Coimbatore, India

Rashmi Singel Manav Rachna International Institute of Research and Studies, Faridabad, Haryana, India

Manisha Singh VIT Business School, VIT Bhopal University, Bhopal, India

Rashmi Singh GLA University, Mathura, India;
Rajiv Academy for Technology and Management, Mathura, India

Sachin Prakash Srivastava School of Business, Dr Vishwanath Karad MIT World Peace University, Pune, India

Sonali Srivastava Jaypee Business School, Jaypee Institute of Information Technology, Noida, India

Akhil Suresh School of Business and Management, Christ University, Bangalore, India

Suresh Suryawanshi Bharati Vidyapeeth (Deemed to be University), Institute of Management, Kolhapur, India

Mohammed Ahmar Uddin Dohfar University, Salalah, Oman

Gitanjali Upadhaya Central University of Himachal Pradesh, Dharamshala, Himachal Pradesh, India

Manish Upadhyay Rajiv Academy for Technology and Management, Mathura, India

N. Usha Devi Vivekananda Degree College, Bangalore, India

B. Veeralakshmi College of Business Studies, COER University, Roorkee, India

K. Venkatesh Bharati Vidyapeeth (Deemed to be University), Institute of Management, Kolhapur, India

C. S. Vinit Sikka School of Leadership and Management, Manav Rachna International Institute of Research and Studies, Faridabad, India

Ankita Yadav NIMS Institute of Business Studies, NIMS University, Rajasthan, Jaipur, India

Leveraging Capital Markets: Strategies for Empowering Women Through Capital Markets for Sustainable Development in Saudi Arabia

Mohammad Aamir Khan, Mohammed Ahmar Uddin, Fatima Mansour Alkhammash, and Nabiha Khan

Abstract Vision 2030, Saudi Arabia is undergoing significant economic transformation. The role of women's in the economy is boosted by harnessing the capital markets. Women can be more actively engaged in business and investment by tapping the financial resources, leading to a more equitable economic environment. This paper examines how capital markets play a vital role in promoting sustainable development, specifically by empowering women in Saudi Arabia. The aim of the study is to build a financial system that supports women entrepreneurs and contributes to sustainable economic growth and social equity in Saudi Arabia. Barriers faced by women are analyzed in accessing financial resources, like cultural, institutional, and regulatory challenges. The study suggests strategies to increase women's participation in capital markets, like educational initiatives, targeted financial products, and partnerships with private and public sectors. This research highlights the importance of collaboration among stakeholders like government agencies, financial institutions, private sector companies, NGOs, educational institutions, community leaders, investors, etc., to prove that women's empowerment is the key to capital market development.

Keywords Capital markets · Women empowerment · Gender equality · Financial education · Stakeholder collaboration

M. A. Khan (✉) · F. M. Alkhammash
Yanbu Industrial College, Yanbu, Saudi Arabia
e-mail: aamir@rcjy.edu.sa

F. M. Alkhammash
e-mail: Alkhammashf@rcjy.edu.sa

M. A. Uddin
Dohfar University, Salalah, Oman
e-mail: ahmar@du.edu.om

N. Khan
Arab Open University, Riyadh, Saudi Arabia
e-mail: n.khan@arabou.edu.sa

K. Khanna et al. (eds.), *From Poverty to Prosperity*, Information Systems Engineering and Management 64, https://doi.org/10.1007/978-3-031-99166-0_1

1 Introduction

Women earn only 10% of their income despite the fact that they produce 60% of the world's food and household goods and invest 90% of the earnings back into their families and communities [1].

According to the International Labor Organization, in 2022, 61.4% of women participate in the world's workforce, against more than 90% of men. Women have fewer career development opportunities than men. According to a survey conducted by McKinsey's Women and Workplace for every 100 men promoted to managers positions 87 women achieved similar promotions [2]. In 2023, women occupied only 18% of C-suite roles, 30% as board members, 22% of executives and 27% of senior management positions in companies' worldwide [3]. So, more opportunities need to be created for women in the financial industry as decision-making employees in the financial institutions.

In a study by IMF, bringing women into the workforce will increase the GDP [4] up to 35% in countries, where the gender inequality is highest as they bring distinct skills and enhance productivity.

An estimated $3.22 trillion [5] of additional investment can be unlocked if women invest at the same rate as men. Women investors earn 1.8% better [6] than men and while making investments they are twice [7] as likely to consider environmental, social, and governance (ESG) factors. Women donate [8] more as their income grows that contributes more to the community.

The labor force participation rate for Saudi women has increased from 20% in 2018 to 35% in 2022 [9]. The increase is because of government policies, higher level of education of Saudi women and greater demand for female workers in the private sector. In 2021 the women's overall participation in the labor market was 36% this was because of increased women's economic participation [10]. The female force participation in 2023 was 34.5% and 79.9% for males [11].

Since 1990, female labor force participation has risen, but the gender gap is still larger in Saudi Arabia compared to other high-income countries. The vulnerable employment was 3.4% for women and 1.4% for men in Saudi Arabia and these data has been stable for women since 1991. Saudi Arabia's half of the GDP makes up of oil revenues. Vision 2030 program aims to boost non-oil sector and reduce the reliance on oil, this will help in increasing the female workforce participation.

The S&P Global Ratings Economics estimates the Saudi Economy will be $39 billion (3.5% larger) if the labor force continues to grow at its current rate over the next 10 years. Saudi Arabia's female labor force participation rate increased from 19% in 2016 to nearly 36% in 2022. However, it still falls behind the global averages, this may be due to the influence of conservative culture and education levels etc. The fertility rate per woman has dropped from nearly 4 in 2000 to 2.4 in 2022. A woman with fewer children can join the workforce more conveniently. There is a rise in the share of the working women and this may be attributable to cultural liberalization, better education and lower fertility rate. The Saudi female labor force participation is only 20% of the employed population, although the overall unemployment rate is low at 5.6% but still they make up the majority of unemployed [12].

Women play a significant role in Saudi workforce and its economic growth. Female entrepreneurs can bring new perspectives and ideas that will help in fostering sustainable development and economic resilience. The Saudi female graduate population is more than 50%, Vision 2030 aims at empowering women and provide equal opportunities in the labor market [13]. Saudi Arabia had 1.27 million small and medium sized-enterprises by the third quarter of 2023 and women led 45% of them [14].

Prince Stephanie Nour who advocates and supports efforts to boost women entrepreneurship, in an interview with Arab News said that despite the advancements the journey toward full empowerment is far from complete and we have barely scratched the surface [15].

1.1 The Role of Capital Markets in Empowering Women

Women can be economically empowered through financial instruments, stock exchanges for financial independence. They have less access to banking system than men and there is a need for conscious effort to involve women in planning and designing of capital needs and its accessibility to women entrepreneurs. Vision 2030 initiatives has created an environment that supports women's economic participation and leadership that will ultimately contribute to a more diversified economy.

1.2 Women Economic Empowerment and Financial Inclusion

Approximately, 1.1 billion women representing 55% of unbanked adults are deprived of financial facilities [16]. The estimated finance gap for micro, small, or medium enterprise (MSME) for women is $1.7 trillion, 740 million are unbanked and 2.4 million lack the same economic rights as men [17]. In many countries, women have less access to verification documents to open even a bank account. The world is designed by and for men. Comfort and practicality are often burdens for women that are wrongly seen as a niche group instead of 50% of the market. A women is denied a loan because of her husband's debt or her loan is not backed by a man. Thus the cycle of inequality is reinforced by failing not to design products with women at the center. The capital accessibility process should be women-centered so that it is designed as per the women's need. This means that the women's needs, ideating solutions, problems and opportunities all should be considered from the women's perspectives. Financiera Confianza the microfinance partners in Peru developed a tailor made loan product for women entrepreneurs. This loan designed for women required no credit history, it had a repayment term to match business seasons and included cancer insurance for breast cancer screenings. The loan amount was initially small, so that the women can build credit and qualify for larger loans later. Despite the global pandemic, the scheme was an instant success with 15,000 loans disbursed within six months [18].

The Saudi government Vision 2030 [19] has made provision to enhance the women participation in the economy by improving accessibility to financial services, creating opportunities for women entrepreneurship and involving them in decision-making roles both in private and public sectors.

Through its Regulatory Approval System (RAS) and technical assistance from FCI (Regulatory Approval System), amended the banking regulations to provide equality in access to credit for women's economic empowerment [20].

The government established the Family Affairs Council that represents women's interests in international bodies and made significant strides in eliminating discrimination and empowered women by aligning with United Nations Convention on the Elimination of All Forms of Discrimination Against Women (CEDAW) and the Sustainable Development Goals (SDGs) [21]. These reforms has improved the Kingdoms progress in women's empowerment score increased from 25.63 in 2019 to 80 in 2022 [22].

With the aim of empowering women under Vision 2030, the Ministry of Commerce and Investment has announced that the women can start their own business without seeking approval from the guardians [23].

It is commendable that 45% of registered companies are owned by women but it is important to equip them with financial knowledge to bring balance between work and entrepreneurship [24]. A Saudi woman has access to bank accounts and is guaranteed property ownership by Sharia law. Alsharq al-Awsat in its report stated $53 billion about 20% of all bank deposits are owned by women [25]. But this also reveals that women with limited financial literacy is unaware of the benefits of money management.

As outlined in the National Savings and Financial Literacy Strategy, the goal is to ensure long-term financial independence and overall economic development by enhancing financial literacy [26]. Government has introduced successful programs like Khazna by Al-Nahdha in Riyadh that focus on to address challenges by underserved women [27]. These initiatives are helping women to make better financial decisions and understand savings, investment options, credit management etc.

2 Women as Entrepreneurs and Business Owners

The future of women in business is promising as Vision 2030 is unfolding the potential of female entrepreneurs. Lubna Olayan an advocate for women in business has played a key role in leading the Olayan Financing Company [28]. Maha Shirah founder of SheWorks has been instrumental in fostering a community of women business owners in Saudi Arabia [29]. Reem Asaad known for her work in empowering women in business is another prominent Saudi entrepreneur [30].

Capital markets can influence the gender diversity by encouraging companies to promote more women to senior roles that can drive business growth and create more opportunities for women. Saudi women are holding 20% of Saudi Shoura Council seats[31] and occupying 44% of middle and senior management positions seats [32].

Capital markets can direct funds toward those sectors that address issues affecting women, such as, fashion, healthcare, education, caregiving, and women's products and services. Such initiatives will directly benefit the women as they will create opportunities for women. Monsha'at is one such initiative by the Saudi Government.

Women can grow their savings and accumulate wealth by investing in stocks, bonds or mutual funds. Capital markets provide access to financial products and services to women. As of 2022 of the 5.8 million individual investors 1.5 million or just 26% are women investors [33].

Capital markets can foster societal changes that benefit women. Socially Responsible Investing (SRI) can break barriers for women both at workplace and society by focusing on companies and funds that prioritize gender equality. They channelize investments into businesses through empowering women at the workplace by equal pay, diversity in leadership, etc.

3 Barriers to Women's Participation in Capital Markets

There is a global growing recognition of the importance of increasing individuals' participation in the capital market. Fostering the participation of individuals in the capital market will promote the creation of a sustainable economic environment in the long term [34]. Women worldwide face greater challenges than men in participating in the capital market. The impact of these challenges strongly restrains their financial independence and economic empowerment. The gender gap and low number of women participating in the capital market are due to several barriers which include cultural, institutional, and regulatory barriers.

The main cultural barriers are gender roles and societal expectations. This refers to how men and women should behave according to their gender. For instance, in developing countries, many young women marry early and drop out of education, as societal expectations often dictate that they stay at home and care for their families. Such an expectation will have an impact on reducing financial confidence and discouraging women from making financial decisions, participating in entrepreneurial activities, or engaging in investments. Furthermore, there is a strong relationship between education and financial literacy. Women's financial literacy is affected by their level of education. The lower educational level leads to lower financial knowledge and negatively affects women's opportunities in investment and economic empowerment [35].

In terms of institutional barriers, despite offering financial infrastructure such as banking services and credit facilities, lacking financial services through the Internet and mobile applications is considered one of the main challenges, especially for women. It will lead significantly to low participation levels in the capital market. Limited financial literacy among women poses a significant barrier to their participation in digital financial systems [36].

Furthermore, the absence of adequate regulatory frameworks that specifically support women's financial inclusion will affect women's accessibility to financial resources. For instance, some financial firms lack interest in women who have no credit history or request them to provide guarantors [37].

Vision 2030 of Saudi Arabia has brought significant changes in terms of new policies and regulations for empowering women. These changes have reshaped Saudi society and expanded greater opportunities for Saudi women [38]. According to recent statistics, over 50% of total university graduates in Saudi Arabia are women. Recognizing the potential of this highly educated demographic, the Saudi government has introduced various regulations and policies to support and empower women [39].

Despite all the efforts and the major transformation that the country of Saudi Arabia is going through, women are still facing similar barriers and challenges that affect negatively their level of participation in the capital market.

Saudi women face societal, organizational, structural, and attitudinal barriers that reduce their chances of advancement in employment. These challenges include gender discrimination in the workplace, limited career advancement, and grow [40]. Gender equality is affected significantly by the culture. For instance, in terms of wages, there are huge variations between genders, industries, and occupations. Also, educated women's employment hider their performance [41].

Institutional barriers remain a significant challenge for women's financial activities in Saudi Arabia. It often fails to provide adequate training or support for women in accessing financial services that are custom-made to their needs.

Thus, the economic empowerment of women through capital markets is a key driver of broader gender equality in both the workforce and society [42]. The level of women's participation varies significantly across different countries and regions in the financial markets, investments, and leadership positions within the capital markets [43].

Studies suggest that women represent less than 20% of all investors in capital markets worldwide [44]. However, U.S. and Europe market trends show that women are more actively engaging in investment decisions, especially in areas such as impact investing, sustainable finance, and environmental, social, and governance (ESG) investing [45]. In the U.S., there has been an increase in women investing in individual stocks, exchange-traded funds (ETFs), and mutual funds [46].

Globally, women's representation in leadership positions in capital markets remains relatively low although it is improving gradually [47]. In the U.S., women hold about 20–25% of board seats in financial services companies, although this is higher in sectors like fintech [48]. The European Union have been calling for gender diversity in corporate boards, supported by regulatory frameworks to ensure that women are included in top leadership positions [49]. Gender quotas have helped increase women on corporate boards in Norway although still there is under representation in leadership in capital markets [50]. Many Western countries have included gender-responsive financial education in their public policies to teach young women, girls, and adult women to manage their personal finances and invest effectively

[51]. Fintech innovations have offered women worldwide to invest and trade with small amounts of capital making capital markets more accessible [52]. There is a growing focus of organizations and institutions to improve financial literacy for women. Although women are more financially literate in many countries, gender gaps still exist especially in developing regions [53]. Women face barriers in participating in capital markets, like investing less than men, biased investment advice, and lack of trust in financial markets [54]. They also struggle to access venture capital and private equity, as the majority of decision-makers are male [55]. Structural biases and stereotypes about women's financial behavior further limit their economic empowerment in global capital markets [56]. Countries like Iceland, France, and Norway have implemented gender quotas for boards that have positively impacted women's leadership in financial institutions [57]. International organizations such as the World Bank and United Nations have launched initiatives to improve women's access to financial services and capital markets [58].

In the past, Saudi women's participation in capital markets was limited due to cultural and social barriers [59]. The Kingdom is encouraging greater female participation in the financial markets by improving opportunities for women to participate in the economy and capital markets. The Saudi Stock Exchange (Tadawul) in 2019 implemented new regulations to enable women to invest and trade in stocks and securities [60]. As of 2021, Saudi women made up around 10% of investors in the Tadawul market, and the figure has been steadily increasing. Their participation in the Saudi labor market is increasing that correlates with greater investment activity [61]. They are more involved in entrepreneurial activities, the numbers of women-led businesses entering the market is on the rise and are actively participating in initial public offerings.

Saudi women have made notable strides in terms of leadership roles, like Sarah Al-Suhaimi the Chairperson of the Saudi Stock Exchange [62]. Government policies like Vision 2030 aims to boost women's representation in decision-making positions across sectors, including finance, investment, asset management, and corporate governance [63]. They are also supported by government-backed educational campaigns and tailored programs to address the lack of access to financial knowledge [64].

Despite reforms, Saudi women still face cultural barriers and social norms that limit their full participation in capital markets, like restricted access to networks, mentorship, and finance, as well as traditional gender roles that hinder business activities. While progress has been made, women still struggle to access loans and financing compared to men [65]. Saudi Arabia's Vision 2030 addresses gender disparities in the workforce and capital markets by introducing gender-neutral labor laws, as well as supporting women's entrepreneurship and better access to finance. The government has also introduced quotas for women in leadership positions in both public and private sectors including in financial institutions and the stock market [66].

Thus a comparison shows that women in many parts of the world and Saudi women face barriers such as gender biases, lower financial literacy, and unequal access to capital despite having greater access to capital markets and leadership roles. However, both Saudi Arabia and global economies are on a positive path in closing these gaps, but still a significant work needs to be done to fully realize gender uniformity in capital markets [67].

4 Strategies to Increase Women's Participation in Capital Markets

Education, innovative financial products, and collaborative efforts between the public and private sectors will enhance women's participation.

For empowering women and increasing their engagement in the capital market, it is crucial to have financial literacy. UN Women is a global entity that has many initiatives that seek gender equity and women empowerment. To empower women, the UN offers women special programs and training to increase their investment opportunities and financial decisions [68]. Lusardi et al. [69] claim that enhanced financial literacy will not only result in reducing the gender gap in market participation but also enhance women's financial decision-making. It will increase women's confidence in making financial decisions.

Advanced technologies such as online financial services, banking platforms, and mobile applications will help overcome the barriers to women's entry into the capital market. Offering such a digital technology will have a direct impact on economy and economy actors including women by providing easy and convince access to financial services. Not having these options will reduce women's ability to access financial services·

Policymakers in government and public sectors must collaborate and develop policies that benefit women and will lead to sustainable economic growth. The policies should focus on promoting, supporting, and encouraging practices and activities to reduce the gender gap and increase financial literacy and accessibility to financial services. One example of a successful policy is introducing a microfinance policy, which will increase women's independence and empowerment.

Saudi Arabia has implemented many strategies to create a sustained economy that does not depend on oil production and its products only. Most of these strategies came under Vision 2030, which focuses on multiple areas such as economic diversification, employment, and women empowerment [70]. These strategies focused on key areas that have a direct impact on increasing participation in the capital market. This includes enhancing financial knowledge, use of digital platforms for financial services, offering women customized financial products, promoting gender equity, and involvement of policymakers in both sectors, public and private.

Under financial education and awareness programs, various initiatives have been made to achieve the goal of improving financial literacy. Alinma Bank [71] held a training session for women titled "Financial Habits for Women" which aims to help aspiring women have financial independence and best practices for saving and investment. Also, Saudi Awwal Bank, in collaboration with SABB and Alnahdhah Society, worked on Khazna project with a total fund of SAR 650,000. This project aims to give a series of workshops that teach Saudi women crucial money management skills [72]. Monsha'at which is The Small and Medium Enterprises General Authority, has many initiatives that strongly contribute to the Saudi economy by empowering Saudi women. The percentage of women-owned businesses increased significantly from 21% in 2016 to 40% in 2023. That reflects government and national efforts in this matter. Monsha'at has a program called "Empowering Female Entrepreneurs" which focuses on increasing women's engagement and enhancing their economic contribution [73].

Furthermore, introducing customized financial products and services that target women will increase their participation in the capital market and it will enhance their financial confidence and meet their needs. One of the Saudi initiatives under this category is Kanaf Loan which is a social financing loan. It is only for Saudi women especially aimed at supporting widows and divorced women. This initiative provides loans to help women manage their finances more effectively [74].

Partnership and collaboration between public and private firms is an effective tool that could help overcome entry challenges. It is encouraged by the Saudi Vision 2030 and has been implemented by some firms. A successful example of this partnership is KAUST and Monshaat partnership. This partnership will formalize continuing collaborations that will provide a basis for advancing the entrepreneurship and SME ecosystem in Saudi Arabia [75].

One of the key objectives of Vision 2030 is to offer equal opportunity and promote gender diversity in the Saudi labor market. The Ministry of Finance has made huge efforts to increase women's participation levels and give them equal opportunities. This effort has extended to Human Resources, and Social Development Ministry's initiatives to promote equal opportunity, engage and empower women. That has led to remarkable results, which show that Saudi Arabia was able to increase women's participation in the labor market by up to 30% in 2020. These initiatives include creating anti-discrimination policies and a women's empowerment strategy. That could be achieved by setting policies that prohibit discrimination in hiring and promotions and ensure an increased number of women in leadership positions.

Digital financial services with the use of AI have improved banking services in Saudi and have encouraged more women to use and participate in the capital market. It simplifies the investment process by collecting data, analyzing it, and offering personalized customer services based on their needs and preferences [76].

5 Conclusion

Saudi Arabia has made great progress in encouraging women's economic empowerment that has been evidenced by increased number of women participation in workplace and in led businesses under Vision 2030. The Saudi capital market holds enormous potential of increasing women empowerment by providing access to differing financial resources, increase investment opportunities, and support saving plans. That leads to unlock new possibilities for sustainable economic growth.

Vision 2030 is catalyst in women's economic empowerment by prioritizing economic diversification, gender equality, and creating opportunities for women. It has boosted the women's economic empowerment that has resulted in an increase in female labor force and growth in women lead businesses however there are challenges to fully realize there economic potential.

As gender equity is prioritized in financial policies and products, Saudi capital market is empowering women by providing them access to investment opportunities, financial resources, and wealth-accumulation options.

Government is conducting training programs and workshops to improve women's knowledge of money management, investment strategies and financial products as financial literacy is important to empower them economically.

The government is ensuring that women have equal access to opportunities and financial resources by promoting policies on gender diversity, minimizing discrimination and creating a supportive environment for women.

Thus, Saudi Arabia has made progress in empowering women participation in the capital market, the government is working hard to eliminate barriers by improving financial literacy, designing customized financial products thereby, and ensuring continued and sustained economic growth.

Despite these efforts by government to empower women economically their access to the capital market is hindered by cultural institutional and regulatory barriers, like, low financial literacy, lack of customized financial products, and inadequate regulatory framework. This requires a joint effort by different stakeholders to improve women's financial literacy through training programs and workshops, designing customized financial products like social financial options, better collaboration between public and private sectors as seen in initiatives by Monsha'at, Alinma Bank, and government ministries.

6 Suggestions for Policymakers

The policy makers can considered the following strategies to promote Saudi Arabia to a more sustainable economy by empowering women through capital markets:

Policymakers should expand financial literacy programs like the Khazna project, especially for women in rural areas. Government can conduct online programs on financial planning, risk management, investment strategies, and the capital markets for women by collaborating with universities banks and financial institutions. A nationwide campaign on financial empowerment of women can be launched to reach a wider group of female audiences. Savings an investment products with low entry thresholds can be developed by the financial is institutions for women. Financial products can be designed in a way that woman without a Credit history can get low interest loans they should be provided with investment options that emphasize on environmental social and governance factors.

Government should strengthen regulations and reforms to remove challenges women face in assessing capital by approving credit without the male guarantor or collateral that is not accessible to them. Regulatory measures should be designed to promote gender diversity and introduce quotes in Companies to have gender equality in leadership roles.

Programs that support women's financial inclusion and entrepreneurial endeavors should be designed by the public and private sectors. Collaborations between public institutions like Monsha'at and private-sector financial institutions should be strengthened to improve access to capital for women wound businesses.

Access to venture capital should be improved by providing government bank loans and reduced borrowing cost for women. Capital market should be made more accessible by simplifying the investment process and offering incentives to encourage female investors. Government-backed loan schemes for women entrepreneurs like Kanaf loan should be implemented more vigorously and special investment schemes or saving plans should be designed to encourage more female participation in the capital markets.

Policy makers must ensure that women's are compensated equally to men in the financial and corporate sector especially in leadership positions to strengthen their economic security and independence.

The government should ensure that the financial literacy programs that target women provide more than basic budgeting knowledge and cover more complex topics including financial planning, risk management, diverse investment opportunities, and strategies. These programs would be most effective if offered by collaborating with universities, banks, and financial institutions.

Financial products can be designed in a way that women without a Credit history can get low interest loans and they should be provided with investment options that emphasize environmental social and governance factors.

Women should be offered repayment terms that align with the business cycle, like Peru's Financiera Confianza, which enables women entrepreneurs to manage cash flows more effectively and grow their businesses. Financial institutions should also provide collateral-free loans and insurance that covers health concerns like breast cancer.

Saudi Arabia can invest and create secure user-friendly digital platforms and mobile banking services that can increase and improve women's access to financial services regardless of their location. Especially for those that are in remote areas. Banks and tech companies should collaborate to ensure that mobile financial services are user-friendly, accessible and secure for women. They should design these services keeping in mind the women's unique needs taking into account factors like literacy level, security concerns and family responsibilities.

The government should strengthen regulations to eliminate discrimination in hiring practices. Clear guidelines should be established for quotas for women across boards and corporations, this will encourage aspiring younger women to look for leadership positions in the organization. Policy makers must ensure that women are compensated equally to men in the financial and corporate sector especially in leadership positions to strengthen their economic security and independence.

The government can provide financial and regulatory incentives like tax rebates or preferential access to government contracts to companies that promote gender diversity. This will help in increasing the number of companies that promote a gender equality environment that supports women and encourages them for leadership positions.

Government should emphasize socially responsible investing (SRI) by encouraging capital markets to direct funds to those sectors that directly benefit women like healthcare, education, fashion, technology and services tackling social inequalities.

Policy makers should encourage partnerships between public and private sectors. These partnerships could be programs that support women's financial inclusion and entrepreneurial endeavors. Successful collaborations between public institutions like Monsha'at and private-sector financial institutions should be strengthened to improve access to capital for women wound businesses. Also, this partnership could include sharing resources that will provide comprehensive financial services and support for women.

Policymakers should create regular monitoring and evaluation systems to track the progress of women's participation in the capital market. They should monitor participation rates, investment returns and the performance of women lead businesses. This will help in identifying gaps major progress and insure that policies are effectively addressing the challenges faced by women. The government can established channels for women investors entrepreneurs and business leaders to share their concerns challenges and needs, There feedback will help the government policymakers to shape future policies and initiatives focused on promoting greater financial independence for women, e.g., they can launch a government backed report on women's financial inclusion and capital market participation with key matrix that tracks women access to financial products investment returns and entrepreneurial real success rates.

By implementing these suggestions, Saudi Arabia will not only achieve its vision 2030 objectives but will also build a more inclusive and diversified economy where women can flourish as entrepreneurs, investors and leaders in the capital markets.

References

1. Klein, K. (2018). *Why women must play a bigger role in capital markets*, from https://knowledge.wharton.upenn.edu/article/women-must-play-bigger-role-capital-markets/
2. Cha, L. M., & World Economic Forum. (2024). *What role can capital markets play in boosting gender diversity at work—and beyond?* from https://www.weforum.org/stories/2024/03/capital-markets-gender-diversity/
3. Choi, S., Yu, C., Ukonu, C., & World Economic Forum. (2024). *Women are poised to reshape financial markets—as investors and financial decision-makers*, from https://www.weforum.org/stories/2024/06/women-shape-influence-revolutionize-financial-markets/
4. Lagarde, C., & Ostry, J. D., (2018). *Economic gains from gender inclusion: Even greater than you thought*, from https://www.imf.org/en/Blogs/Articles/2018/11/28/blog-economic-gains-from-gender-inclusion-even-greater-than-you-thought

5. BNY. (2022). *BNY Mellon Investment Management: It's time to create a more inclusive investment world*, from https://www.bny.com/corporate/global/en/about-us/newsroom/press-release/bny-mellon-investment-managementpercent3a-itpercent27s-time-to-create-a-more-inclusive-investment-world-130254.html#:~:text=It%20also%20shows%20that%20women,flowing%20to%20more%20responsible%20investments.
6. WBS. (2018). *Are women better investors than men?* from https://www.wbs.ac.uk/news/are-women-better-investors-than-men/
7. RBC Wealth Management. (2018). *As women gain power, interest in ESG investing grows*, from https://www.rbcwealthmanagement.com/en-us/insights/as-women-gain-power-interest-in-esg-investing-grows#:~:text=And%20increasingly%2C%20studies%20show%20that,into%20their%20policies%20and%20decisions.
8. Yahoo News. (2021). *Giving while female: Women are more likely to donate to charities than men of equal means*, from https://www.yahoo.com/news/giving-while-female-women-more-132535498.html?guce_referrer=aHR0cHM6Ly93d3cuZ29vZ2xlLmNvbS8&guce_referrer_sig=AQAAAJzb5wRL_GNPmRAQKc5S4IN76edqJ5kmW91aXh-nB-GJJOoPuRJAUWLj1gUC7y3oGtMHnUo60s1XIaL4RPeKBcUDHJ9RX7ecZx8HMfXrbwXFaFPBzYhQLWyIkjk7y6x5EzLTZscFkOpM7aoGf-qaTeBPd2nofkEuhhmIMz6gCuAT
9. Atlantic Council. (2023). *The rising female workforce in Saudi Arabia and its impact on the private sector*, from https://www.atlanticcouncil.org/event/the-rising-female-workforce-in-saudi-arabia-and-its-impact-on-the-private-sector/
10. GASTAT. (2023). *GASTAT Issues Saudi Women's Report 2022*, from https://www.stats.gov.sa/en/news/472
11. World Bank Group. (2024). *Saudi Arabia*, from https://genderdata.worldbank.org/en/economies/saudi-arabia
12. S&P Global. (2023). *Economic research: Greater share of working women bolster Saudi Arabia's economic growth, improving productivity will entrench it*, from https://www.spglobal.com/ratings/en/research/articles/230920-economic-research-greater-share-of-working-women-bolster-saudi-arabia-s-economic-growth-improving-productiv-12848378#:~:text=Saudi%20Arabia's%20overall%20female%20labor,2016%20(see%20chart%201).
13. Riyadh Behavioral Insights Center. (2022). *Achieving gender equality at work in Saudi Arabia*, from https://www.hrsd.gov.sa/sites/default/files/2024-03/20220520_RBIC%20-%20KSA%20gender%20norms%20and%20BI%20report_3.pdf
14. Monsha'at. (2023). *Monsha'at quarterly report*, from https://www.monshaat.gov.sa/sites/default/files/2024-02/SME%20Monitor%20-%20Q4%202023%20EN_0.pdf
15. Arab News. (2024). *Saudi Arabia a beacon for female entrepreneurship, according to industry leaders*, from https://www.arabnews.com/node/2491916/business-economy
16. World Economic Forum. (2015). *Why are women in developing economies excluded from banking?* from https://www.weforum.org/stories/2015/09/why-are-women-in-developing-economies-excluded-from-banking/
17. World Economic Forum. (2023). *The finance gap for women entrepreneurs is $1.7 trillion. Here's how to close it*, from https://www.weforum.org/stories/2023/10/women-entrepreneurs-finance-banking/
18. World Economic Forum. (2022). *Why putting women in charge of their own financial security pays dividends*, from https://www.weforum.org/stories/2022/03/women-centred-design-finances/
19. Vision 2030. (2016). *Kingdom of Saudi Arabia Vision 2030*, from https://www.vision2030.gov.sa/media/rc0b5oy1/saudi_vision203.pdf
20. The World Bank. (2023). *Achieving results for women in the Middle East and North Africa Region*, from https://documents1.worldbank.org/curated/en/099408005082326000/pdf/IDU030ef0a3c0ae6e04e5d0af0508e31a51af154.pdf
21. Ministry of Finance. *Women empowerment initiative within the Saudi Annual Budget*, from https://www.mof.gov.sa/en/budget/2023/Documents/Women%20Empowerment.pdf
22. Arab News. (2024). *Saudi Arabia jumps 25 places in World Bank's national statistics indicator*, from https://www.arabnews.com/node/2446701/business-economy

23. World Bank. (2024). *Women, business and the law 2024*, from https://wbl.worldbank.org/content/dam/documents/wbl/2024/snapshots/Saudi-arabia.pdf
24. Arab News. (2024). *Saudi females are driving the SME boom in the Kingdom*, from https://www.arabnews.com/node/2334436/business-economy
25. Asharq Al-Awsat. (2019). *Saudi Banks: 20% of deposits for women amounting to $53 billion*, from https://aawsat.com/home/article/1526961/
26. Financial Sector Development Program. (2021). *Enhance financial literacy*, from https://www.vision2030.gov.sa/media/ud5micju/fsdp_eng.pdf
27. Arab News. (2019). *Alawwal, SABB partner with Al-Nahda to support women*, from https://www.arabnews.com/node/1583811/spa/aggregate
28. Forbes. (2023). *Alawwal, the world's most powerful women*, from https://www.forbes.com/profile/lubna-s-olayan/
29. The Gulf Entrepreneur. (2022). *Maha Shirah: an outstanding personality with a passion to serve others*, from https://thegulfentrepreneur.com/maha-shirah-an-outstanding-personality-with-a-passion-to-serve-others/
30. Woman Interwoven. (2021). *Reem Asaad, or the key role of lingerie in Saudi Women entering the workforce*, from https://www.womaninterwoven.com/post/reem-asaad-or-the-key-role-of-lingerie-in-saudi-women-entering-the-workforce
31. Human Rights Commission. (2023). *Saudi women can*, from https://www.hrc.gov.sa/storage/reports-attachment/2leIi2eu35vfTfKHoSxOnRlnNz8F7Wrs80muzKr6.pdf
32. Business Start Up Saudi Arabia. (2024). *Saudi women hold over 1,700 key posts in employment market*, from https://www.businessstartupsaudiarabia.com/news/saudi-women-hold-1700-posts-employment/
33. Maaal.com. (2023). *In the Saudi stock market, women make up 26% of investors. What are the causes of this low proportion, and what suggestions are there to raise it?* from https://en.maaal.com/2023/03/in-the-saudi-stock-market-women-make-up-26-of-investors
34. Kazemikhasragh, A., Cicchiello, A. F., Monferrá, S., & Girón, A. (2022). Gender inequality in financial inclusion: An exploratory analysis of the Middle East and North Africa. *Journal of Economic Issues, 56*(3), 770–781. https://drive.google.com/file/d/1QQdPmQ2txyo-QOMiSsS4Sb8QvBt432y_/view
35. OECD. (2013). Barriers to women's financial empowerment and factors affecting gender differences in financial literacy. In *Women and financial education: Evidence, policy responses and guidance*. OECD Publishing. https://doi.org/10.1787/9789264202733-6-en
36. Le Quoc, D. (2024). The relationship between digital financial inclusion, gender inequality, and economic growth: Dynamics from financial development. *Journal of Business and Socio-economic Development*. https://www.emerald.com/insight/content/https://doi.org/10.1108/JBSED-12-2023-0101/full/pdf
37. Saluja, O. B., Singh, P., & Kumar, H. (2023). Barriers and interventions on the way to empower women through financial inclusion: A 2 decades systematic review (2000–2020). *Humanities and Social Sciences Communications, 10*(1), 1–14. https://www.nature.com/articles/s41599-023-01640-y.pdf
38. Vision 2030. (2020). *National transformation program delivery plan 2020–2025*. Retrieved from https://www.vision2030.gov.sa/media/nhyo0lix/ntp_eng_opt.pdf
39. Human Resources and Social Development. (2022). *Achieving gender equality at work in Saudi Arabia*. Retrieved from https://www.hrsd.gov.sa/sites/default/files/2024-03/20220520_RBIC%20-%20KSA%20gender%20norms%20and%20BI%20report_3.pdf
40. Al-Asfour, A., Tlaiss, H. A., Khan, S. A., & Rajasekar, J. (2017). Saudi women's work challenges and barriers to career advancement. *Career Development International, 22*(2), 184–199. https://doi.org/10.1108/CDI-11-2016-0200
41. Al-Rashdi, N. A. S., & Abdelwahed, N. A. A. (2022). The empowerment of Saudi Arabian women through a multidimensional approach: The mediating roles of self-efficacy and family support. *Sustainability, 14*(24), 16349. https://www.mdpi.com/2071-1050/14/24/16349/pdf
42. IMF. (2018). *Pursuing women's economic empowerment*, from https://www.imf.org/-/media/Files/Publications/PP/2018/pp053118pursuing-womens-economic-empowerment.ashx

43. Abudy, M. M., Mugerman, Y., & Wiener, Z. (2021). Stock markets and female participation in the labor force. *Journal of International Financial Markets, Institutions and Money*. https://doi.org/10.1016/j.intfin.2021.101297
44. Sahay, R., Cihak, M., et al. (2018). *Women in finance: A case for closing gaps*, from https://www.imf.org/-/media/Files/Publications/SDN/2018/SDN1805.ashx
45. Morgan Stanlay. (2019). *Rise of the SHEconomy*, from https://www.morganstanley.com/ideas/womens-impact-on-the-economy
46. AMF. (2023). *Women and investment savings and investment barometer*, from https://www.amf-france.org/sites/institutionnel/files/private/2023-03/Barom%C3%A8treAMF_Les%20femmes%20et%20l%27investissement%20en%20bourse_Mars2023_VEN.pdf
47. World Economic Forum. (2024). *Global gender gap report 2024*, from https://es.weforum.org/publications/global-gender-gap-report-2024/in-full/economic-and-leadership-gaps-constraining-growth-and-skewing-transitions-7b05a512cb/
48. Rogish, A., Shemluck, N., Danielecki, P., & Hauria, S. (2022). *Advancing more women leaders in financial services: A global report*, from https://www2.deloitte.com/us/en/insights/industry/financial-services/gender-diversity-in-global-financial-services.html
49. European Commission. (2025). *New EU rules to improve gender balance in corporate boards enter into application*, from https://ec.europa.eu/commission/presscorner/detail/en/ip_25_22
50. Wang, M., & Kelan, E. (2013). The gender quota and female leadership: Effects of the Norwegian gender quota on board chairs and CEOs. *Journal of Business Ethics, 117*(3), https://doi.org/10.1007/s10551-012-1546-5
51. Sundarasen, S., Rajagopalan, U., Kanapathy, M., & Kamaludin, K. (2023). Women's financial literacy: A bibliometric study on current research and future directions. *Heliyon, 9*(12), https://doi.org/10.1016/j.heliyon.2023.e21379
52. IFC. (2024). *Her Fintech Edge*, from https://www.ifc.org/content/dam/ifc/doc/2024/her-fintech-edge.pdf
53. OECD. (2013). *Women and financial education*, from https://www.oecd.org/content/dam/oecd/en/publications/reports/2013/10/women-and-financial-education_g1g30fb5/9789264202733-en.pdf
54. The World Bank. *Using digital solutions to address barriers to female entrepreneurship*, from https://digitalforwomen.worldbank.org/access-finance
55. Gray. C. (2024). *Why do more men invest in venture capital than women?* from https://www.linkedin.com/pulse/why-do-more-men-invest-venture-capital-than-women-catherine-gray-qyjec/
56. The World Bank. (2018). *Women economic empowerment study*, from https://documents1.worldbank.org/curated/en/861491551113547855/pdf/134846-WP-PUBLIC-march-2-WB-Women-Study-EN.pdf
57. Storvik, A. (2024). *Women on boards—experience from the Norwegian quota reform*, from https://www.ifo.de/DocDL/dicereport111-rm2.pdf
58. World Bank Group. (2014). *Expanding women's access to financial services*, from https://www.worldbank.org/en/results/2013/04/01/banking-on-women-extending-womens-access-to-financial-services
59. Al-Amir, E. A., Othman, J., & Qureshi, N. I. (2020). Women and stock market participation: Empirical evidence from the Saudi stock market. *Asian Social Science, 16*(5), https://doi.org/10.5539/ass.v16n5p29
60. Felimban, R., Floros, C., & Nguyen, A. N. (2018). The impact of dividend announcements on share price and trading volume: empirical evidence from the Gulf Cooperation Council (GCC) countries. *Journal of Economic Studies, 45*(2), https://doi.org/10.1108/JES-03-2017-0069/full/html
61. Ministry of Finance. (2024). *Budget statement*, from https://www.mof.gov.sa/en/budget/2024/Documents/Bud-E%202024%20F4.pdf
62. ARABISK. (2024). *Sarah Al-Suhaimi is the Saudi financial market's first female leader*, from https://www.arabisklondon.com/12090

63. Alafeef. (2024). The role of vision 2030 in supporting Saudi women to participate in marketing activities in the Kingdom of Saudi Arabia. *Migration Letters, 21*(2). ISSN: 1741-8984
64. Almoaidbed, H. (2024). *Saudi women are learning financial literacy and it's helping the country grow*, from https://www.atlanticcouncil.org/blogs/menasource/saudi-arabia-financial-literacy-women-vision-2030/
65. Morsy, H. (2020). *Access to finance: Why aren't women leaning in?*, from https://www.imf.org/-/media/Files/Publications/Fandd/Article/2020/March/africa-gender-gap-access-to-finance-morsy.ashx#:~:text=Worldwide%2C%20women's%20access%20to%20finance,between%20women's%20and%20men's%20access
66. Parveen, M. (2023). Reflection of Saudi Women's participation and leadership: A study on the gender differences in leadership and structural barriers. *Rupkatha Journal, 15*(2). https://doi.org/10.21659/rupkatha.v15n2.16
67. Polok, B., & Akeel, M. (2024). Vision 2030 has done wonders for women. In *But there's still room to enhance their economic roles in Saudi Arabia*, from https://www.atlanticcouncil.org/blogs/menasource/vision-2030-women-economy-saudi-arabia/
68. UN Women. (n.d.). *Empowering women through financial inclusion.* Retrieved from https://www.unwomen.org
69. Lusardi, A., Mitchell, O. S., & Curto, V. (2010). Financial literacy among the young. *Journal of Consumer Affairs, 44*(2), 358–380. https://www.nber.org/system/files/working_papers/w15352/w15352.pdf
70. Topal, A. (2019, September). Economic reforms and women's empowerment in Saudi Arabia. *Women's Studies International Forum, 76*, 102253. https://doi.org/10.1016/j.wsif.2019.102253
71. Alinma Bank. (2024). *Alinma Bank financial awareness and trust program.* Retrieved November 30, 2024, from https://www.alinma.com/en/about-the-bank/amad/alinma-bank-financial-awareness-and-trust-program
72. Saudi Awwal Bank. (2019). *Alawwal Bank, SABB join forces with Alnahda Society to increase financial literacy among Saudi women.* Retrieved from https://www.sab.com/en/about-us/newsroom/news-year-2019/alawwal-bank-sab-join-forces-with-alnahda-society-to-increase-financial-literacy-among-saudi-women/
73. Monshaat. (2024). *Empowering female entrepreneurs program.* Retrieved November 30, 2024, from https://www.monshaat.gov.sa/en/node/12845
74. Saudi National Portal. (2024). *Kanaf Loan from Social Development Bank.* Retrieved from https://www.my.gov.sa/wps/portal/snp/servicesDirectory/servicedetails/14308
75. King Abdullah University of Science and Technology (KAUST). (2022). *KAUST and Monsha'at sign partnership to advance Saudi economy by creating opportunities for SMEs n.* Retrieved from https://innovation.kaust.edu.sa/kaust-and-monshaat-to-create-opportunities-for-smes/
76. Arab News. (2024). How AI and fintech are redefining the banking landscape in Saudi Arabia. *Arab News.* Retrieved from https://www.arabnews.com/node/2577599/saudi-arabia

Analysis of the Digital Lending Procedures Utilized by Banks and MSMEs in India

Mohd Farhan

Abstract In the recent times, digital lending reports have emerged as a fundamental resource for individuals seeking to acquire knowledge regarding both the quantity of cutting-edge fintech start-ups and the emergent technologies that are utilized. This study aims to identify the challenges in the procedure for complicated applications, prolonged processing times, opportunities and challenges encountered by banks and micro, small, and medium enterprises (MSMEs) in the NCR region of Delhi during the digital lending process including restricted capital and inadequate banking knowledge, opaqueness in the loan process, unavailability of suitable technology, and the need to re-innovate and re-direct efforts. Technological advancements, an innovative operating model, cost efficiency, credit acquisition, reduction of the credit gap, and easy banking are also considered. The data was obtained from a sample of 600 respondents. It covers the demographic variables, challenges, opportunities, and the utilization of descriptive and analytical statistical tools to examine the correlation between these factors and the digital lending process. According to the findings of this research, the digital lending process is beneficial for both banks and MSMEs. The digital lending process is an emerging technology to reduce processing time and deliver economical, user-friendly banking solutions.

Keywords Challenges · Opportunities · Digital lending · Banks · MSMEs

1 Introduction

The digital lending process comprises a sequence of activities that financial service providers execute to extend credit. This process yields numerous advantages for both banks and micro, small, and medium-scale enterprises (MSMEs), including enhanced customer experiences, increased operational efficiencies, greater cross-selling opportunities, cost savings, and improved decision-making. However, banks and MSMEs in India face substantial challenges due to the high cost of credit,

M. Farhan (✉)
Mittal School of Business, Lovely Professional University, Phagwara, Punjab, India
e-mail: mohdfarhan7@gmail.com

K. Khanna et al. (eds.), *From Poverty to Prosperity*, Information Systems Engineering and Management 64, https://doi.org/10.1007/978-3-031-99166-0_2

which hampers their ability to maintain competitive strengths and protect intellectual property without adequate financial support.

Existing literature extensively discusses the challenges and opportunities banks and MSMEs face, along with the resources available to support MSME development in growing economies. In India, specific challenges include complex application procedures, long processing times, limited capital, inadequate banking knowledge, lack of transparency in the loan process, and unavailability of suitable technology. Conversely, opportunities lie in re-innovation and redirection, technological advancements, innovative operating models, and cost efficiency.

2 Literature Review

Curtin and Reynolds [1] explore the foundational elements of the Panel Study of Entrepreneurial Dynamics (PSED), highlighting that the deterioration in a unit's ability to adapt to environmental changes should be examined internally rather than distinguishing between external and internal causes. Ramdani et al. [2] Identify marketing as a critical issue affecting the growth of small-scale enterprises, emphasizing the necessity of proper organization to overcome marketing challenges. Walsh and Lipinski [3] assert that marketing plays a central role in decision-making, enhancing a firm's entrepreneurial orientation and value. Irwin and Scott [4] investigate the persistent financial challenges in banks, including money scarcity and defaults on repayments. Mallick et al. [5] analyse non-performing assets (NPAs) in banks, concluding that NPAs are significantly sticky over time, with larger advances correlating with larger advances. Rao [6] highlight the importance of the small-scale industry (SSI) in achieving accelerated industrial growth and creating employment opportunities in India. Dey [7] examine the effective tasks for connecting manufacturers to new markets, emphasizing the practical evolution of business enterprises. Wang et al. [8] study the online lending system, finding that trust and information asymmetry significantly influence the loan process. Khan and Dalu [9] discussed the advancement of SSIs through financial resources, while Ghatak [10] explored the importance of small-scale industries in national development. Rajeshekar [11] note the exponential expansion of SSIs in India. Biswas [12] investigated the strategies and issues related to small-scale industries, emphasizing the importance of competitive factors in market selection. Bidja and Mandizvidza [13] analyse the relationship between competitive needs and upper hands in small-scale ventures. Ozili [14] examine the implications of digital finance for financial stability and inclusion during economic crises. Subrahmanya [15] highlight the benefits of maintaining bookkeeping for small-scale industries, noting that the lack of bookkeeping records can lead to a perception of poor business performance. This comprehensive review underscores the multifaceted challenges and opportunities within the digital lending landscape for banks and MSMEs, offering valuable insights for further research and policy development.

3 Research Methodology

This study adopts a descriptive and analytical approach, utilizing empirical analysis based on survey data. A structured interview schedule was chosen as the method for data collection, encompassing aspects of the digital lending process, respondents' perceived opportunities and challenges, and sustainability. The sample size comprises 600 respondents from the NCR region. The primary data collected were analysed using Statistical Package for the Social Sciences (SPSS-21), employing descriptive statistics and simple linear regression models. Research objectives are as follows:

(i) To investigate the challenges encountered by banks and MSMEs in the realm of digital lending.
(ii) To examine the factors influencing the digital lending procedures of banks and MSMEs.

4 Interpretation and Analysis of Data

Table 1 presents the regression analysis for the relationship between the complexity of the application procedure and digital lending outcomes. The model shows a weak positive correlation ($R = 0.070$) and a very low R^2 (0.007), indicating that only 0.7% of the variance in digital lending is explained by the complexity of the application procedure. The adjusted R^2 is slightly lower at 0.005, with a standard error of 1.3452. The ANOVA results show an F-statistic of 2.564 with degrees of freedom (df1 = 1, df2 = 596) and a significance level of 0.02, suggesting that the relationship is statistically significant, though the practical impact of the predictor is minimal given the low R^2 value.

Table 2 presents the regression analysis results for the impact of complex application procedures on digital lending. The unstandardized coefficient (B) for the constant is 2.703 with a standard error of 0.139, and for the complex application procedure, the coefficient is 0.081 with a standard error of 0.033. The standardized coefficient (Beta) for the complex application procedure is 0.080, indicating a small effect size. The *t*-values for the constant and the complex application procedure are 18.217 ($p < 0.001$) and 3.207 ($p = 0.034$), respectively, suggesting that both the constant and the predictor are statistically significant. This implies that as the complexity of

Table 1 Regression model for the intricate digital lending process (Model Summary)[b]

Model	*R*	R^2	Adjusted R^2	Std. error of the estimate	ANOVA results			
					F	df1	df2	Sig.
1	0.070[a]	0.007	0.005	1.3452	2.564	1	596	0.02

[a] Predictors: (constant), complex application procedure
[b] Dependent variable: digital lending
Source Authors' compilation

the application procedure increases, digital lending outcomes increase slightly but significantly.

Table 3 presents the regression analysis for the effect of ineffective marketing strategies on long processing times in the digital lending process. The model shows a weak positive correlation ($R = 0.179$) and a low R^2 (0.012), indicating that only 1.2% of the variance in long processing times can be explained by ineffective marketing strategies. The adjusted R^2 is slightly higher at 0.024. Although the relationship between the predictor and the dependent variable is weak, the model provides some insight into the minimal impact of ineffective marketing strategies on processing times in digital lending.

Table 4 provides the regression analysis results for the effect of long processing times on digital lending. The unstandardized coefficient (B) for the constant is 1.453 with a standard error of 0.238, and for long processing times, the coefficient is 0.171 with a standard error of 0.031. The standardized coefficient (Beta) for long processing times is 0.141, indicating a modest effect size. The *t*-values for the constant and long processing times are 18.723 ($p < 0.001$) and 2.979 ($p = 0.004$), respectively, both showing statistical significance. This suggests that longer processing times are significantly associated with higher digital lending outcomes, though the effect size is relatively small.

Tables 5 and 6 present the regression analysis results for the impact of limited capital and improper banking knowledge on digital lending. The unstandardized coefficient (B) for the constant is 1.245 with a standard error of 0.133, and for insufficient appetite for credit, the coefficient is 0.122 with a standard error of 0.042. The

Table 2 Unstandardized beta estimates and predictor effects for digital lending with a complex application procedure (Coefficients)[a]

Model		Unstandardized coefficients		Standardized coefficients	*t*	Sig.
		B	Std. error	Beta		
1	(Constant)	2.703	0.139	–	18.217	0.000
	Complex application procedure	0.081	0.033	0.080	3.207	0.034

[a] Dependent variable: digital lending

Table 3 Regression model for long processing times in the digital lending process(Model Summary)[b]

Model	*R*	R^2	Adjusted R^2	Std. error of the estimate	ANOVA results			
					F	df1	df2	Sig.
1	0.179[a]	0.012	0.024	2.134	6.234	1	596	0.004

[a] Predictors: (constant), ineffective marketing strategy
[b] Dependent variable: digital lending
Source Authors' compilation

Table 4 Unstandardized beta estimates and predictor effects for digital lending with prolonged processing delays (Coefficients)[a]

Model		Unstandardized coefficients		Standardized coefficients	t	Sig.
		B	Std. error	Beta		
1	(Constant)	1.453	0.238	–	18.723	0.000
	Long processing times	0.171	0.031	0.141	2.979	0.004

[a] Dependent variable: digital lending
Source Authors' compilation

standardized coefficient (Beta) for insufficient appetite for credit is 0.231, indicating a moderate effect size. The t-values for the constant and insufficient appetite for credit are 15.910 ($p < 0.001$) and 3.213 ($p = 0.002$), respectively, both of which are statistically significant. This suggests that insufficient appetite for credit significantly influences digital lending outcomes, with a moderate effect size.

Table 7 provides the regression analysis for the effect of an enabling regulatory environment on the digital lending process. The model shows a very weak positive correlation ($R = 0.045$) and an R^2 of 0.005, indicating that only 0.5% of the variance in digital lending can be explained by the regulatory environment. The adjusted R^2 is slightly higher at 0.006, with a standard error of the estimate at 1.398. The ANOVA results indicate an F-statistic of 2.654 with degrees of freedom 1 and 596

Table 5 Regression model for the improper banking knowledge and limited capital in digital(Model Summary)[b]

Model	R	R^2	Adjusted R^2	Std. error of the estimate	ANOVA results			
					F	df1	df2	Sig.
1	0.564[a]	0.012	0.017	1.657	11.453	1	596	0.007

[a] Predictors: (constant), no sufficient appetite for credit
[b] Dependent variable: digital lending
Source Authors' compilation

Table 6 Unstandardized beta estimates and predictor effects of "no sufficient appetite for credit" for digital lending (Coefficients)[a]

Model		Unstandardized coefficients		Standardized coefficients	t	Sig.
		B	Std. error	Beta		
1	(Constant)	1.245	0.133	–	15.910	0.000
	No sufficient appetite for credit	0.122	0.042	0.231	3.213	0.002

[a] Dependent variable: digital lending
Source Authors' compilation

and a significance level of 0.021. This suggests that while the relationship is statistically significant, the regulatory environment has a minimal impact on digital lending outcomes.

Table 8 presents the regression analysis results for the impact of the enabling regulatory environment on digital lending, focusing on the lack of transparency in the loan process. The unstandardized coefficient (B) for the constant is 2.567 with a standard error of 0.143, and for the lack of transparency, the coefficient is 0.067 with a standard error of 0.035. The standardized coefficient (Beta) for the lack of transparency is 0.077, indicating a small effect size. The *t*-values for the constant and lack of transparency are 12.543 ($p < 0.001$) and 2.786 ($p = 0.031$), respectively, both of which are statistically significant. This suggests that while the lack of transparency in the loan process has a statistically significant impact on digital lending, the effect size is small.

Table 9 details the regression analysis examining the impact of the non-availability of suitable technology on the digital lending process. The model indicates a very weak positive correlation ($R = 0.040$) and an R^2 of 0.001, suggesting that only 0.1% of the variance in digital lending is explained by the non-availability of suitable technology. The adjusted R^2 is slightly higher at 0.005, and the standard error of the estimate is 1.356. The ANOVA results show an F-statistic of 2.567 with degrees of freedom 1 and 597 and a significance level of 0.018. This implies that while the relationship between the non-availability of suitable technology and digital lending is statistically significant, the impact is minimal given the low R^2 value.

Table 7 Regression model for the regulatory environment that facilitates the digital lending process (Model Summary)[b]

Model	R	R^2	Adjusted R^2	Std. error of the estimate	ANOVA results			
					F	df1	df2	Sig.
1	0.045[a]	0.005	0.006	1.398	2.654	1	596	0.021

[a] Predictors: (constant), enabling regulatory environment
[b] Dependent variable: digital lending
Source Authors' compilation

Table 8 Regression analysis results for the impact of the enabling regulatory environment on digital lending (Coefficients)[a]

Model		Unstandardized coefficients		Standardized coefficients	t	Sig.
		B	Std. error	Beta		
1	(Constant)	2.567	0.143	–	12.543	0.000
	Lack of transparency in loan process	0.067	0.035	0.077	2.786	0.031

[a] Dependent variable: digital lending
Source Authors' compilation

Table 9 Regression model for the lack of appropriate technology in digital lending procedure (Model Summary)[b]

Model	R	R^2	Adjusted R^2	Std. error of the estimate	ANOVA results			
					F	df1	df2	Sig.
1	0.040[a]	0.001	0.005	1.356	2.567	1	597	0.018

[a] Predictors: (constant), non-availability of suitable technology
[b] Dependent variable: digital lending
Source Authors' compilation

Table 10 presents the regression analysis results for the impact of the non-availability of suitable technology on digital lending outcomes. The unstandardized coefficient (B) for the constant is 2.4646 with a standard error of 0.187, and for the non-availability of suitable technology, the coefficient is 0.456 with a standard error of 0.345. The standardized coefficient (Beta) for the non-availability of suitable technology is 0.789, indicating a moderate effect size. The t-value for non-availability of suitable technology is 2.678 ($p = 0.017$), suggesting statistical significance. This indicates that the non-availability of suitable technology significantly influences digital lending outcomes, with a moderate effect size, despite the overall weak explanatory power of the model as indicated by the low R^2 value.

Table 11 presents the regression analysis examining the impact of re-innovating and redirecting digital lending processes. The model shows a moderate positive correlation ($R = 0.435$) and an R^2 of 0.021, indicating that 2.1% of the variance in digital lending is explained by these efforts. The adjusted R^2 is slightly higher at 0.034, with a standard error of the estimate of 1.4567. The ANOVA results show an F-statistic of 6.112 with degrees of freedom 1 and 596, and a significance level of 0.005, suggesting that the regression model is statistically significant. This indicates that re-innovating and redirecting digital lending processes have a measurable impact on digital lending outcomes, although the effect size is moderate.

Table 12 provides the regression analysis results for the impact of re-innovating and redirecting digital lending processes on digital lending outcomes. The unstandardized coefficient (B) for the constant is 2.786 with a standard error of 0.134, and for re-innovating and redirecting, the coefficient is 0.233 with a standard error of

Table 10 Beta estimates and predictor effects (unstandardized) for digital lending associated with the lack of suitable technology (Coefficients)[a]

Model		Unstandardized coefficients		Standardized coefficients	t	Sig.
		B	Std. error	Beta		
1	(Constant)	2.4646	0.187	–	13.567	0.000
	Non availability of suitable technology	0.456	0.345	0.789	2.678	0.017

[a] Dependent variable: digital lending
Source Authors' compilation

Table 11 Regression analysis examining the impact of re-innovating and redirecting digital lending processes (Model Summary)[b]

Model	R	R^2	Adjusted R^2	Std. error of the estimate	ANOVA results			
					F	df1	df2	Sig.
1	0.435[a]	0.021	0.034	1.4567	6.112	1	596	0.005

[a] Predictors: (constant), re-innovate and re-direct
[b] Dependent variable: digital lending
Source Authors' compilation

0.056. The standardized coefficient (Beta) for re-innovating and redirecting is 0.111, indicating a small effect size. The *t*-value for re-innovating and redirecting is 2.521 ($p = 0.006$), suggesting statistical significance. This implies that re-innovating and redirecting digital lending processes significantly influence digital lending outcomes, although the effect size is relatively small based on the standardized coefficient.

Table 13 presents the regression analysis assessing the impact of technological advancements on digital lending processes. The model shows a weak positive correlation ($R = 0.067$) and an R^2 of 0.007, indicating that technological advancements explain 0.7% of the variance in digital lending outcomes. The adjusted R^2 is slightly lower at 0.005, with a standard error of the estimate of 1.6745. The ANOVA results reveal an F-statistic of 2.71 with degrees of freedom 1 and 596, and a significance level of 0.023, suggesting statistical significance. This indicates that while technological advancements have a statistically significant impact on digital lending outcomes, the effect size is minimal, reflecting the limited explanatory power of the model.

Table 14 provides the regression analysis results for the impact of technological advancements on digital lending outcomes. The unstandardized coefficient (B) for the constant is 2.675 with a standard error of 0.3645, and for technological advancements, the coefficient is 0.083 with a standard error of 0.0431. The standardized coefficient (Beta) for technological advancements is 0.0871, indicating a small effect size. The *t*-value for technological advancements is 3.561 ($p = 0.066$), suggesting marginal statistical significance at the conventional 0.05 level. This suggests that while there may be some influence of technological advancements on digital lending outcomes,

Table 12 Beta estimates and predictor effects (Unstandardized) for digital lending associated with re-innovate and re-direct (Coefficients)[a]

Model		Unstandardized coefficients		Standardized coefficients	*t*	Sig.
		B	Std error	Beta		
1	(Constant)	2.786	0.134	–	12.456	0.000
	Re-innovate and re-direct	0.233	0.056	0.111	2.521	0.006

[a] Dependent variable: digital lending
Source Authors' compilation

Table 13 Regression model for technological advancements in the digital lending process (Model Summary)[b]

Model	R	R^2	Adjusted R^2	Std. error of the estimate	ANOVA results			
					F	df1	df2	Sig.
1	0.067[a]	0.007	0.005	1.6745	2.71	1	596	0.023

[a] Predictors: (constant), technological advancements
[b] Dependent variable: digital lending
Source Authors' compilation

the effect size is small and the statistical significance is borderline, implying caution in interpreting the results.

Table 15 presents the regression analysis exploring the impact of an innovative operating model on digital lending processes. The model indicates a modest positive correlation ($R = 0.114$) and an R^2 of 0.022, suggesting that 2.2% of the variance in digital lending outcomes can be explained by the innovative operating model. The adjusted R^2 is slightly higher at 0.023, with a standard error of the estimate of 1.3478. The ANOVA results show an F-statistic of 6.0213 with degrees of freedom 1 and 595, and a significance level of 0.006, indicating statistical significance. This suggests that the innovative operating model has a measurable impact on digital lending outcomes, although the effect size remains modest, highlighting the complex nature of factors influencing digital lending processes.

Table 14 Unstandardized beta estimates and predictor effects of technological advancements for digital lending (Coefficients)[a]

Model		Unstandardized coefficients		Standardized coefficients	t	Sig.
		B	Std. error	Beta		
1	(Constant)	2.675	0.3645	–	23.453	0.000
	Technological advancements	0.083	0.0431	0.0871	3.561	0.066

[a] Dependent variable: digital lending
Source Authors' compilation

Table 15 Regression model for the innovative operating model in the digital lending process (Model Summary)[b]

Model	R	R^2	Adjusted R^2	Std. error of the estimate	ANOVA results			
					F	df1	df2	Sig.
1	0.114[a]	0.022	0.023	1.3478	6.0213	1	595	0.006

[a] Predictors: (constant), innovative operating model
[b] Dependent variable: digital lending
Source Authors' compilation

Table 16 provides the regression analysis results for the impact of an innovative operating model on digital lending outcomes. The unstandardized coefficient (B) for the constant is 3.498 with a standard error of 0.129, and for the innovative operating model, the coefficient is 0.492 with a standard error of 0.044. The standardized coefficient (Beta) for the innovative operating model is 0.117, indicating a small effect size. The *t*-value for the innovative operating model is 2.674 ($p = 0.006$), suggesting statistical significance. This indicates that the innovative operating model significantly influences digital lending outcomes, although the effect size is relatively modest based on the standardized coefficient.

Table 17 presents the regression analysis examining the relationship between cost efficiency, credit acquisition, and digital lending outcomes. The model shows a moderate positive correlation ($R = 0.173$) and an R^2 of 0.021, indicating that 2.1% of the variance in digital lending can be explained by these factors. The adjusted R^2 is slightly lower at 0.020, with a standard error of the estimate of 1.2510. The ANOVA results indicate an F-statistic of 4.235 with degrees of freedom 1 and 597, and a significance level of 0.006, suggesting that the regression model is statistically significant. This implies that both cost efficiency and credit acquisition significantly impact digital lending outcomes, though the effect size remains modest, underscoring the multifaceted nature of influences on digital lending processes.

Table 18 provides the regression analysis results for the impact of cost efficiency and credit acquisition on digital lending outcomes. The unstandardized coefficient (B) for the constant is 1.320 with a standard error of 0.130, and for cost efficiency and credit acquisition, the coefficient is 0.211 with a standard error of 0.033. The

Table 16 Predictor effects and beta estimates (unstandardized) for digital lending that are associated with an innovative operating mechanism (Coefficients)[a]

Model		Unstandardized coefficients		Standardized coefficients	*t*	Sig.
		B	Std. error	Beta		
1	(Constant)	3.498	0.129	–	17.487	0.000
	Innovative operating model	0.492	0.044	0.117	2.674	0.006

[a] Dependent variable: digital lending
Source Authors' compilation

Table 17 Regression model for the cost efficiency and credit acquisition in the digital lending process (Model Summary)[b]

Model	R	R^2	Adjusted R^2	Std. error of the estimate	ANOVA results			
					F	df1	df2	Sig.
1	0.173[a]	0.021	0.020	1.2510	4.235	1	597	0.006

[a] Predictors: (constant), cost efficiency and credit acquisition
[b] Dependent variable: digital lending
Source Authors' compilation

standardized coefficient (Beta) for cost efficiency and credit acquisition is 0.103, indicating a small effect size. The *t*-value for cost efficiency and credit acquisition is 2.779 ($p = 0.008$), suggesting statistical significance. This indicates that improving cost efficiency and enhancing credit acquisition processes significantly influence digital lending outcomes, although the effect size is relatively modest based on the standardized coefficient.

Table 19 presents the regression analysis investigating the impact of reducing the credit gap on digital lending outcomes. The model indicates a moderate positive correlation ($R = 0.137$) and an R^2 of 0.031, suggesting that 3.1% of the variance in digital lending can be explained by efforts to reduce the credit gap. The adjusted R^2 is lower at 0.014, with a standard error of the estimate of 1.3356. The ANOVA results reveal an F-statistic of 6.849 with degrees of freedom 1 and 595, and a significance level of 0.002, indicating statistical significance. This implies that reducing the credit gap significantly influences digital lending outcomes, although the effect size remains modest, emphasizing the complex interplay of factors impacting digital lending processes.

Table 20 provides the regression analysis results for the impact of reducing the credit gap on digital lending outcomes. The unstandardized coefficient (B) for the constant is 2.197 with a standard error of 0.135, and for the reduction of the credit gap, the coefficient is 0.132 with a standard error of 0.041. The standardized coefficient (Beta) for the reduction of the credit gap is 0.158, indicating a moderate effect size. The *t*-value for the reduction of the credit gap is 3.060 ($p = 0.003$), suggesting statistical significance. This indicates that efforts to reduce the credit gap significantly

Table 18 Beta estimates and predictor effects (Unstandardized) for digital lending about cost efficiency and credit acquisition (Coefficients)[a]

Model		Unstandardized coefficients		Standardized coefficients	*t*	Sig.
		B	Std error	Beta		
1	(Constant)	1.320	0.130	–	14.675	0.000
	Cost efficiency and credit acquisition	0.211	0.033	0.103	2.779	0.008

[a] Dependent variable: digital lending
Source Authors' compilation

Table 19 Regression model for the mitigation of credit gaps in the digital lending process model *R*, R^2 Adjusted R^2 standard error of the estimate ANOVA results (Model Summary)[b]

Model	*R*	R^2	Adjusted R^2	Std. error of the estimate	ANOVA results			
					F	df1	df2	Sig.
1	0.137[a]	0.031	0.014	1.3356	6.849	1	595	0.002

[a] Predictors: (constant), reduction of credit gap
[b] Dependent variable: digital lending
Source Authors' compilation

influence digital lending outcomes, with a moderate effect size based on the standardized coefficient, underscoring the importance of addressing credit accessibility in digital lending processes.

Table 21 presents the regression analysis examining the impact of easy banking on digital lending processes. The model shows a weak positive correlation ($R = 0.070$) and an R^2 of 0.059, indicating that easy banking practices explain 5.9% of the variance in digital lending outcomes. The adjusted R^2 is very low at 0.004, with a standard error of the estimate of 1.2945. The ANOVA results reveal an F-statistic of 5.720 with degrees of freedom 1 and 596, indicating statistical significance at a significance level of 0.05. This suggests that while easy banking practices have a statistically significant impact on digital lending outcomes, the effect size is minimal, highlighting the need for further investigation into additional factors influencing digital lending processes.

Table 22 provides the regression analysis results for the impact of easy banking practices on digital lending outcomes. The unstandardized coefficient (B) for the constant is 1.256 with a standard error of 0.145, and for easy banking, the coefficient is 0.070 with a standard error of 0.036. The standardized coefficient (Beta) for easy banking is 0.050, indicating a small effect size. The *t*-value for easy banking is 2.6754 ($p = 0.023$), suggesting statistical significance. This implies that easy banking practices significantly influence digital lending outcomes, although the effect size is relatively small based on the standardized coefficient. Further exploration into additional variables affecting digital lending processes would be beneficial to better understand the overall impact of easy banking practices.

Table 20 Beta estimates and predictor effects (Unstandardized) for digital lending associated with credit gap reduction (Coefficients)[a]

Model		Unstandardized coefficients		Standardized coefficients	*t*	Sig.
		B	Std error	Beta		
1	(Constant)	2.197	0.135	–	31.123	0.001
	Reduction of the credit gap	0.132	0.041	0.158	3.060	0.003

[a] Dependent variable: digital lending
Source Authors' compilation

Table 21 Regression model for the digital lending process in easy banking model *R*, R^2 Adjusted R^2 standard error of the estimate ANOVA results (Model Summary)[b]

Model	*R*	R^2	Adjusted R^2	Std. error of the estimate	ANOVA results			
					F	df1	df2	Sig.
1	0.070[a]	0.059	0.004	1.2945	5.720	1	598	0.014

[a] Predictors: (constant), easy banking
[b] Dependent variable: digital lending
Source Authors' compilation

Table 22 Beta estimates and predictor effects (unstandardized) for digital lending associated with easy (Coefficients)[a]

Model		Unstandardized coefficients		Standardized coefficients	*t*	Sig.
		B	Std. error	Beta		
1	(Constant)	1.256	0.145	–	14.345	0.000
	Easy banking	0.070	0.036	0.050	2.6754	0.023

a. Dependent variable: digital lending
Source Authors' compilation

5 Conclusion

The research analysis indicates that the challenges (complex application procedure, long processing times, limited capital and improper banking knowledge, lack of transparency in the loan process, non-availability of suitable technology) and opportunities (re-innovate and re-direct, technological advancements, innovative operating model, cost efficiency and credit acquisition, reduction of credit gap, Easy Banking) have a moderate impact on digital lending. The Micro, Small and Medium Enterprises (MSMED) Act, which is the backbone of the Indian economy, was introduced by the Indian government to protect the interests of entrepreneurs. The MSME sector in the country is supported by a variety of programmes and schemes. The study's analysis reveals that the opportunities (complex application procedure, long processing times, limited capital and improper banking knowledge, lack of transparency in the loan process, non-availability of suitable technology) and challenges (re-innovate and re-direct, technological advancements, innovative operating model, cost efficiency and credit acquisition, reduction of credit gap, Easy Banking) faced by banks and MSMEs are significantly correlated with the digital lending process. The digital lending process is profitable for both banks and MSMEs, and it is anticipated that traditional money lenders will adopt the digital lending process.

6 Recommendations

1. It is recommended that banks and financial institutions simplify the application procedures involved in the digital lending process.
2. It is recommended that banks and financial institutions conduct a significant number of orientation programmes to raise awareness. Regarding the diverse financial programmes available to entrepreneurs in the micro, small, and medium-sized enterprise sectors.

3. It is recommended that banks and financial institutions implement an easy banking system in the digital lending process, which is a technological advancement.
4. Industrial policies should be periodically evaluated and reviewed.

7 Managerial Implications

The results of this research indicate that the digital lending process is profitable for both banks and SMEs. The study offers analysed inputs for banks, while simultaneously demonstrating that the digital lending process is profitable for SMEs. Developing Digital Lending Products and Services for MSMEs. This study provides an understanding of the relative factors and their effects on the digital landing process. The results of this study can be utilized by Financial Institutions and Digital Lending departments to improve the design, development, and implementation of digital lending services for MSMEs.

References

1. Curtin, R. T., & Reynolds, P. D. (2009). *New firm creation in the United States: Initial explorations with the PSED II data set*. Springer.
2. Ramdani, B., Kawalek, P., & Lorenzo, O. (2009). Predicting SMEs' adoption of enterprise systems. *Journal of Enterprise Information Management, 22*, 10–24
3. Walsh, M. F., & Lipinski, J. (2009). The role of the marketing function in small and medium-sized enterprises. *Journal of Small Business, 16*, 569–585
4. Irwin, D., & Scott, J. M. (2010). Barriers faced by SMEs in raising bank finance. *International Journal of Entrepreneurial Behavior & Research, 16*, 245–259
5. Mallick, S. K., Sarkar, A., Roy, K. K., Duttachaudhuri, T., & Chakrabarti, A. (2010). Dynamics of emerging India's banking sector assets: A simple model. *Journal of Asset Management, 11*(1), 62–70.
6. Rao, G. J. (2010). A study on socioeconomic background of entrepreneurs in small scale industries. *Indian Journal of Marketing, 40*(9), 55–60.
7. Dey, S. K. (2014). MSMEs in India: Its growth and prospects. *Abhinav National Monthly Refereed Journal of Research in Commerce & Management, 3*(8), 26–33.
8. Wang, P., Zheng, H., Chen, D., & Ding, L. (2015). Exploring the critical factors influencing online lending intentions. *Financial Innovation, 1*(1), 1–11.
9. Khan, J. G., & Dalu, R. S. (2015). Role of small & medium enterprises in industrial development of Vidarbha Region. *International Journal of Engineering Technology, Management and Applied Sciences, 3*(7), 78–84.
10. Ghatak, S. (2016). Micro, small and medium enterprises (MSMEs) in India: An appraisal. *Journal of Technology Management & Innovation, 6*(1), 66–76.
11. Rajeshekar, D. (2016). Role of Banks in empowering MSMEs. *Airo International Research Journal*. MSME Annual Report for the year 2015–16, Govt. of India. Official Website of Reserve Bank of India, pp. 2015–2016.

12. Biswas, I. (2016). *Analysis of technology mix in clusters of micro small and medium enterprises (MSMEs) in West Bengal: Identification of adaptation gaps and policy prescriptions* (Published Ph. D Thesis). TERI University, Centre for Regulatory and Policy Research, New Delhi.
13. Bidja, A. B., & Mandizvidza, K. (2017). Strengthening enterprise's growth and effectiveness in developing economies. A case of very Small, Small and Medium-sized Enterprises in Cameroon. *IOSR Journal of Business and Management (IOSR-JBM), 19*(3), 19–27.
14. Ozili, P. K. (2018). Impact of digital finance on financial inclusion and stability. *Borsa Istanbul Review, 18*(4), 329–340.
15. Subrahmanya, M. B. (2019). Small-scale industry development for export promotion: India's experience under liberalization. In *Micro and small enterprises in India* (pp. 67–87). Routledge.

Analyzing the Relationship Between Mutual Fund Flows and the Magnitude of NAV Corrections: Evidence from Historical Data

Deepali M. Gala, K. Venkatesh, Bhakti Ranjit Pawar, Suresh Suryawanshi, and Hiten N. Shah

Abstract Mutual funds play a pivotal role in modern financial systems, offering investors diversified portfolios and professional management. Net Asset Value (NAV) serves as a key performance metric, reflecting fund valuation and investor confidence. This study examines the relationship between mutual fund flows, representing investor sentiments, and the magnitude of NAV corrections, which often mirror market responses to inflows and outflows. Using a comprehensive dataset of fund performance across categories, statistical and econometric techniques such as correlation analysis, regression modeling, and time-series analysis are employed to quantify this relationship. The findings reveal a significant positive relationship between fund flows and NAV corrections, with larger outflows associated with steeper NAV declines. Time-lagged effects indicate that changes in fund flows impact NAV corrections over a 2–3 week period. These insights offer valuable implications for fund managers, policymakers, and investors, highlighting the importance of fund flows as a predictive tool for market behavior and aiding in strategic decision-making.

Keywords Mutual funds · Net Asset Value (NAV) · Fund flows · Investor sentiment · Market corrections · Financial analysis · Asset Under Management (AUM) · Statistical modeling

D. M. Gala (✉) · K. Venkatesh · S. Suryawanshi
Bharati Vidyapeeth (Deemed to be University), Institute of Management, Kolhapur, India
e-mail: deepali.gala@bharatividyapeeth.edu

K. Venkatesh
e-mail: kanthaswamy.venkatesh@bharatividyapeeth.edu

S. Suryawanshi
e-mail: Suresh.Suryawanshi@bharatividyapeeth.edu

B. R. Pawar
School of Leadership and Management, Manav Rachna Educational Institutions, MRIIRS, Faridabad, Haryana, India
e-mail: bhaktijit5eco@gmail.com

H. N. Shah
Research Scholar, Bharathiyar University, Coimbatore, India
e-mail: hiten65@gmail.com

K. Khanna et al. (eds.), *From Poverty to Prosperity*, Information Systems Engineering and Management 64, https://doi.org/10.1007/978-3-031-99166-0_3

1 Introduction

Mutual funds are essential components of financial markets, which enables individuals to pool resources for diversified investment opportunities. These funds democratize access to capital markets, allowing retail investors to benefit from professional management and risk diversification [1]. Performance metrics like NAV, serves as a critical indicators of fund health and investor confidence. NAV corrections often signal market sentiment shifts were fund inflows and outflows act as catalysts [2].

In an increasingly volatile market environment, it is evident to understand the interplay between mutual fund flows and NAV corrections. Fund flows are reflective of investor confidence and market sentiments. Large inflows often signify optimism, whereas outflows indicate uncertainty or dissatisfaction [3]. NAV corrections are the perfect indicator that mirror the market's response to these inflows and outflows. This study investigates these dynamics, focusing on the relationship between fund flows and NAV corrections to unravel patterns critical for strategic decision-making.

2 Review of Literature

The literature extensively explores mutual funds' role in shaping modern financial systems. Kopsch et al. [4] studied the determinants of mutual fund flows, emphasizing the influence of macroeconomic indicators and investor behavior. Previous research highlights that NAV serves as a reflection of a fund's market performance, incorporating the impact of market fluctuations and internal fund management efficiency [2].

The relationship between mutual fund flows and the magnitude of NAV corrections is complex and influenced by various factors. Study highlights the impact of constant NAVs on fund redemptions, particularly during financial stress periods [5]. It was observed that the funds which had a constant NAV have accomplished negative net flows during the run on the reserve primary fund, suggesting that constant NAVs may be associated with higher fund redemptions during crises1. This indicates that the structure of NAVs can significantly affect investor behavior and fund stability.

Another critical aspect is the first-mover advantage by Capponi (2020), where investors who redeem their shares early can do so at more favorable prices, potentially leading to a cycle of falling prices and increasing redemptions [6]. This dynamic introduces a nonlinear relationship between market shocks and impact on a fund's NAV. Swing pricing is proposed as a solution to mitigate this issue by transferring liquidation costs to redeeming investors, thus reducing the risk of fund failure. Properly designed swing pricing can prevent the adverse effects of the first-mover advantage and stabilize fund flows [6].

In addition to structural and policy considerations, predictive models for NAV can also play a role in understanding and managing mutual fund flows. Advanced models, such as those using Support Vector Machines (SVM), have been developed to predict NAVs by incorporating various parameters, including macroeconomic factors [7]. These models aim to provide more accurate predictions of NAV changes, which can help in managing investor expectations and fund strategies. NAV corrections, often triggered by market downturns, signify adjustments in the market's valuation of underlying assets. Researchers have identified correlations between NAV corrections and external factors such as macroeconomic changes, regulatory developments, and sector-specific performance [8]. However, the nuanced relationship between fund flows and NAV corrections remains underexplored, warranting further investigation. Approach adjusts the net asset value (NAV) of funds to allocate trading costs directly to transacting shareholders, effectively mitigating the competitive advantage inherent in traditional pricing structures. Research using unique datasets, such as investor transactions in U.K. corporate bond funds, has demonstrated that swing pricing reduces redemption pressures during periods of market stress. This stabilizing effect is notably more pronounced among institutional investors and those with longer investment skyline [9].

The literature suggests that the relationship between mutual fund flows and NAV corrections is influenced by the structure of NAVs, investor behavior, and predictive modeling. Constant NAVs may lead to higher redemptions during crises, while swing pricing can mitigate the risks associated with the first-mover advantage [10]. Additionally, advanced predictive models can enhance the understanding and management of NAV changes, contributing to more stable mutual fund operations [11]. Investor sentiment is a pivotal factor influencing fund flows. Studies have shown that fund inflows often correlate with market optimism, while significant outflows are observed during periods of economic downturn. Behavioral finance literature suggests that investor decisions are often driven by herd mentality, amplifying fund flow volatility [12].

The role of ESG investments in reducing information asymmetry and improving transparency in financial markets. Also highlights the issue of "greenwashing" in sustainable finance. The study has explored the effectiveness of ESG investments in mitigating financial risks, particularly during economic downturns such as the COVID-19 pandemic [12]. A systematic literature review of green finance analyzing 978 studies across 59 countries [13]. Highlights the growth of green finance research, the role of interdisciplinary studies, and key areas such as renewable energy and sustainable investing.

3 Objectives

- Quantify the relationship between mutual fund inflows/outflows and NAV corrections.
- Identify key trends and statistical patterns in NAV movements.
- Analyze variations in fund performance across categories.

4 Hypothesis

- **Ho**—There is no significant relationship between mutual fund flows and the magnitude of NAV corrections.
- **Ha**—There is a significant relationship between mutual fund flows and the magnitude of NAV corrections.

5 Methodology

This study examines the relationship between mutual fund flows and Net Asset Value (NAV) corrections across 478 mutual fund schemes with 23 different types of mutual funds, categorized by fund types such as Small Cap, Mid Cap, Multi Cap, Banking, ELSS, Transport and Logistics, Pharma, Infrastructure, Consumption, Dividend Yield, Energy, Focused Equity, Flexi Cap, International Equity, PSU, and ESG funds. The dataset includes NAV movements and percentage declines from all-time highs, providing insights into historical volatility and valuation trends.

Key performance metrics include NAV, representing the latest fund valuations and 52-week highs, and the percentage decline from all-time highs, highlighting volatility patterns. Fund flow analysis, expressed as net inflows or outflows relative to Assets Under Management (AUM), reflects investor sentiment and fund popularity. AUM, representing the total capital managed by the fund, serves as a critical indicator of market presence and fund stability. Data sources such as AMFI and Morningstar ensure reliability in assessing fund performance. By integrating these metrics, the study provides a comprehensive perspective on mutual fund behavior and the extent to which fund flows influence NAV corrections.

During data preprocessing phase, normalization was applied to address inconsistencies across datasets, ensuring that all data points were on a similar scale. Outliers were identified and treated using statistical thresholds, helping to minimize their impact on the analysis. Additionally, time-lagged variables were introduced to capture delayed effects, allowing the model to account for temporal dependencies and provide more accurate predictions. These steps were essential in preparing the data for further analysis, ensuring its quality and relevance.

In the analysis phase, descriptive statistics were used to summarize key metrics, providing a clear overview of the data's central tendency and variability. Correlation analysis was employed to explore the linear relationships between variables, helping to identify potential connections. Regression analysis was then conducted to quantify the impact of fund flows on NAV corrections, allowing for a deeper understanding of how changes in fund flows influence NAV movements. Lastly, time-series analysis, specifically using ARIMA models, was applied to examine lagged effects, capturing temporal dependencies and enhancing the predictive accuracy of the analysis. These techniques collectively provided a robust framework for understanding the dynamics of mutual fund performance.

6 Data Analysis and Interpretation

Table 1 represents data on mutual fund performance across various fund categories. It focuses on two key metrics as sum of fall (%) from 52-Week High which represents the cumulative percentage decline from the highest Net Asset Value (NAV) observed in the past 52 weeks for each fund category. The second is sum of 52-week high NAV which indicates the total highest NAV for all funds within each category over the past year.

The Flexi Cap Fund has the highest 52-week high NAV (8219.406), followed by ELSS (8008.679), Large and Midcap Fund (7920.239), and Midcap Fund (14,463.687). These values indicate the overall strong market presence and asset

Table 1 Mutual fund performance across various fund categories

S. No.	Row labels	Sum of fall (%) from 52-week high	Sum of 52-week high NAV
1	Banking fund	– 178.92	1691
2	Business cycle fund	– 139.1	273.8
3	Consumption fund	– 195.08	1276.3
4	Dividend yield fund	– 98.16	1146.6
5	ELSS	– 315.5	8008.679
6	Energy fund	– 47.8	168.4
7	ESG fund	– 90	446.6
8	Flexi cap fund	– 358.8	8219.406
9	Focused equity fund	– 262.4	1977.7
10	Infra fund	– 202.38	2405.4
11	International equity fund	– 30.5	201.8
12	International equity fund	– 265.4	998.9
13	Large and midcap fund	– 282.1	7920.239
14	Large cap fund	– 209.9	4381.5
15	Manufacturing fund	– 116.8	202.2
16	Midcap fund	– 265.44	14463.687
17	MNC fund	– 49.4	2349.17
18	Multi cap fund	– 238.41	3098.7
19	Pharma fund	– 77.95	1545.3
20	PSU equity fund	– 75.3	179.5
21	Small cap fund	– 226.26	2642.6
22	Thematic fund	– 454	2669.2
23	Transport and logistics fund	– 67.06	365.7
	Grand total	– 4246.66	66632.381

Source Author compilation

Table 2 Top mutual funds by assets under management (AUM)

Fund name	AUM
Aditya Birla sun life equity advantage fund–regular plan	960.8
Aditya Birla sun life mid cap fund–regular plan	842.2
ICICI prudential multi cap fund	833.7
HDFC capital builder value fund	772.6
Templeton India value fund	771.7

Source Author compilation

accumulation of these fund types. The International Equity Fund has a relatively low 52-week high NAV (201.8), suggesting a smaller scale compared to domestic funds. Thematic Fund (– 454%) recorded the highest percentage fall, indicating extreme volatility. Flexi Cap Fund (– 358.8%), ELSS (– 315.5%), and Large and Midcap Fund (– 282.1%) also experienced significant declines. Funds like Energy Fund (– 47.8%) and MNC Fund (– 49.4%) exhibited relatively lower declines, suggesting greater stability.

The descriptive statistics for the dataset reveal key insights into mutual fund performance and size. The average NAV stands at ₹28.49, while the mean fall percentage is 8.96%, indicating moderate corrections across most funds. The fund flow range spans from ₹6.56 to ₹854.85, and the AUM range varies from ₹7.71 to ₹960.8, highlighting significant variation in both fund size and investor activity. This variability points to the importance of careful fund selection and diversification across categories, as most funds showed corrections consistent with broader market trends. Investors can mitigate risk and optimize returns by considering these factors in their decision-making.

The initial correlation analysis revealed key relationships between the variables. There is a slight positive correlation between NAV and Fall Percentage ($r \approx 0.25$), suggesting that as NAV increases, the fall percentage tends to rise slightly, although the relationship is not strong. On the other hand, a strong positive correlation was found between AUM and Fund Flow ($r \approx 0.7$), indicating that larger funds generally experience higher net inflows or outflows, highlighting a clear relationship between the size of the fund and investor activity. These findings provide valuable insights into how fund size and performance metrics are interconnected.

The linear regression analysis revealed a significant positive relationship between fund flows and NAV corrections ($p < 0.05$), indicating that larger outflows were associated with steeper declines in NAV. This suggests that when funds experience higher net outflows, their NAV tends to drop more significantly. The regression equation that captures this relationship as NAV Correction $= \beta 0 + \beta 1 \times$ Fund Flow $+ \epsilon$.

In addition, the time-series analysis using ARIMA models revealed a lagged effect, where fund outflows impacted NAV corrections over a 2–3 week period. This indicates that changes in fund flows do not immediately affect NAV but rather show their impact after a short delay, highlighting the importance of considering temporal dependencies in mutual fund performance analysis.

The distribution of NAV is right-skewed, with the majority of values concentrated between ₹20 and ₹40, indicating that most funds are valued within this range, but a few funds have significantly higher NAVs. The fall percentage distribution is approximately normal, with most values clustered around 9%. This suggests that most funds experience moderate corrections, with fewer funds showing extreme falls. A slight positive correlation is observed, where higher NAVs tend to coincide with marginally larger fall percentages. This relationship, while not strong, suggests that larger funds may experience slightly steeper declines when corrections occur.

There is a clear trend where larger funds with higher AUM experience proportionately greater fund flows, highlighting that funds with a larger asset base tend to attract more investor activity, either through inflows or outflows. These insights from the visualizations offer a better understanding of the underlying patterns and relationships within the mutual fund data, aiding investors in making more informed decisions (Fig. 1).

Hypothesis Testing

To test the hypothesis, various steps were performed starting with correlation analysis, wherein it revealed a weak positive relationship ($r \approx 0.25$) between NAV and fall percentage, suggesting a mild association between the two variables. This initial analysis provided an insight into the potential connection but did not establish a strong link.

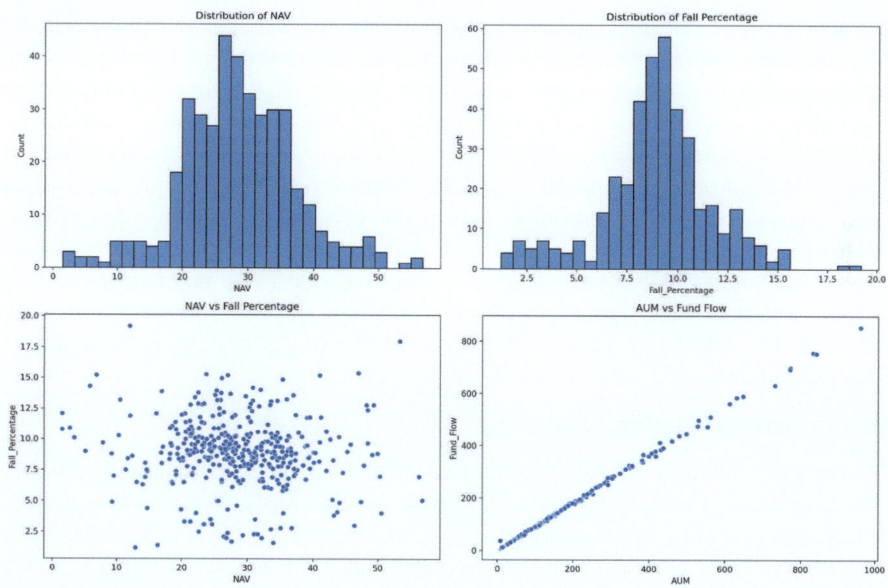

Fig. 1 Distribution of NAV and fall percentage, Author calculation

The regression model confirmed that fund flows significantly influence NAV corrections, with a p-value less than 0.05. This indicates that larger outflows are associated with steeper declines in NAV, providing evidence of a statistically significant relationship between the two variables.

When time-series analysis was applied, ARIMA models were applied to examine the lagged effects, and the results confirmed that fund outflows have a delayed impact on NAV corrections over a 2–3 week period. This finding highlights the temporal nature of the relationship, suggesting that immediate reactions in NAV are not observed, but rather, a delayed effect occurs.

The F-test was performed to test the overall significance of the regression model. The null hypothesis was rejected ($p < 0.05$), further supporting the conclusion that fund flows are indeed a significant predictor of NAV corrections (Table 2).

Funds with higher AUM generally exhibited resilience, with lower fall percentages. Notable examples include Aditya Birla Sun Life Equity Advantage Fund, AUM ₹960.8 crores, Fall 9.0%. HDFC Capital Builder Value Fund, AUM ₹772.6 crores, Fall 9.4%. These funds have shown stability and moderate corrections despite broader market trends, indicating that larger funds are less susceptible to extreme NAV declines.

7 Findings and Suggestions

Outliers and Robustness Checks

A crucial aspect of the analysis involved identifying and managing outliers to ensure data reliability and the robustness of statistical results. Extreme fund flow values were detected using Winsorization and Interquartile Range (IQR) filtering, minimizing their influence on regression estimates. Subsample Analysis is used to examine the different fund categories (e.g., Small Cap vs. Large Cap) to assess variations in NAV correction trends. Alternative Lag Structures for testing multiple lag periods (e.g., 1-week vs. 4-week lags) to confirm the persistence of fund flow effects. Heteroskedasticity and Serial Correlation Adjustments were performed utilizing Newey-West standard errors to account for heteroskedasticity and autocorrelation in fund flow data. The robustness checks confirmed that fund flows significantly influence NAV corrections, reinforcing the reliability of the findings across different model specifications.

Implications for Financial Stability

The study also highlights significant risks associated with large mutual fund outflows, which can amplify market instability. Large-scale redemptions force fund managers to sell assets quickly, often at distressed prices, exacerbating NAV declines. Fund outflows and NAV corrections create feedback loops, further accelerating sell-offs and market instability.

Fund Categories—Different fund categories displayed varying levels of volatility:

- **Small Cap Funds**: These funds experienced higher volatility, with fall percentages ranging from 7.0 to 15.2%. They tend to be more sensitive to market fluctuations.
- **Large Cap Funds:** Moderate corrections were observed in large-cap funds, with average falls around 9.1%. These funds are generally more stable, reflecting their larger and more established market positions.
- **Thematic Funds:** These funds, particularly those focused on niche sectors like energy, showed significant declines, with fall percentages reaching up to 16.1%. The concentrated nature of thematic funds exposes them to higher risks.

Implications for Investors

There is a clear correlation between large fund outflows and steeper NAV corrections. When investors redeem their shares in large quantities, it can cause more significant declines in NAV, indicating a potential risk for investors in times of heavy outflows. To mitigate risks, it is recommended that investors diversify their portfolios across different fund categories. This strategy can help reduce the impact of volatility and mitigate risks associated with any single sector or fund type. The lagged effects observed in the analysis offer opportunities for strategic market timing. Understanding the delayed relationships between fund flows and NAV corrections can help investors identify optimal entry or exit points, potentially maximizing returns or minimizing losses. These insights emphasize the importance of understanding fund characteristics and market timing to make informed investment decisions.

8 Conclusion

The analysis supports the alternative hypothesis (H_1), confirming that there is a significant relationship between mutual fund flows and the magnitude of NAV corrections. Fund outflows exacerbate NAV declines, with the effects becoming more pronounced over a lag period. This evidence underscores the critical role of fund flows in driving NAV changes, particularly during periods of high outflows, which can lead to more severe declines. This study highlights the intricate relationship between mutual fund flows and NAV corrections, demonstrating that the structure of fund flows plays a crucial role in shaping NAV movements. Funds with higher AUM and diversified portfolios tend to exhibit resilience, experiencing moderate corrections even in adverse market conditions. These funds are more susceptible to significant volatility, with larger declines in NAV observed, especially in more specialized sectors like energy or emerging markets.

The study's insights are valuable for fund managers and policymakers, emphasizing the need for continuous monitoring of fund flows and the implementation of strategic decision-making practices. Effective management of outflows can help mitigate NAV declines, especially in times of market stress. Higher AUM funds exhibit greater resilience, experiencing lower NAV corrections, while smaller and sector-specific funds remain highly volatile. Investors should adopt a diversified approach, balancing exposure to both large-cap and thematic funds to mitigate risks. Regulatory policies should focus on stress-testing fund liquidity, enhancing disclosure norms, and implementing risk-adjusted pricing mechanisms.

Future research should explore real-time fund flow data and machine learning models to refine predictive strategies for NAV fluctuations. The integration of AI-driven risk assessment models could provide early warning indicators for fund managers and policymakers, enhancing market stability and investor confidence.

References

1. Merton, R. C., & Bodie, Z. (2004). *Financial infrastructure and economic growth.* Federal Reserve Bank of Kansas City.
2. Campbell, J. Y., & Viceira, L. M. (2002). *Strategic asset allocation: Portfolio choice for long-term investors.* Oxford University Press.
3. Ferris, S. R., & Chance, D. B. (2003). The effects of mutual fund flows on stock market liquidity and volatility. *Journal of Banking & Finance, 27*(4), 615–636.
4. Kopsch, C., Liesenfeld, R. G., & Reif, M. (2015). Macroeconomic determinants of mutual fund flows. *Journal of International Financial Markets, Institutions & Money, 34*, 145–160.
5. Witmer, J. (2016). The run on the reserve primary fund: evidence from the 2008 financial crisis. *Financial Review, 51*(3), 401–420.
6. Capponi, A., & Glasserman, C. (2020). Swing pricing for open-end mutual funds. *Management Science, 66*(6), 2318–2338.
7. Yu, L., & Wang, S. (2009). Forecasting mutual fund NAV using support vector machines. *Expert Systems with Applications, 36*(4), 7311–7315.
8. Agarwal, A., Daniel, N. D., & Naik, N. Y. (2009). Flows, performance, and managerial incentives in hedge funds. *The Journal of Finance, 64*(2), 985–1030.
9. Cetorelli, N. (2021). Swing pricing and financial stability. In *Federal Reserve Bank of New York Staff Report no. 977.*
10. N. Barberis and R. Thaler (2003). A survey of behavioral finance. In G. M. Constantinides, M. Harris, & R. M. Stulz (Eds.), *Handbook of the economics of finance* (Vol. 1, pp. 1053–1128). Elsevier.
11. Liu, Y., Zhang, X., & Huang, J. (2023). ESG investments, transparency, and greenwashing: Evidence from global financial markets. *Journal of Sustainable Finance & Investment, 13*(1), 45–67.
12. Yi, Y., Li, M., & Wu, H. (2021). ESG performance and firm resilience during the COVID-19 crisis: Evidence from global markets. *Finance Research Letters, 41*, 101871.
13. Mudalige, H. M. N. K. (2023). Green finance: A systematic literature review. *Future Business Journal, 9*(1), Article 29.

Green Bonds as Catalysts for Sustainable Development: Strategies and Implications for India

C. S. Amit Kumar and C. S. Vinit Sikka

Abstract This chapter looks at the emerging world of green bonds in India, one of the very important financial instruments lying at the intersection of sustainable development and economic growth. As the world tilts toward green investments, India is extending a case that is in tune with rapid economic development and commitment to sustainable practices. The text covers the global movement of green bonds, setting the development and present importance of such bonds within the framework of a climate-conscious economy. This section also comprises the analysis of the projects that amass finance from green bonds in India, starting from renewable energy to sustainable forms of infrastructure. This evaluation demonstrates various forms of investment and simultaneously emphasizes the role of those projects in ensuring sustainable development goals for India. The chapter critically analyzes the challenges and opportunities in the Indian green bond market, including issues of transparency and standardization, the role of investors, and that of green bonds in mobilizing additional investments in sustainable systems and operations. Strategic recommendations to enhance efficiency and attractiveness toward green bonds in India are provided in the end. Examples of such approaches are providing policy recommendations, supporting market development programs, and educating through campaigns for raising awareness and involvement in green finance.

Keywords Green bonds · Green technologies · Green finance · Investment

C. S. Amit Kumar (✉) · C. S. Vinit Sikka
School of Leadership and Management, Manav Rachna International Institute of Research and Studies, Faridabad, India
e-mail: amitkumar.slm@mriu.edu.in

C. S. Vinit Sikka
e-mail: Vinit.slm@mriu.edu.in

K. Khanna et al. (eds.), *From Poverty to Prosperity*, Information Systems Engineering and Management 64, https://doi.org/10.1007/978-3-031-99166-0_4

1 Introduction

1.1 Background and Importance of Green Bonds

Over these years, the capital market has undergone a transitional shift in the path of sustainable development. Green bonds have emerged as one of the vital financial instruments at the forefront of sustainable development. Green bonds are a form of sustainable finance, a debt instrument that raises capital to be used specifically for projects that offer environmental or climate benefits. They are akin to regular bonds in form but differ in their pledge to finance projects that add to environmental sustainability, including renewable energy developments, energy efficiency, sustainable agriculture, and conservation.

Green bonds have a particular prominence in today's global financial environment. This is an instrument that mobilizes private capital markets for activities, projects, or programs addressing critical environmental challenges such as climate change, biodiversity loss, and pollution. Green bonds therefore provide a viable mechanism for channeling the required financial resources into sustainable projects as countries around the world strive to meet their commitments under international agreements such as the Paris Agreement. This has precipitated exponential growth in the green bond market, with issuances reaching record levels as governments, corporations, and financial institutions realize the dual benefit of these instruments in supporting environmental goals and offering competitive returns to investors.

The emergence of the green bond market comes at a time when the global economy needs to pursue full speed toward a low-carbon economy. According to the IEA, for a probable achievement of the climate objectives of the Paris Agreement, there would be a need for more investment in clean energy and associated infrastructure by about $4 trillion annually up to 2030; green bonds are bound to play a key role in filling this financing gap. Also, they support financial markets in being compatible with sustainability goals and incentivize changing investor behavior toward long-term value creation from a focus on short-term profits [1].

This makes India one of the most important players in the green bond market, given its fast-growing economy and serious environmental issues. Ambitious targets have been set for renewable energy, including achieving 500 GW of nonfossil fuel capacity by 2030, which reflects the country's commitment to sustainable development. Green bonds have played an instrumental role in financing projects that contribute to such targets, positioning India as one of the leading issuers of green bonds among emerging markets.

1.2 Objectives of the Chapter

The present chapter focuses on an in-depth analysis of the development and existing state of the green bond market in India, placing the country within a wider global context. Based on the above observations, the focus areas of this chapter would be as follows:

- **To Understand the Global Evolution of Green Bonds**: This encompasses the historical beginning, major drivers, institutional agents, and positioning within the global financial market. This chapter will compare green bonds with more traditional instruments of finance, underlining their distinctive features and advantages.
- **For the Green Bond Market Development Analysis in India**: The chapter, therefore, critically examines the development of green bonds in India from a historical and regulatory point of view, the influence of the key player or regulator in the market dynamics of green bonds, notably the SEBI supported by other regulatory bodies, and recent reforms and policy initiatives that have influenced the issuance and performance of green bonds.
- **Impact Analysis of Green Bonds on Sustainable Development in India**: This chapter discusses the kind of projects that green bonds have financed in India. Therefore, it undertakes an analysis of these projects regarding their contribution toward the fulfillment of SDGs. It also presents case studies of successfully implemented projects under the financing of green bonds to depict their tangible benefits.
- **Evaluation of Problems and Opportunities Regarding the Indian Green Bond Market**: The chapter will critically assess a few major challenges for the Indian green bond market, such as issues related to transparency, standardization, and market depth, while exploring strategic opportunities that will make green bonds more attractive and effective in the country.
- **To Suggest Strategic Recommendations for Future Growth in the Green Bond Market of India**: The chapter, based on the analysis, shall put forward policy suggestions, market development programs, and education programs targeted at the growth of India's green bond market. The suggestions would be addressed to the policymakers, investors, and other stakeholders associated with sustainable finance.

1.3 Methodology and Structure of the Chapter

Employing a mixed-methods approach, this chapter combines quantitative data analysis of green bond issuances between 2017 and 2024 with qualitative insights from the regulatory landscape as well as investor sentiment. Data are sourced from SEBI, publications by the Climate Bonds Initiative, and credible financial news platforms.

Literature Review: The chapter starts with the literature review, presenting a summary of relevant studies, reports, and writings, which serves as a backdrop for the subsequent exploration of sustainable finance.

Empirical and Case Study Analysis: An analysis of India's green bond issuance trends with data from October 2023, and case studies from projects successfully financed through green bond issuance providing the analysis with context and evidence such as economic and environmental impact.

Market-Expanding Challenges and Opportunities: The chapter assesses impediments to strengthening the market such as secretive problems, regulatory vacuums, and investor anxieties, while signposting potential sectors of growth, including policy reform and new outlets for capital.

Strategic Recommendation: Harmonize policy apparatus to enhance investor confidence, regulatory clarity, and increase market penetration through policy and market-based recommendations. These are aimed at policymakers, regulators, and financial institutions.

2 Global Context of Green Bonds

2.1 Evolution and Growth of the Global Green Bond Market

Over the past twelve years, since its inception, the green bond market has only continued to show incredible growth and turn from a niche area of sustainable finance to a mainstream instrument for funding environmentally friendly projects. Green bonds are a type of bond instrument that first entered the market as early as 2007 under the name "Climate Awareness Bond" via the European Investment Bank. Basically, this was meant to start laying the framework for developing a market in debt instruments targeted at financing projects that could have impacts on the environment, such as projects on renewable energy, energy efficiency, and climate adaptation projects [2].

The green bond market grew at a tepid pace in those early years, with little interest owing to but a lack of standardized definitions and frameworks about what made a bond "green." This market began to take off after the first green bond issued by the World Bank in 2008. In 2008, the World Bank issued the first green bond to finance climate-related projects in developing countries. This landmark issuance demonstrated that green bonds could tap capital from institutional investors whose ESG standards were become more exacting.

In the past decade, the green bond market has expanded enormously in terms of issuance volume and geographic coverage. As of 2021, the global market size in green bonds was \$500 billion + of new issuances per year according to estimates by the Climate Bonds Initiative, and current trends could now even imply a possibility of \$1 trillion already in 2023. This development is fueled by an increasing recognition of climate risks, growing demand for sustainable investment opportunities, and regulatory support in key markets like the European Union, the USA, and China [3]. The other important milestone, of great importance, was the adoption of the Green Bond Principles by the International Capital Market Association in 2014. The GBP sets guidelines for voluntary issuers regarding the use of proceeds,

Table 1 Evolution and growth of the green bond market: key developments and milestones

Aspect	Details
Introduction	Green bonds have evolved from a niche sustainable finance tool to a mainstream funding instrument for green projects. The European Investment Bank (EIB) launched the first green bond in 2007 under the Climate Awareness Bond
Early challenges	The market initially struggled due to a lack of standardized definitions and frameworks. The World Bank issued the first official green bond in 2008 to support climate-focused projects in developing nations
Market growth	Over the past decade, green bond issuances have increased significantly, exceeding $500 billion annually by 2021. Projections estimated the market could surpass $1 trillion in annual issuances by 2023 [7]
Key regulatory milestone	Developed by the International Capital Market Association (ICMA), the Green Bond Principles (GBP) are a set of voluntary guidelines established in 2014 for issuers of green bonds regarding the use of proceeds, project evaluation, and reporting. This helped to standardization and increase investor confidence [8]
Diversification of Issuers	The market has expanded beyond financial institutions to corporations, municipalities, and sovereign governments. Countries like France, Poland, and the UK are issuing sovereign green bonds to finance their national climate strategies [9]

Source Authors' compilation

project evaluation, and, consequently, project selection, management of proceeds, and reporting. The principles contributed to market harmonization and an increase in investor confidence and can be observed in Table 1 summarization.

The green bond market in India has grown significantly, and there is a cumulative issuance of $21 billion as of Feb 2023 led by the private sector (84%) [4]. Nonetheless, about $600,000 was raised through green bonds in the first half of 2024, which pales in comparison with figures in China [5]. In the meantime, there has been innovation in bond structures present in the market, such as green securitizations and most recently, sustainability-linked bonds, where the financial terms of the bond are linked to the attainment of certain environmental objectives set by the issuer [6].

2.1.1 Comparison with Traditional Financial Instruments

A green bond differs from other financial instruments in a number of ways. Oriented to the environmental aspect and the purpose of using the proceeds, it also differs from traditional bonds in that the latter are issued for capital raising to meet general corporate or governmental purposes, whereas a green bond is specifically used for projects that bring forth clear environmental benefits. This specific use of proceeds is what differentiates green bonds and makes them desirable for investors who want to make their investments compatible with their sustainability objectives [10].

As of 2021, India's green bond market was valued at $19.5 billion, accounting for the major part of the existing $24 billion green, social, and sustainability (GSS) bond market in the country. But a look at China shows a vast difference in the scale of the market and growth paths [11]. By 2024, the total amount of green bonds issued by China cumulatively had far surpassed India's total. This included an extraordinary issuance of more green bonds in the first quarter of 2024 than India had cumulatively issued to date, illustrating the difference between the two nations in terms of market growth and investor interest [5]. The difference highlights how much faster China has accelerated its green finance sector—driven, perhaps, by greater policy incentives, greater institutional demand, and a more established base of investors, yet for India, it is just an emerging market, relatively earlier in the life cycle [12].

3 Green Bonds in India: Development and Current Scenario

3.1 Historical Development of Green Bonds in India

Though very recent, India's entry in the green bond market has been very rapid, making the country a front-runner among emerging markets in sustainable finance. The year 2015 saw Yes Bank releasing the first-ever green bond in India and that was the beginning of one of the significant milestones in the financing of ecologically viable projects in the country. This bond issuance worth INR 1,000 crore (~USD 160 million) was mainly to raise capital for renewable energy projects and was relevant to India's overall nationally determined contributions to increase the renewable energy capacity [2].

Furthermore, the green bond market in India grew by government emphasis put on sustainability and climate action. The articulation of some ambitious targets in the renewable energy sector—like 175 GW by 2022 to 500 GW by 2030—gave a strong fillip to green bond issuance in India. Together with growing investor interest in sustainable finance, this is gradually ramping up the issuance of green bonds over the years [4].

3.2 Role of the Securities and Exchange Board of India (SEBI)

SEBI played an instrumental role in the development of the regulatory framework governing green bonds in the country. Since the SEBI recognized that sustainable finance was gaining momentum, it laid out guidelines for green bonds in 2017, thus giving a formal shape to the issuance of green bonds in India. At the time, the guidelines were an important milestone because they defined the eligible category of green bonds and disclosure and reporting requirements of issuers [13].

3.3 Recent Reforms and Policies

Recent reforms also relate to developing an effective regulatory regime that can make the green bond market work in a way that is transparent and efficient. For example, SEBI has from time to time issued amendments to its guidelines with a view to meeting new challenges and best international practices. In 2021, SEBI added more disclosures related to green bond issuance, such as the requirement for the issuer to specify environmental objectives for projects and expected climate impact [14].

In February 2023, SEBI broadened the category of green debt securities to include blue bonds, yellow bonds, and transition bonds to align with global sustainability norms [15]. SEBI also required the appointment of independent third-party reviewers, both prior to and after issuance, to improve transparency and investor confidence [13].

It was very proactive in advocating for green bonds as a means of sustainable development in international forums. The membership of various global initiatives by India, such as the International Platform on Sustainable Finance, gives sufficient proof that it is putting effort into integrating sustainability into its financial markets [16].

3.4 Current Market Dynamics and Key Stakeholders

Over time, the Indian green bond market evolved and grew in terms of more number of issuers with a diversified investor base in the last couple of years. In 2023, the green bond market in India would be dynamic, comprising a set of different types of actors: issuers, investors, regulators, and external reviewers.

Despite an initial wave of excitement, investors' demand for 10-year green bonds has been cooling in recent auctions, due to a lack of incentives and market-driven expectations about yield [17]. On the other hand, 30-year green bonds face more demand from insurance companies with longer time horizons [18].

The Indian green bond market scale has also been broadened from just banks to contain corporates, public sector, and state governments financing a variety of projects including solar parks and green buildings. Institutional investors such as pension funds, insurance companies, and mutual funds are increasingly investing in green bonds for their ESG portfolios, and there has been growing retail participation [19]. SEBI acts as the market watchdog, ensuring regulatory compliance and transparency, while the RBI furthers the growth of sustainable banking. External reviewers, notably CRISIL and Moody's, validate use of proceeds and environmental benefits, increasing market credibility. Together, this ecosystem mobilizes capital for green initiatives to help India transition to a sustainable economy [20].

4 Projects Financed by Green Bonds

4.1 Types of Projects Supported

Green bonds have become one of the most effective financial instruments for projects ranging from high environmental sustainability. In India, the majority of green bonds have been issued to refinance renewable energy projects which are aligned with India's commitment to reduce its carbon footprint and meet the international targets on renewable energy. But slowly, the green bonds coverage has grown to include projects increasingly useful to the environment. A number of various key areas in which the whole project can be allocated are as follows:

- **Renewable Energy**: This is, till now, the most prevalent application for the proceeds of green bonds in India. The projects cover solar power plants, wind farms, and small hydropower projects. With India committed to installing 500 GW of renewable energy by the year 2030 [21], green bonds have played a stellar role indeed in mobilizing funds for these commitments.
- **Energy Efficiency**: Green bonds finance energy efficiency projects, which are performed by the upgrading of the already existing infrastructure with the core aim of reducing energy consumption. This includes retrofitting buildings with energy-efficient technologies such as LED lighting, heating ventilation, and air-conditioning systems with high energy efficiency besides the upgrading of industrial processes to reduce the wastage of energy.
- **Sustainable Transportation**: Another very critical area in which green bonds have contributed much is sustainable transportation. This, in turn, includes financing mass transit systems, electric vehicle infrastructure, and low-carbon transportation solutions. A particular example of the use of green bonds in financing metro rail projects across various Indian cities with a view to overcoming urban congestion while reducing GHG emissions [22].
- **Water Management and Sustainable Infrastructure**: Green bonds in India have been used to finance projects related to water management and the development of sustainable infrastructure. It covers construction in sewage treatment plants, water purification systems, and other infrastructural projects that help raise resilience due to climate change–flood control systems and sustainable urban drainage [23].
- **Waste Management**: Green bonds have also gained applications in waste management for recycling projects, waste-to-energy initiatives, and the management of hazardous waste in a secure manner. This helps reduce environmental pollution and further furthers the concept of a circular economy [24].
- **Biodiversity Conservation and Land Use**: A little less common, but some green bonds have been used to finance activities such as biodiversity conservation and/or the sustainable use of land, for instance, the protection of natural habitat, land restoration, and sustainable agriculture [25].

4.2 *Case Studies of Successful Green Bond Financed Projects*

- **ReNew Power Green Bond Issue (2021)**: It is one of the largest independent renewable energy powers in India, which had issued green bonds aggregating to USD 585 million during 2021. The proceeds had been utilized for financing construction of solar and wind power projects across the country that contribute to the renewable energy capacity of India. Not only was it large in size, but also it had drawn encumbering demand from international investors—showing their increasing faith in India's green energy sector [26].
- **NTPC's Green Masala Bond (2017)**: In a first-of-its-kind issue, NTPC, India's largest power utility, issued a green masala bond worth INR 20 billion (~USD 300 million) on the London Stock Exchange. Proceeds were subsequently utilized to finance renewable energy projects, including solar and wind farms, as part of NTPC's strategy in diversifying its energy portfolio and reducing its reliance on coal. The deal was one of the first major green bond issuances by an Indian public sector enterprise, reinforcing the potential for green bonds to open up international investment flows into the Indian clean energy sector [27].
- **Ghaziabad Municipal Corporation Green Bond**: First-ever green bond of Ghaziabad Municipal Corporation made it the first ever by an urban local body in the country. It mobilized INR 150 crore, approximately equivalent to USD 20 million, which would go to fund a tertiary water treatment plant, distributing treated water for industrial use, thus reducing the load on fresh water supply in the city. This was done with the objective of presenting a promise for municipal green bonds in the funding of sustainable urban infrastructure projects in India [28].
- **DMRC Green Bonds**: DMRC issued several green bonds over these years to finance the continuously expanding metro network in Delhi, one of the most polluted cities in the world. These have typically financed new metro lines that help reduce traffic congestion and, consequently, vehicle emissions in the capital [29]. DMRC's use of green bonds underlines the role of sustainable transport in India's urban development strategy.

4.3 *Impact on Sustainable Development Goals (SDGs)*

It is due in large part to this fact that projects financed through green bonds have been able to contribute significantly so far to the attainment of United Nations Sustainable Development Goals, particularly those related to climate action, clean energy, and sustainable cities and communities can be observed in Table 2.

As conclusion, green bonds have emerged as a key source of financing toward meeting the larger share of the project cost of various development projects in India. From renewable energy to sustainable urban infrastructure, the range of projects they

Table 2 Impact of green bonds with SDGs

Sustainable development goal (SDG)	Impact of green bonds
SDG 7: Affordable and clean energy	Funds solar, wind, and renewable energy projects, increasing clean energy share and reducing fossil fuel dependency
SDG 11: Sustainable cities and communities	Supports metro rail, water treatment plants, and sustainable urban infrastructure, reducing pollution and enhancing resilience
SDG 13: Climate action	Encourages carbon reduction, energy efficiency, and sustainable transport to mitigate climate change impacts
SDG 6: Clean water and sanitation	Finances water treatment and sanitation projects, improving water management in urban centers
SDG 12: Responsible consumption and production	Funds waste management and recycling projects, promoting sustainable consumption and reducing environmental waste

Source Authors' Compilation

fund illustrates both their flexibility and their potential to work in the varied contexts of environmental challenges in the country. As India embarks on ambitious climate and sustainability commitments, the role of green bonds in mobilizing necessary financial resources will also become more critical.

5 Strategic Recommendations

5.1 Policy Recommendations for Enhancing Effectiveness

Several key policy recommendations can be implemented to further enhance the effectiveness of the green bond market in India. The development of these recommendations will be based on existing challenges and lessons learned in order to unlock the full potential of green bonds as a means of financing sustainable development.

- **Strengthening Regulatory Frameworks**: Even though SEBI has already notified the guidelines with respect to green bond issuance, these need further refinement to bring in consistency and transparency. A necessary certification process for the green bonds needs to be brought into place, pertaining to internationally accepted standards like Climate Bonds Standard. This will avoid the issue of "greenwashing" and make sure that only genuinely green projects get financed.
- **Tax Incentives and Subsidies**: Government could consider introducing tax incentives on green bonds both for the issuer and investor. For the issuers, this may be in the form of reduced tax rates on interest income earned by the green bonds, or even a subsidy on the projects that have been funded through green bonds. In

the case of investors, tax exemption returns from investment in green bonds can make these instruments more attractive. Incentives of this kind, fiscal in nature, would help to increase market supply and demand for the green bond market.

- **Linkage with National Climate Objectives**: Green bonds should be more clearly integrated with India's national objectives on climate change and international commitments, such as the Paris Agreement, which may include specific targets for green bond issues within greater national strategies for renewable energy, sustainable infrastructure, and climate resilience. Additionally, the government should accord priority to projects which contribute directly to the achievement of those targets.
- **Improved Reporting and Disclosure**: There is a requirement to further work upon the reporting and disclosure related to green bonds to help inspire better confidence among investors. This could include the development of a centralized platform for tracking the usage of proceeds and environmental performance of projects financed by green bonds. The platform can be managed by a regulatory body like SEBI or an independent third-party provider who can provide investors with credible and easy-to-access information.
- **To the Local Government and Municipality**: Encouraging local governments and municipalities to increasingly issue green bonds could be one way to partly bridge the sustainable urban infrastructure funding gap. In this regard, the central government should provide capacity-building programs, financial guarantees, and other forms of technical support to municipal green bond issuances.

5.2 Market Development Programs

The implementation of targeted market development programs is essential to ensure the sustained growth of the green bond market in India. These initiatives should address structural challenges and promote investor confidence. Increasing awareness can enhance market liquidity and transparency, enabling investors to trade efficiently via a dedicated green bond exchange or trading platform. Issuance is also expected to be more efficient with standardized templates for documentation and disclosure, which will lower transaction costs.

Encouraging corporate participation, especially from high carbon industries that are moving toward more sustainable practices, corporate issuance can be spurred by government-led incentives, green bond challenges, and partial credit guarantees, leading to further engagement with sustainable financing. Symbiotic with this are financial innovation like sustainability-linked bonds and green sukuk that can attract a wide array of investors, matching bond terms to environmental targets.

Promoting PPPs in green bond-financed projects has the potential to leverage private sectors know-how and funds in the public infrastructure. By appropriately structuring a PPP framework with clear guidelines around risk-sharing and performance metrics, the financing gap for large-scale sustainable projects can be closed.

Global cooperation is just as important. India must become active in the global green finance community, working with organizations like the International Finance Corporation and Green Climate Fund to access co-financing/technical support. Solidifying international partnerships will allow India to have a prominent seat at the table within the green finance ecosystem, attracting investment and expertise.

5.3 Educational Campaigns and Stakeholder Engagement

Education and stakeholder engagement will be key to growing India's green bond market through increased awareness, capacity building, and sustainability.

Investor Awareness: The introduction of green bonds can be accompanied by targeted campaigns, such as workshops, webinars, and other informational materials, that can help to educate investors on the benefits of green bonds, associated risks, and how green bonds fit into their portfolios. Particular efforts should target retail investors who are less familiar with sustainable finance.

Capacity Building for Issuers: A number of public sector and smaller firms lack capacity to issue green bonds. Issuers can benefit from training programs, technical assistance, and partnerships with industry experts in project selection, structuring, and reporting (both qualitative and quantitative) requirements.

Market Trends Discussion: Corporates, financial institutions, and NGO(s) collaborate through forums and roundtables to discuss the trends in different markets.

Academics: Schools of business and universities should offer courses on sustainable finance, ESG investing, and green bond issuance to help develop a workforce proficient in green finance.

Public Awareness Campaigns: Media-driven campaigns, through social media outreach and case studies, can highlight successful green bond projects, promoting wider acceptance.

These initiatives will enhance India's green bond ecosystem, attract investments, and further climate and sustainability objectives, establishing India as an international leader in green finance.

6 Future Prospects and Conclusions

Future Trends in Green Bond Market in India

An enabling regulatory environment, together with fast-growing investor interest and ambitious climate change targets, will combine to create a sudden exponential growth in the Indian green bond market. A set of emerging trends is likely to shape the future course for this market, enabling India to keep pace at the forefront of world sustainable finance.

1. **Diversification of Issuer Base**: The issuer base for green bonds in India is likely to go beyond the traditional players like banks and large corporations into a broader set of entities. This would include many more public sector undertakings, state governments, and urban local bodies issuing green bonds to finance infrastructure projects. Besides, SMEs could also increasingly tap the green bond market as awareness and capacity-building efforts gather steam.
2. **Integration with Digital Finance**: The application of green bonds combined with new digital finance technologies like blockchain will gain speed. Blockchain technology could add value to green bonds by offering traceability of proceeds, ensuring it is indeed spent for the designated purpose and reducing greenwashing. This will most probably attract more tech-savvy investors into this market and enhance its reputation in general.
3. **Emphasis on Sustainable Infrastructure**: With rapid urbanization in India, the concentration would turn to ensuring sustainable infrastructure through green bonds in respect of transportation that is clean, renewable energy, and water management. There would be a focus on building resilient cities which are adaptable to the effects of climate change while economically viable.
4. **Thematic Bonds Emerge**: Not only this, but the thematic bonds focused on the specific environmental challenge that is beyond traditional green bonds will also mark their presence in the market, such as blue bonds for ocean conservation and biodiversity bonds. These thematic bonds will be provided toward niche investor segments and contribute to India's broader environmental goals like the protection of marine ecosystems and preservation of biodiversity.
5. **International Cooperation and Investment**: India is likely to enhance its collaboration with international financial institutions and investors for the scaling up of its green bond market. Involvement in global green finance initiatives and platforms will contribute to foreign investment, bringing much-needed capital for the country's climate and development goals. In this way, the sovereign green bonds could also contribute significantly and attract funds for big national projects
6. **Strengthening of Policy and Regulatory Framework**: The regulatory environment for green bonds is bound to evolve in India, with possible enhancements targeted to converge toward best practices globally. This may include stringent disclosure-related requirements, incentivizing the issuers, and national green taxonomy development. Consequently, this would lead to greater market standardization, appealing both to domestic and international investors.

7 Concluding Thoughts on Green Bonds as a Tool for Sustainable Development

The green bond has gained momentum as one of the useful financial instruments to mobilize capital for projects supporting sustainable development. In India, they have emerged as a key source for renewable energy, sustainable infrastructure, and other environmentally relevant projects that parallel the country's commitments to addressing climate change and attaining the SDGs.

The green bond market would continue to build up in India; there was, however, a dire need to pay strong attention to transparency, accountability, and the measurement of impacts for environmental deliveries. Further innovation, enabling regulatory frameworks, and active stakeholder engagement will be required to successfully bring about its expansion.

In other words, the green bonds project is a unique opportunity for having the financial markets aligned with the sustainable development goals by creating already social and economic win–win scenarios. Interlinking India's status as a nascent but growing green bond market with the objective of creating and sustaining a durable and inclusive economy is important from a financial perspective as well as from the perspective of fostering an environmentally sustainable growth trajectory in the economy. You will be financing by the market in future, which will depend more and more on its role in facing the challenges of climate change and environmental degradation that have now confronted the whole world.

References

1. Khurram, M. U., Xie, W., Mirza, S. S., & Tong, H. (2023). Green bonds issuance, innovation performance, and corporate value: Empirical evidence from China. *Heliyon, 9*(4), e14895. https://doi.org/10.1016/j.heliyon.2023.e14895
2. Aneja, R., Kappil, S. R., Das, N., & Banday, U. J. (2023). Does the green finance initiatives transform the world into a green economy? A study of green bond issuing countries. *Environmental Science and Pollution Research, 30*(14), 42214–42222. https://doi.org/10.1007/s11356-023-25317-w
3. Lin, L., & Hong, Y. (2022). Developing a green bonds market: Lessons from China. *European Business Organization Law Review, 23*(1), 143–185. https://doi.org/10.1007/s40804-021-00231-1
4. Hussain, F., & Dill, H. (2024). *India incorporates green bonds into its climate finance strategy*. World Bank Blogs. https://blogs.worldbank.org/en/climatechange/india-incorporates-green-bonds-its-climate-finance-strategy
5. Bharucha, Mishra, P.-V., & Nangru, P. (2024, December 9). *Slow growth of green deposits and green bonds in India*. Lexology. https://www.lexology.com/library/detail.aspx?g=3b859fca-b9e0-405a-9004-e8e3ac693123
6. Khurram, M. U., Xie, W., Mirza, S. S., & Tong, H. (2023). Green bonds issuance, innovation performance, and corporate value: Empirical evidence from China. *Heliyon*, 9(4), e14895. https://doi.org/10.1016/j.heliyon.2023.e14895

7. Chen, X., Weber, O., & Saravade, V. (2022). Does it pay to issue green? An institutional comparison of Mainland China and Hong Kong's stock markets toward green bonds. *Frontiers in Psychology, 13*, 833847. https://doi.org/10.3389/fpsyg.2022.833847
8. Piñeiro-Chousa, J., López-Cabarcos, M. Á., Caby, J., & Šević, A. (2021). The influence of investor sentiment on the green bond market. *Technological Forecasting and Social Change, 162*, 120351. https://doi.org/10.1016/j.techfore.2020.120351
9. Datta, S. (2024). An empirical assessment of India's position in global sustainable bond market. *Journal of Applied Economic Sciences (JAES), 19*(16), 369. https://doi.org/10.57017/jaes.v19.4(86).02
10. Bansal, S., Mani, S. P., Gupta, H., & Maurya, S. (2023). Sustainable development of the green bond markets in India: Challenges and strategies. *Sustainable Development, 31*(1), 237–252. https://doi.org/10.1002/sd.2386
11. Statista. (2024). *India: Green, social, and sustainability bond market size by type*. Statista. https://www.statista.com/statistics/1376192/india-green-social-and-sustainability-bond-market-size-by-type/
12. Kumar, S. (2022). Strategy and corporate governance for risk management. *SSRN Electronic Journal*. https://doi.org/10.2139/ssrn.4135582
13. Srivastava, S. (2024). *Indian capital market regulator's updated green debt guidelines: Unlocking the potential of sustainable finance*. https://ieefa.org/resources/indian-capital-market-regulators-updated-green-debt-guidelines-unlocking-potential
14. Paez. (2023). *Setting standards in the sustainable bond market: From guidelines to regulation, on the shoulders of giant principles?* https://insights.issgovernance.com/posts/setting-standards-in-the-sustainable-bond-market-from-guidelines-to-regulation-on-the-shoulders-of-giant-principles/
15. juriscorp.in. (2023, February 13). *NCS regulations amended by SEBI*. Juris Corp | Advocates & Solicitors. https://www.juriscorp.in/ncs-regulations-amended-by-sebi/
16. Nader Naifar, A. E. (Ed.). (2023). *Green finance instruments, FinTech, and investment strategies, sustainable portfolio management in the post-COVID Era*. https://doi.org/10.1007/978-3-031-29031-2
17. Dhutia, D. (2024, November 29). India cenbank devolves 70% of new green bonds, cutoff below 10-year note. *Reuters*. https://www.reuters.com/world/india/india-cenbank-devolves-70-new-green-bonds-cutoff-below-10-year-note-2024-11-29/
18. Dhutia, D. (2025, January 29). India bond investors eye small rise in government borrowing, expect yields to decline. *Reuters*. https://www.reuters.com/world/india/india-bond-investors-eye-small-rise-government-borrowing-expect-yields-decline-2025-01-29/
19. Manglunia, A. (2023). *Unlocking full potential of green bonds in India—The Economic Times*. https://economictimes.indiatimes.com/markets/bonds/unlocking-full-potential-of-green-bonds-in-india/articleshow/101541808.cms?from=mdr
20. Lehner, O. M., Harrer, T., Silvola, H., & Weber, O. (Eds.). (2023). *The Routledge handbook of green finance*. Routledge. https://doi.org/10.4324/9781003345497
21. PIB, D. (2023). *Steps taken to increase production of renewable energy and for setting up pumped storage projects*. https://pib.gov.in/pib.gov.in/Pressreleaseshare.aspx?PRID=1989807
22. Kwasniok, R. (1920, December 30). Green bonds. *Changing transport*. https://changing-transport.org/green-bonds/
23. Brears, R. (2024). *Financing water security: How green bonds and loans propel sustainable water management*. https://www.linkedin.com/pulse/financing-water-security-how-green-bonds-loans-propel-robert-brears-vcixf
24. Liu, Y. (2024). *Does the issuance of green bonds nudge environmental responsibility engagements? Evidence from the Chinese green bond market | Financial Innovation*. https://jfin-swufe.springeropen.com/articles/https://doi.org/10.1186/s40854-024-00620-8
25. Thompson, B. (2022). *Impact investing in biodiversity conservation with bonds: An analysis of financial and environmental risk—Thompson—2023—Business Strategy and the Environment*. Wiley Online Library. https://onlinelibrary.wiley.com/https://doi.org/10.1002/bse.3135

26. Nair, R. (2021). *ReNew power to raise $585 million through dollar green bonds*. https://www.mercomindia.com/renew-power-raise-585-million-green-bonds
27. NTPC. (2016). *NTPC lists world's first green Masala bond by an Indian issuer on London Stock Exchange*. https://www.londonstockexchange.com/discover/news-and-insights/about:blank
28. ET. (2021). GMC green bonds: Ghaziabad Municipal Corp lists first green bonds. *The Economic Times*. https://economictimes.indiatimes.com/markets/stocks/news/ghaziabad-municipal-corp-lists-first-green-bonds/articleshow/81974055.cms?from=mdr
29. Kapoor, A., & Kumar, N. (2023). Delhi metro: Going the electric extra mile: Delhi Metro's last-mile connectivity. *The Economic Times*. https://economictimes.indiatimes.com/industry/renewables/going-the-electric-extra-mile-delhi-metros-last-mile-connectivity/articleshow/97133952.cms?from=mdr

Entrepreneurship for Engineers: Supply Chain, Strategies of Circular Economy and Innovation as a Competitive Advantage for Structural Engineering Consultants, Road and Railway Companies (Construction Industry)

Omid Reza Baghchesaraei, Alireza Baghchesaraei, and Amirali Khademi

Abstract Today, due to technological advancements and changes in the level of living, people's needs have changed in all areas. Structural engineers are equipped to develop and integrate their designs with those of other professionals, ensuring cohesive and functional outcomes. It aims to provide opportunities for the rapid and organized growth of innovation in this field by analyzing the gap between the current and desired levels. The circular economy system has been proposed as an alternative for linear economy; this circular system can reduce the economic, environmental, and social burden of the projects. This reduction is achieved several times with the triple principles of reduction, reuse, and recycle, which the road and construction industry, as one of the largest industries in the world, can accept these principles. The road and construction industry, by consuming about 30% of the total primary resources and producing more than 25% of the total solid residual in the world, plays a prominent role in creating all kinds of effects and burdens, and applying the circular economic principles in this industry can reduce these burdens and increase the productivity.

Keywords Circular economy · Construction industry · Innovation · Sustainable development · Entrepreneurship

O. R. Baghchesaraei (✉)
Centre for Infrastructure Engineering, Sydney, Australia
e-mail: o.baghchesaraei@westernsydney.edu.au

A. Baghchesaraei
Infrastructure Management (Engineering), Construction Industry, Sydney, Australia
e-mail: baghchesaraeialireza@gmail.com

A. Khademi
Department of Civil Engineering, Sharif University of Technology, Tehran, Iran
e-mail: amirali.khademi81@sharif.edu

K. Khanna et al. (eds.), *From Poverty to Prosperity*, Information Systems Engineering and Management 64, https://doi.org/10.1007/978-3-031-99166-0_5

1 Introduction

To succeed in structural engineering under today's conditions, it is essential to transform organizational activities, particularly in the leadership and management practices [1]. Currently, one of the concerns for organizations is to implement strategies guaranteeing their success and survival in evolving environmental conditions. The actual conflict between countries in the next decades will not be on energy nor on markets, but it will be on human resources. Therefore, countries more than ever need people whose talents and inherent capabilities are developed so that they well possess knowledge, skill, and more importantly, the ability to creative thinking. Maintaining competitive advantage in structuring, research, and industrial companies depends on innovation [2]. So, it can be argued that based on the fact that the majority of the third millennium organizations and research-oriented, what the structural engineering desperately needs at the present age to be able to have a place in today's dynamic and complex markets and obtain competitive advantage of itself is the ability to producing structures and new, unique, and competitive services in today's changing markets. This will not be achieved unless by innovation and creativity of the structural engineers, as well as their effect on boosting their own team's creativity, so that they try together toward the growth and development of their company. The business strategy of any organization, which is directed in this context, has an important role in guiding the organization to reach the desired position among the competitors. As previously noted, creativity and innovation are crucial for organizational improvement, long-term survival, and competitive advantage. Therefore, effectively institutionalizing and aligning them with organizational goals to formulate and implement competitive strategies is essential. This study aims to identify strategies for increasing creativity and innovation within structural engineering, fostering entrepreneurship, evolution, and dynamism in this competitive field. Evaluating productivity capabilities reveals growth opportunities, enhances innovation, directs resources toward advanced technology, and informs strategic planning.

2 Literature Review

As mentioned above, structural engineers are tasked with the creative and efficient utilization of funds, structural components, and materials to achieve these objectives [3]. Structural engineers can operate in various fields, referred to below:

- Consulting companies, key players in building design, significantly influence customer demand. Structural engineers can leverage science and technology to improve designs, incorporating new technologies to enhance construction speed, quality, and project applications. The Burj Khalifa, for example, considered by many an "engineering" masterpiece and once deemed impossible, demonstrates how innovative design and analysis software can lead to successful completion, even for complex projects.

- Contractors, as primary construction project executors, utilize management techniques and technical specialists. The tunnel template method, initially chosen by a contractor for mass-build projects, is now a highly desirable approach due to its speed.
- Manufacturers are crucial to civil engineering. Structural engineers help meet evolving building industry needs by improving materials, the core of construction. Material suppliers must stay attuned to market demand, as human needs and the need for innovative materials constantly change. Structural engineers, key players in building production, must leverage existing materials to create new ones, such as the increasingly important prefabricated structures and panels.
- Many structural engineers now work in sales, providing construction materials and technologies. This involves identifying market demand, sourcing domestic or imported products, and even exporting. Crucially, their main role is offering technical support for these products.
- Education is another sector, and large number of structural engineers are working in this field, which represents training programs for students and researchers who are trying to educate the engineering community. These engineers can provide students with a variety of strategies by applying the methods, technologies, and modern materials of the world. Conducting and participating in seminars and conferences can be a key factor in improving scientific knowledge in this field [3].
- Creativity is producing something both novel and valuable, whether tangible (e.g., a painting) or intangible (e.g., an idea). Sternberg and O'Hara's research highlights six influencing factors: deepening knowledge, intellectual flexibility to connect ideas, an innovative thinking style, intrinsic motivation to realize ideas, a persistent and independent personality, and a supportive environment.

3 Sustainable Development

Sustainable development is a process aimed at societal stability in resource utilization, ensuring replenishment. It encompasses economic growth and human development, seeking continuous progress beyond basic economic gains [4]. It prioritizes the sustainability of nonrenewable resources for future generations, creating a positive future by integrating economic, social, and environmental factors (Fig. 1).

4 Innovation

Innovation is often defined as a "new idea, device, or method." Innovation can be described as something original and more effective that, as a result, introduces itself into the market or society.

Fig. 1 Sustainable development in construction industry. *Source* Created by the author

Afuah believes that the purpose of innovation in using the new technology tools, and the market knowledge is to deliver a new product or service to the customers [5]. Tarek Khalil also believes that innovation means delivering a new product, service, or process to the market by creating new applications for the existing technology or creation and commercialization of new technologies [6]. According to Shompiz, innovation is not even limited to technical aspects; it includes organizational aspects as well [7]. Innovation refers to the creativity that has been manifested and come into practice. In other words, innovation means the realized creative idea; it is delivering new products, processes, and services to the market; it is utilizing intellectual abilities to create a new thought or concept [8]. Innovation is the process of turning creativity into practical application, from initial idea to commercialization. Entrepreneurs use innovation to transform opportunities into marketable products or services, driving change. Innovation is realizing and implementing new ideas. It takes many forms, including fundamental, process, product, production, service, and marketing innovation.

According to Shompiz, different kinds in innovation in structural engineering are [9] as follows:

1. Innovation in materials.
2. Innovation in structure and building.
3. Innovation in methods, processes, execution, analysis, design, and building the structure.
4. Innovation in executive management of the structure.
5. Innovation in the market.

Innovation is becoming more and more a key competence in the competitive environment of construction. Several studies have already focused on the barriers to and enablers of innovation in construction firms and projects (Tatum [10]; Slaughter [11]; Ling [12]; Bossink [13]; Blayse and Manley [14]). Van de Ven [15] defines the process of innovation as "the development and implementation of new ideas by people who over time engage in transactions with others within an institutional context." Within this process, Van de Ven recognizes four central problems in the management of innovation: drawing attention to new needs and opportunities, managing ideas into good currency, managing relationships, and institutional leadership.

4.1 Creativity and Innovation in Structural Engineering

Structural engineers, vital for building stability, also contribute significantly beyond basic requirements. Skilled professionals add value, enhance functionality, mitigate project risks (cost and time), and expand architects' creative options. Innovative engineers are creators, exploring new ways to use materials. Their expertise is invaluable for meeting client needs and realizing architectural visions. Creativity and innovation are intertwined: Creativity generates ideas, and innovation implements them. Essentially, creativity is ideation, and innovation is the realization of that idea [16]. According to management experts, creativity process in structural engineering, the process of turning thought into action, can be divided into three stages: (1) creating an idea, (2) developing the idea, and (3) implementing the idea. Practical creativity in structural engineering, however, involves five stages: (1) information acquisition, (2) inspiration, (3) testing, (4) treating, and (5) delivering.

4.2 Creativity Elements in Structural Engineering

Creativity consists of different elements and components. Amabile has identified three elements as the main components of creativity:

- The skills related to the domain or the subject of skills related to the domain are: knowledge and perception of the subject, facts, principles, opinions, and ideas emended in that subject.
- The skills related to creativity: If a person possessed the highest level of skills, but lacks the creativity and innovation, they will never be able to accomplish a

Table 1 Elements of creativity in structural engineering

Motivation	Creativity skills	Subject skills
Attitude toward working is one's perception of their motivation to perform the task, which depends on: – The initial amount of internal motivation to work – The presence/absence of certain external factors in the social environment – The ability of individuals to reduce environmental restrictions	The methodology proportional to recreational and implicit information of exploration method in creating new ideas The working procedure of the manufacturer depends on: – Training – Expertise in creating new ideas – Personality characteristics	Understanding the subject of necessary technical skills to perform a task. The talent of that subject depends on: – Cognitive abilities – Movement, perception, and innate skills – Formal and informal education

Source Authors compilation

creative task. The skills related to creativity utilize the subject skills in a new way by crossing the frameworks.

Giving up a habit by the structural engineer.

Postponing ideas evaluations to prevent the initial loss of structural ideas.

Perception of the complexities based on the structures with more complex architecture and their involvement.

Looking at the structures and building from different perspectives.

Breadth of thought and making relationships among different structural ideas.

- The skills related to motivation: Motivation is one of the essential and most important components in this context. A structural engineer cannot have new and creative thoughts and ideas without internal and external motivations.

Table 1 describes the elements of creativity in structural engineering.

4.3 Entrepreneurship

Entrepreneurship is the process of developing and implementing new ideas, leveraging available resources, and taking risks. Entrepreneurs gather the necessary resources (land, labor, materials, capital) and use their skills and decision-making power to design, organize, launch, and manage a new venture, generating profit. Entrepreneurs are not necessarily capitalists but effectively utilize existing capital. They may be inventors, experts, or possess other skills.

Jeffrey Timmons believes that the entrepreneur creates a valuable thing from nothing. Bill Gates states that most solutions are discovered in another place, so they have to be identified and transferred from their discovery domain to our own domain. This is what we call entrepreneurship. Factors affecting structural engineering entrepreneurship include: (1) economic and market conditions, (2) the building industry's structure and dynamism, (3) Engineering Council regulations, (4) social capital, and (5) entrepreneur-related factors.

Entrepreneurship is actually the commercialized invention. In fact, entrepreneurship = creativity + innovation. This kind of entrepreneurship is a process in which creativity starts with a technology idea in the structural engineering and reaches a technology business. Factors affecting this process are structural engineer, the Engineering Council, society.

5 Economy in Construction

Sustainability's core principles—economic, environmental, and social—mean all engineering projects impact the environment. However, circular economy principles (reduce, reuse, recycle) can minimize environmental damage and operational costs [17]. It can help the sustainability of projects by using the link between economy and environment and reducing the consumption of raw materials and can be used to achieve sustainability goals. These triple principles are defined as follows [18]:

- Reduce: Reducing the amount of input of primary raw materials into production systems and reducing energy consumption. This is achieved through increasing productivity in production.
- Reuse: Using wastage and products with multiple applications in different stages of production. This is achieved through the maximum use of the capacity of a product.
- Recycle: Making changes in a used material and creating conditions for its use as a substitute for non-recycled raw materials.

The circular economy analyzes systems encompassing people, materials, and science, examining material input, production, consumption, and disposal networks. It replaces the linear "take-make-dispose" model with a circular system based on reducing, reusing, and recycling materials [19]. Of course, in some studies, circular economy principles are categorized into four principles of reuse, recycling, repair, and replacement, in which the repair principle is integrated into the rest of the principles in triple categories.

Beyond its core principles, the circular economy is a sustainable development model based on productive use and resource cycling. It emphasizes minimizing raw material needs and consumption, reducing greenhouse gas emissions, efficient resource use (materials, water, energy), and maximizing renewable resource utilization [20].

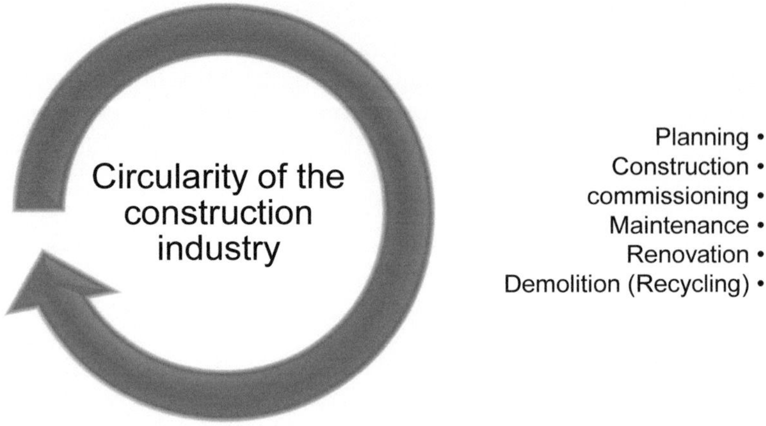

Fig. 2 The process of circularity of the construction industry. *Source* Created by author

5.1 Linear Economics in the Construction Industry

The construction industry currently operates on a linear economy model, consuming raw materials traditionally. This model starts with raw material extraction, proceeds through production and supply, creating products that are often difficult or costly to disassemble or repurpose at end of life. These products are assembled into buildings with a useful lifespan, generating waste during installation. Even at a building's end of life, many components retain functionality, but their full potential is not utilized [21].

5.2 Circular Economy in Construction Industry

The circular economy aims to lessen the environmental impact of buildings, from raw material production to construction, maintenance, and demolition. Recently, especially in the EU, policies have promoted increased reuse and recycling of construction and demolition waste [22]. Figure 2 shows the path toward circularization of the construction industry.

5.3 Assemble and Disassemble

Assembling and disassembling is a new approach in the construction industry, which is inspired by the automobile industry. Just as you can scrap a car at the end of its life and reuse the parts that still function, this is the case for buildings [23]. Designing buildings that can be assembled and disassembled at the end of their life has economic, environmental, and social benefits. It is to reduce the amount of wastage produced after demolition, as well as ease repairs and reduce maintenance costs [24].

In the mentioned design, the components of the building are considered as layers separable for replacement, reuse, or repair, without significant damage to the remaining layers [25]. Simple separation of components allows them to be recycled more easily or transported to their reuse site more quickly. This allows quick, convenient, and low-cost change of uses. NEXT 21 residential complex in Osaka, Japan, is an example of this type of design. This residential complex has the ability to accept all kinds of changes and new configurations. It is an integrated system including a variety of independent subsystems.

Contrary to the mentioned benefits, consumers may not have enough trust in opened components of buildings and may not show interest in using such components. This led to some challenges. Also, according to their defined value, the opened components cannot be moved for long distances due to the high cost of transportation, so they must be recycled or reused near the previous building site [26].

5.4 Adaptability and Flexibility

A key factor in architectural design is considering residents' and users' needs, which often evolve over time, necessitating building repairs [27]. Flexible initial designs that adapt to changing tastes and needs significantly reduce renovation costs by minimizing energy and material consumption, thus aligning with circular economic principles. In this regard. Table 2 describes the characteristics of an adaptable building.

Due to the reduction of energy consumption and raw materials as well as the recycling and reuse of structural components, the flexible and adaptive design of buildings helps to achieve the goals of sustainable development and provides the following facilities [28].

- Adapting the building spaces to the individual needs and preferences of the users.
- Adapting a single space in the building to the needs of different users.
- Adapting the building to the life cycle of a family or user.
- Adapting the building to the needs of the disabled and the elderly.
- Change in the function of the building and using the spaces.
- Reducing the usable space in the building by introducing operators with the ability to change the shape and use of spaces.
- Autonomy, independence, and individualization of building spaces in order to reduce the consumption of energy, raw materials, etc.
- Applying the principles of prefabricated and modular components in the construction of the structure.

Table 2 Characteristics of an adaptable building

Component	Characteristic	
Capacity to change	Changing the size and use of spaces	
	Changing the capacity, performance, and efficiency of the building	
	Failure to implement frequent changes and early implementation of external changes	
	Corrected, moved	
Ability to remain consistent with a defined goal	Reducing inconsistencies between building uses and the goals of its users	
Value	Maximizing constructive and productive uses	
	Adapting the use of the building to the demands of the owners	
	Minimizing the costs	
Time	Speed of change	Quick change of shape and arrangement
		Easy response to change
	Changes during useful life	Changes in future
		Long-term changes
		Use development

Source Authors compilation

To increase the productivity in versatility and flexibility, it is suggested to study new approaches regarding raw materials, technological innovations, and implementation methods. Also, environmental and economic factors should also be studied, because some adaptable and flexible building components have a high price and may not be optimal from an environmental point of view. From a social point of view, it should be pointed out whether societies accept the individualization and independence of building components and the implementation of an independent design in their work and home spaces [29].

5.5 *Circular Economy in the Road Construction Industry*

Road infrastructure is crucial and costly. Due to high maintenance expenses and the environmental impact of road development (due to energy and material consumption), countries are working to reduce lifecycle costs and environmental effects [30]. Developed countries are adopting circular economy principles in road construction to reduce maintenance and development costs while minimizing environmental impact. Since paving represents a significant portion of these costs, efforts focus on circularizing road paving to improve lifecycle cost-effectiveness [31]. Figure 3 presents the lifecycle assessment of a pavement.

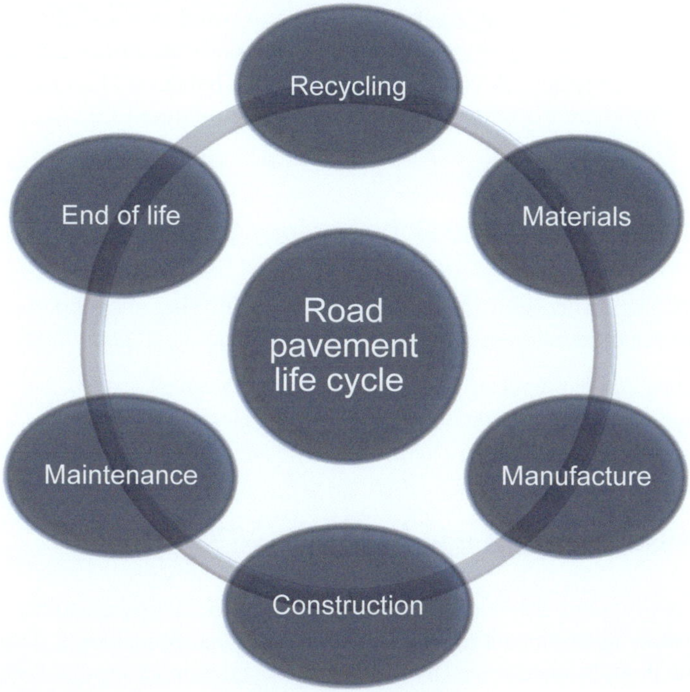

Fig. 3 Road pavement life cycle. *Source* Created by the author

To optimize road paving life cycles, Germany is legislating the use and recycling of waste in paving, pursuing circular economy goals. Belgium, Britain, and Denmark prioritize preventative maintenance to extend road lifespan and avoid costly repairs. Lithuania is testing waste utilization as a road material, while Slovenia and Slovakia are removing legal barriers to asphalt recycling. The Netherlands uses recycled materials in asphalt mixes for frequent preventative repairs, minimizing the need for expensive overhauls and reducing lifecycle costs [32].

Generally, the approaches of these European countries are summarized in the following cases:

- Removal of legal barriers to asphalt recycling in order to reduce wastage production.
- Increasing the useful life of asphalt pavement by performing preventive repairs.
- Testing the capacities of using wastage as raw materials in asphalt pavements.

Among all the approaches followed to realize the circular economy in the road industry, the use of asphalt chips is the most common approach.

Reclaimed asphalt is processed pavement material collected during road repair, improvement, or reconstruction. Road infrastructure development has various environmental, economic, and social impacts; therefore, using reclaimed asphalt is a key step toward sustainability and circular economy goals. Asphalt chips can be used between 10 and 30% in road paving. This restriction is mostly due to legal restrictions and technical problems caused by excessive use of asphalt chips in the pavement. Its technical problems are due to variable behavior of reclaimed asphalt and uncertainty about the performance of the asphalt mixture [33].

6 The Stages of Structural Technology Entrepreneurship

Preliminary stage: decision-making regarding identifying the opportunities and forming the structure plan, providing a structural commercial plan, finial decision.

Establishment stage: preparing and organizing the resources.

The stage of expanding market penetration, developing a firm.

6.1 Making a Creative and Innovative Structure

Structural engineering thrives with less rigid control, fostering creativity and innovation by allowing engineers to think outside traditional frameworks. This requires independent thought, encouraged by managers. Organizational structure also matters: Organic structures, more adaptable to change and offering staff flexibility, are better suited for creative innovation than mechanical ones.

6.2 Characteristics of Organic Structural Engineering

Organic structural engineering is characterized by: (1) parallel relationships; (2) decentralized decision-making; (3) commitment to objectives; (4) a flat organizational structure (design engineer, architect, executor); (5) close professional relationships; (6) engineer collaboration; and (7) fewer hierarchical command-and-control structures.

6.3 Methodology and Research Model

A combination of quintet approaches was used in this study to evaluate the innovation. This model is schematically shown in Fig. 4, which consists of 11 criteria.

According to the fact that the results of this study can be used in structural engineering, this research is applied regarding the objective, and given the authors are

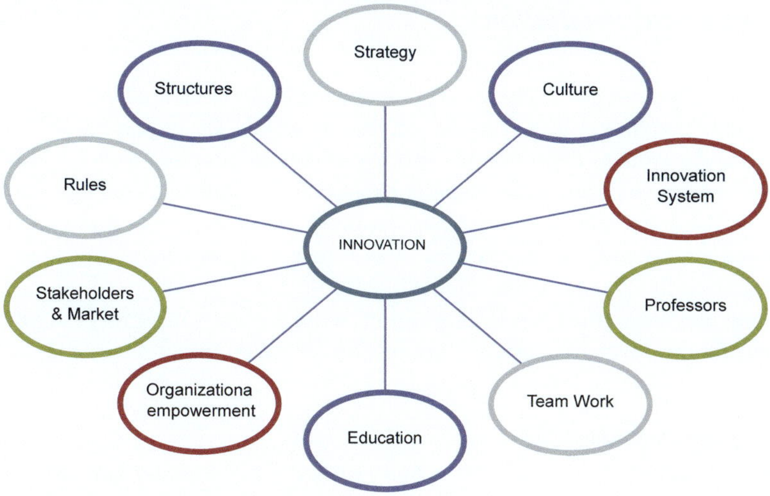

Fig. 4 Conceptual model (authors' work)

structural engineers themselves, collecting the experts' information from the structural engineers using the questionnaire is methodologically a field survey. Furthermore, since the research is carried out in the field of structural engineering, it is considered as a case study.

7 Research Objectives and Questions

Considering the abovementioned issues, creativity and innovation in structural engineering are essential, and therefore, it is necessary to examine more precisely some of the issues to understand the principles of innovation in this field and the differences between these principles and the reality and the ideal conditions. Successful innovation projects in the field of structural engineering need to be identified. Here are three basic questions that can help to better understand the issue:

- What are the innovative indicators in structural engineering and how far is it to the ideal level for high productivity?
- How successful is the use of innovation techniques, and is creativity in structural engineering alone feasible for innovation, or does it require other characteristics?
- What techniques and strategies can be used to create a creativity of the least possible risk of failure?

7.1 Statistical Population

The statistical population includes experienced engineers and students. That is why the research conducted as a general survey, in which the engineers with undergraduate and graduate degrees and working experience higher than 10 years, and PhD and master students as the experts constitute the statistical population of this research. The total number of the population is about 30, and 28 questionnaires distributed among them were returned, whose results are analyzed according to the corresponding figures and tables (Table 3 and Fig. 5).

Second question: third research question: what are the strategies to compensate the gaps and improve any aspects of innovation in structural engineering?

Table 3 Innovation inventory (1)

Aspects	Index	Average index (%)	Standard deviation	Gap
Strategy	How much attention has paid to structural engineering strategy regarding innovation?	34	2	66
	Where do structural engineers stand considering the status of innovation in structure?	34	2	70
	How much different levels of Engineering Council know about the status of innovation in organizational strategies	45	3	65
Structure	How much does the structure of structural engineering help the promotion and development of innovation?	29	2	61
	How much does the structure of structural engineering facilitate the team working and problem-solving groups?	37	3	64
Culture	How much do society and people bear the failures and mistakes of structural engineers regarding innovation?	31	2	63
	How much does the structural engineering field encourage and motivate people to innovate?	32	2	40

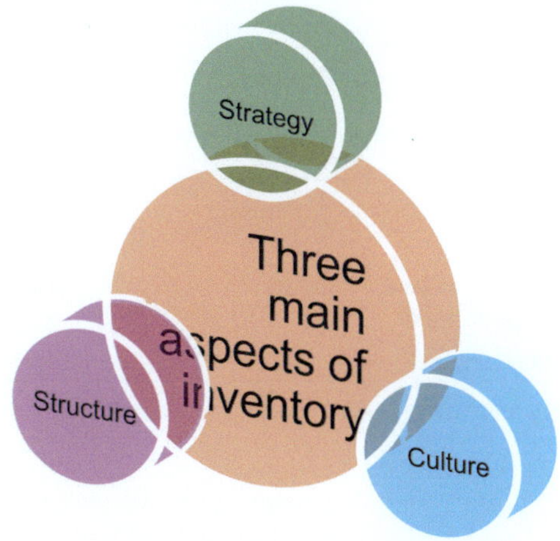

Fig. 5 Three main aspects of inventory, authors' work

According to the scores obtained in structural engineering based on the classifications presented in this field, most aspects of innovations have been poorly developed, and a wide range of capabilities are required to reconstruct them.

The operational strategies to enhance creativity and innovation in competitive strategies can summarized as below:

- The first and most important step is to understand the importance of creativity and innovation in the current era.
- Applying some differences in rewards and material and spiritual benefits among the Engineering Council engineers, who work creatively and wisely, compared to those who adopt common methods of doing things.
- Holding workshops in the context of creativity and its techniques in the Engineering Council and conducting activities to clarify the significance of these courses for engineers.
- Determining some groups to carry put applied research in the context of improving the innovation process.
- Identifying new and creative needs in the business.
- Improving the spirit of scientific courage, criticism, and critical acceptance.

1. Creating a suitable structure for an innovative and creative Engineering Council (organic structure). Target-aligned organizational structure is key to fostering creativity and innovation in structural engineering. Organic (or biological) structures, which adapt quickly to change and offer staff flexibility, are ideal. These structures feature parallel communication, decentralized decision-making, commitment to objectives, a flat hierarchy, and strong collaboration. Larger organizations increasingly adopt these principles, emphasizing teamwork and horizontal communication. Innovative organizations often restructure into smaller units to empower creative teams and encourage entrepreneurship.

Table 4 Innovation inventory (2)

Dimensions	Average index (%)	Dimensions	Average index (%)
Strategy	25	Innovation system	34.7
Culture	22.7	Professors	30.3
Teamwork	25.2	Stakeholders and market	31.2
Education	33.9	Rules	35.3
Organizational empowerment	50.2	Structures	27.3
Mean value		31.7%	

Source Authors' work

2. Innovative organizations cultivate a culture and strategy that supports creativity and entrepreneurship, with managers actively guiding the innovation process. These organizations' strategies, visions, values, and guidelines emphasize innovation, accept failure, and encourage risk-taking. Managers must foster a culture that expects innovation and accepts setbacks, eliminating risk aversion.
3. Training different structural engineers in universities to succeed in the whole stages of the process.

Innovation needs different people, namely thoughtful people who create insights and new opinions, people having information related to the knowledge sources, product experts who advocate adopting new methods, project managers performing technical tasks to maintain an innovative project in the desired path, and leaders who actively encourage, support, and guide others to follow innovation (Table 4).

8 Conclusion

Circular economy with it triple principles of reduction, reuse, and recycling of materials can contribute to the sustainability of projects and the realization of development goals due to the close relationship between the three principles of sustainable, economic, environmental, and social development. Considering the importance of the road and construction industry and the position of this industry in employment, consumption of raw materials and energy, as well as wastage production, by applying circular economy concepts in the construction industry, the economic, environmental, and social burdens of road and construction projects can be reduced. By applying the approaches presented in Table 1, it is possible to realize the goals of the circular economy and reduce the burdens from the projects to an optimized level.

Among the proposed approaches, adaptability and flexibility, prefabrication, and independence of layers can be more operational and effective in the stages of the project life cycle and effective in reducing the economic and environmental costs of the project and inside the country, according to the existing infrastructure and technical and operational experience, by applying the principle of prefabrication in the road and construction industry, the economic and environmental costs of the project and its construction time can be simultaneously reduced by creating an efficient supply chain. Also, in order to move to the different economic sectors to apply circular economy principles in the road and construction industry, the policies can be used in this direction, policies that encourage different economic sectors for circular economy principles. These policies can be for achieving the goals of economy of resistance, and it is possible to create a sustainable supply and production chain in sanctions time. It is worth mentioning that by applying the principles of circular economy in the road and construction industry, it is possible to increase the speed of construction, especially by using prefabrication, adaptability, and flexibility methods to meet the short-term or long-term needs of the country for housing in a shorter time.

The structural engineer should spend all time successfully learning to innovate. While doing so, steps have been taken to establish an innovation process which is sustainable, and which promotes a culture of collaborative working. The findings indicate that the role of client and innovative procurement route can help to promote the culture of innovation. Management of innovation can also be improved by reducing bureaucratic hurdles, feedback to the originators about the progress of the idea, identification of owners who can take the process forward and by rewarding those who have originated the idea. This paper elucidates the findings of a case study examining innovation processes within the construction sector. The study underscores that the success of construction innovations hinges not only on the adept management of their implementation but also on the effective oversight of their generation and development phases. It is imperative that the initial stages of the innovation process establish conditions that facilitate efficient and effective implementation. By referencing Van de Ven's four central problems of managing innovation, the study identifies the pivotal roles of institutional leadership and the management of part-whole relationships. Institutional leadership is crucial for cultivating the commitment and organizational culture necessary for the adoption of innovations. The management of part-whole relationships ensures that the impacts of innovative ideas on both technical and social systems are recognized promptly and accurately. Although the company's innovation process is in its nascent stages and requires benchmarking against best practices from other organizations, it provides valuable insights into how support service providers can establish innovation processes within their organizations to enhance existing systems and processes. This study clearly demonstrates how the appropriate conditions for fostering innovation can be nurtured within a contemporary construction organization.

References

1. Kang, M. J., & Hwang, J. (2016). Structural dynamics of innovation networks funded by the European Union in the context of systemic innovation of the renewable energy sector. *Energy Policy, 96*, 471–490.
2. Garcia-Perez-de-Lema, D., Madrid-Guijarro, A., & Martin, D. P. (2016). Influence of university–firm governance on SMEs innovation and performance levels. *Technological Forecasting and Social Change.*
3. Saouma, V. E. (2007). *Lecture notes in: Structural engineering analysis and design.*
4. Baghchesaraei, O. R., Baghchesaraei, A., & Cheshnasar, S. P. (2020, November). Assessment of forecasting energy consumption in residential buildings based on local climate of major cities in Australia. In *2020 international conference on smart grids and energy systems (SGES)* (pp. 729–733). IEEE.
5. Afuah, A. (2003). *Innovation management: Strategies, implementation and profits.* Oxford University Press.
6. Hosni, Y. A., & Khalil, T. M. (Eds.). (2004). Management of technology: Internet economy-opportunities and challenges for developed and developing regions of the world. In *Selected papers from the eleventh international conference on management of technology.* Elsevier.
7. Kumar, V. (2012). *101 design methods: A structured approach for driving innovation in your organization.* Wiley.
8. Kováč, J., & Rudy, V. (2014). Innovation production structures of small engineering production. *Procedia Engineering, 96*, 252–256.
9. Lewandowska, M. S., Szymura-Tyc, M., & Gołębiowski, T. (2016). Innovation complementarity, cooperation partners, and new product export: Evidence from Poland. *Journal of Business Research, 69*(9), 3673–3681.
10. Tatum, C. B. (1987). Process of innovation in construction firm. *Journal of Construction Engineering and Management, 113*, 648–663.
11. Slaughter, S. (2000). Implementation of construction innovations. *Building Research and Information, 28*, 2–17.
12. Ling, F. Y. Y. (2003). Managing the implementation of construction innovations. *Construction Management and Economics, 21*, 635–649.
13. Bossink, B. (2004). Managing drivers of innovation in construction networks. *Construction Engineering and Management, 130*(3), 337–345.
14. Blayse, A., & Manley, K. (2004). Key influences on construction innovation. *Construction Innovation, 4*, 143–154.
15. van de Van, A. (1986). Central problems in the management of innovation. *Management Science, 32*(5), 590–607.
16. Cebon, P., Newton, P., & Noble, P. (1999). *Innovation in firms: Towards a framework for indicator development.* Melbourne Business School, Working Paper (99-9).
17. Heshmati, A. (2017). A review of the circular economy and its implementation. *International Journal of Green Economics, 11*(3–4), 251–288.
18. Zhu, Q., Geng, Y., & Lai, K. H. (2010). Circular economy practices among Chinese manufacturers varying in environmental-oriented supply chain cooperation and the performance implications. *Journal of environmental management, 91*(6), 1324–1331.
19. Qiao, F., & Qiao, N. (2013). Circular economy: An ethical and sustainable economic development model. *Prakseologia, 154*, 253–272.
20. Su, B., Heshmati, A., Geng, Y., & Yu, X. (2013). A review of the circular economy in China: Moving from rhetoric to implementation. *Journal of cleaner production, 42*, 215–227.
21. Mangialardo, A., & Micelli, E. (2018). Rethinking the construction industry under the circular economy: principles and case studies. In *Smart and sustainable planning for cities and regions: Results of SSPCR 2017* (pp. 333–344). Springer International Publishing.
22. "UI GreenMetric." Accessed: August 21, 2022 [Online]. Available: https://greenmetric.ui.ac.id/rankings/overall-rankings-2021; McCurdy, A. H., Sengupta, A., & Glodis, P. (2004).

Compact measurement of low-PMD optical telecommunication fibers. *Optics Express, 12*(6), 1109–1118.

23. Salama, W. (2017). Design of concrete buildings for disassembly: An explorative review. *International Journal of Sustainable Built Environment, 6*(2), 617–635.
24. Baghchesaraei, A., & Baghchesaraei, O. (2014). The Importance of Infrastructures in the development of modern methods of construction. *International Journal of Applied Engineering Research, 9*(21), 11689–11692.
25. Baghchesaraei, A., & Baghchesaraei, O. R. (2015). Space definition in modern and innovative houses. *Indian Journal of Fundamental and Applied Life Sciences, 5*(S2), 266–269.
26. Munaro, M. R., & Tavares, S. F. (2023). Design for adaptability and disassembly: Guidelines for building deconstruction. Construction Innovation.
27. Estaji, H. (2017). A review of flexibility and adaptability in housing design. *International Journal of Contemporary Architecture, 4*(2), 37–49.
28. Magdziak, M. (2019, February). Flexibility and adaptability of the living space to the changing needs of residents. In *IOP conference series: Materials science and engineering* (Vol. 471, No. 7, p. 072011). IOP Publishing.
29. De Paris, S. R., & Lopes, C. N. L. (2018). Housing flexibility problem: Review of recent limitations and solutions. *Frontiers of Architectural Research, 7*(1), 80–91. https://doi.org/10.1016/j.foar.2017.11.004; Innella, F., Arashpour, M., & Bai, Y. (2019). Lean methodologies and techniques for modular construction: chronological and critical review. *Journal of Construction Engineering and Management, 145*(12), 04019076. https://doi.org/10.1061/(ASCE)CO.1943-7862.0001712
30. Vandewalle, D., Antunes, V., Neves, J., & Freire, A. C. (2020). Assessment of eco-friendly pavement construction and maintenance using multi-recycled rap mixtures. *Recycling, 5*(3), 17.
31. Mantalovas, K., Di Mino, G., Del Barco, J., Carrion, A., Keijzer, E., Kalman, B., Parry, T., & Lo Presti, D. (2020). European national road authorities and circular economy: An insight into their approaches. *Sustainability, 12*(17), 7160.
32. Mantalovas, K., & Di Mino, G. (2019). The sustainability of reclaimed asphalt as a resource for road pavement management through a circular economic model. *Sustainability, 11*(8), 2234.
33. Milios, L. (2018). Advancing to a circular economy: Three essential ingredients for a comprehensive policy mix. *Sustainability Science, 13*(3), 861–878.

Empowering Women and Marginalized Groups Through Capital Markets: Strategies for Inclusive Growth, Financial Access, and Economic Equity

Salu Dsouza and Nikolay Bodrov

Abstract Empowering women and marginalized groups through capital markets is crucial for promoting inclusive growth, enhancing financial access, and fostering economic equity. Despite progress in financial inclusion, significant gaps persist in the participation of underrepresented groups in capital markets, exacerbating socio-economic inequalities. This paper explores strategic interventions to bridge these gaps by leveraging capital markets as instruments of empowerment. Key strategies include enhancing financial literacy, promoting gender-responsive investment products, improving access to credit and investment opportunities, and fostering an enabling regulatory environment. Financial literacy initiatives tailored to the needs of women and marginalized communities are fundamental to equipping them with the knowledge and skills necessary to participate meaningfully in capital markets. Programs that integrate culturally sensitive and context-specific approaches can demystify complex financial instruments, enabling informed investment decisions and long-term wealth accumulation. Developing gender-responsive and socially inclusive financial products can attract diverse investors and channel capital into projects that support social impact. Impact investing, social bonds, and gender bonds represent viable instruments for driving capital toward initiatives that uplift disadvantaged groups. Additionally, the rise of environmental, social, and governance (ESG) investing offers opportunities to align financial goals with broader social objectives, promoting equity and inclusion. Access to credit and investment opportunities remains a significant barrier for marginalized communities. Strengthening microfinance institutions, expanding peer-to-peer lending platforms, and supporting community-driven investment models can provide alternative pathways to financial participation. These mechanisms are particularly effective in underserved regions where traditional financial institutions may not operate effectively. An enabling regulatory framework is critical for sustaining progress. Policymakers should implement

S. Dsouza (✉)
School of Law, Christ University, Bengaluru, India
e-mail: saludsouza@yahoo.com

N. Bodrov
Department of Forensic Science, Kutafin Moscow State Law University, Moscow, Russia
e-mail: bodrovnf@gmail.com

K. Khanna et al. (eds.), *From Poverty to Prosperity*, Information Systems Engineering and Management 64, https://doi.org/10.1007/978-3-031-99166-0_6

regulations that encourage transparency, reduce barriers to entry, and incentivize financial institutions to serve diverse populations. Collaborations between governments, financial institutions, and non-profit organizations can amplify the impact of these policies through targeted programs and incentives. Furthermore, technology-driven solutions, such as fintech and digital finance, offer innovative avenues for enhancing financial access. Digital platforms can reduce transaction costs, simplify access to capital markets, and provide tailored financial services to remote and underserved populations. However, addressing the digital divide and ensuring equitable access to technology remain essential to maximizing these benefits. Ultimately, empowering women and marginalized groups through capital markets not only contributes to social justice but also enhances economic resilience and productivity. By creating opportunities for diverse participants in financial ecosystems, societies can unlock untapped potential, drive innovation, and build more inclusive economies.

Keywords Inclusive growth · Financial access · Economic equity · Capital markets · Financial literacy · Impact investing · Regulatory framework

1 Introduction

Economic equity and financial inclusion are critical components of sustainable development, playing a vital role in reducing poverty, enhancing social cohesion, and promoting economic stability [1]. Financial inclusion involves providing individuals and businesses with access to useful and affordable financial products and services that meet their needs, delivered responsibly and sustainably [2]. When combined with economic equity—the fair distribution of economic resources and opportunities—these concepts contribute to building resilient economies that offer opportunities for all segments of society [3]. Empowering women and marginalized groups through financial inclusion not only uplifts these communities but also stimulates broader economic growth by harnessing untapped potential [4]. However, significant disparities remain in access to capital markets, particularly for women and marginalized groups. According to the Global Findex Database, nearly 1.4 billion adults globally remain unbanked, with women constituting a disproportionately high share of this population [2]. Gender-specific barriers, such as lower financial literacy, limited property ownership, and socio-cultural constraints, often hinder women's access to formal financial systems [5]. Marginalized groups, including ethnic minorities, rural communities, and individuals with disabilities, face additional challenges such as geographic isolation, discrimination, and bureaucratic obstacles [4]. These barriers restrict their ability to access credit, save securely, and invest in growth opportunities, thereby perpetuating cycles of poverty and exclusion [6]. Capital markets, which include equity, debt, derivatives, and commodities markets, hold significant potential for promoting inclusive growth by providing diverse financial instruments that can cater to the needs of underserved populations [7]. By facilitating the efficient allocation of resources, capital markets can support small businesses, encourage entrepreneurship, and enable long-term wealth creation for individuals across socio-economic strata [8]. Moreover, innovative financial instruments such as gender bonds, social impact bonds, and microfinance products have demonstrated success in channeling investments toward social goals, including gender equity and poverty reduction [9]. These developments underscore the transformative capacity of capital markets in bridging financial gaps and driving inclusive economic development.

The role of capital markets in empowering women and marginalized groups extends beyond financial access to fostering broader socio-economic equity. For instance, studies have shown that financial inclusion among women leads to improved household welfare, higher investments in education and health, and increased economic resilience [4]. Similarly, providing marginalized groups with access to financial tools and investment opportunities enhances their economic agency, enabling them to contribute more effectively to economic growth [10]. Capital markets, through mechanisms like equity financing, debt instruments, and community-based investment funds, can offer viable pathways to economic integration for these populations [7]. Despite this potential, several challenges hinder the effective integration of women and marginalized groups into capital markets. Structural barriers such as regulatory constraints, lack of financial literacy, and limited access to digital financial services remain significant [9]. Additionally, market-based challenges, including investor biases and the concentration of financial products in urban areas, often exclude rural and low-income populations [8]. Addressing these challenges requires targeted interventions that not only enhance access to financial services but also ensure that these services are designed to meet the unique needs of diverse groups [6].

The primary objective of this research is to analyze the role of capital markets in empowering women and marginalized groups by enhancing financial access and promoting economic equity. This involves examining how capital markets can serve as a catalyst for inclusive growth by providing tailored financial instruments, fostering investment in social impact initiatives, and facilitating equitable wealth distribution [4]. The research also aims to identify effective strategies for overcoming existing barriers to financial inclusion, focusing on regulatory frameworks, innovative financial products, and the role of technology in bridging access gaps [1]. A critical aspect of this study is evaluating existing models that have successfully leveraged capital markets for inclusive growth. Case studies of initiatives such as microfinance programs in Bangladesh, gender-focused investment funds, and community-based cooperatives in Kenya will provide valuable insights into best practices and scalable models [9]. By comparing these models with less successful initiatives, the research will highlight key factors contributing to the effectiveness of capital market strategies in promoting financial inclusion. This comparative analysis will inform the development of policy recommendations aimed at enhancing the impact of capital markets on underserved populations [2].

The research also seeks to propose policy recommendations that support the creation of a conducive environment for inclusive capital markets. These recommendations will focus on strengthening regulatory frameworks to encourage investments in gender and social equity initiatives, promoting financial literacy programs, and leveraging technology to expand access to financial services [5]. Additionally, the study will advocate for public–private partnerships as a means of mobilizing resources for inclusive growth and ensuring that the benefits of capital markets are distributed equitably across different societal groups [6]. In conclusion, empowering women and marginalized groups through capital markets is not only a matter of social justice but also a strategic economic imperative. By addressing existing disparities in financial access and leveraging the inclusive potential of capital markets, policymakers and financial institutions can unlock new opportunities for growth and equity [4]. This research aims to contribute to this objective by providing evidence-based insights and actionable recommendations for building more inclusive and resilient capital markets that benefit all members of society.

The integration of women and marginalized groups into capital markets is a critical pathway to achieving inclusive growth and economic equity [1]. This research seeks to address three primary questions: How can capital markets be leveraged to promote inclusive growth? What are the barriers faced by women and marginalized groups in accessing capital markets? and Which strategies have proven effective in promoting financial inclusion and equity? These questions are interlinked, as understanding the challenges and opportunities within capital markets is essential for formulating effective strategies that drive inclusive growth. Capital markets, comprising equity, debt, derivatives, and commodities markets, play a vital role in economic development by facilitating resource allocation, supporting entrepreneurship, and enabling wealth creation [8]. When effectively utilized, capital markets can empower underserved communities by providing access to financial instruments that support business growth, housing, education, and other critical needs [7]. The first research question explores the potential of capital markets to create inclusive financial ecosystems that offer opportunities for historically marginalized populations, including women, rural communities, ethnic minorities, and individuals with disabilities [4]. This inquiry will focus on identifying mechanisms through which capital markets can contribute to poverty alleviation, enhance economic participation, and reduce inequalities [2].

The second research question delves into the barriers that hinder women and marginalized groups from fully participating in capital markets. Despite advancements in financial inclusion, gender-specific obstacles such as limited financial literacy, lack of property ownership, and restrictive socio-cultural norms persist [5]. Women often face challenges in accessing credit and investment opportunities due to systemic biases and discriminatory practices within financial institutions [10]. Similarly, marginalized groups encounter barriers related to geographic isolation, economic exclusion, and institutional discrimination, which limit their access to formal financial systems [6]. By examining these challenges, the research aims to highlight structural, cultural, and economic factors that contribute to the financial exclusion of vulnerable populations [4].

Table 1 addresses some barriers to women's financial freedom. The third research question focuses on identifying and evaluating strategies that have successfully empowered women and marginalized groups through capital markets. These strategies include policy interventions such as gender quotas and inclusive financial regulations, innovative financial instruments like gender bonds and social impact funds, and capacity-building initiatives aimed at improving financial literacy and economic agency [9]. Case studies of successful models, such as microfinance programs in Bangladesh and women's cooperatives in Kenya, will provide practical insights into scalable solutions [2]. This analysis will also consider the role of technology, particularly digital finance and fintech innovations, in overcoming traditional barriers to financial access [5].

This research adopts a broad geographical and demographic focus while maintaining clear conceptual boundaries to ensure depth and relevance. Geographically, the study will analyze both global trends and specific regional contexts, with an emphasis on developing economies where financial exclusion is most pronounced [4]. Particular attention will be given to regions such as Sub-Saharan Africa, South Asia, and Latin America, where women and marginalized groups often face significant socio-economic barriers [1]. The research will also draw comparisons with developed economies to understand how established financial markets integrate underserved communities and what lessons can be adapted for emerging markets [8]. Demographically, the study will focus on women, rural populations, ethnic minorities, individuals with disabilities, and economically disadvantaged groups. These populations are chosen due to their historical exclusion from formal financial systems and their potential to contribute to economic growth if provided with adequate financial tools and opportunities [7]. By examining these diverse groups, the research aims to develop a comprehensive understanding of the multifaceted challenges and needs within capital markets [6].

Conceptually, the research will be grounded in theories of financial inclusion, economic equity, and inclusive growth. It will explore the intersection of gender and social equity within economic systems, drawing from frameworks such as Inclusive Capitalism and the Capability Approach [3]. The study will focus on capital markets' role in providing access to financial services, enhancing investment opportunities, and promoting wealth creation among marginalized groups [11]. However, it will not

Table 1 Barriers to financial access for women and marginalized groups

Barrier category	Description	Examples
Legal and regulatory	Discriminatory laws or lack of protective policies	Inheritance laws, lack of gender-specific financial policies
Financial literacy	Limited knowledge of financial products	Low awareness of investment and savings instruments
Socio-cultural challenges	Social norms restricting economic participation	Gender roles, discrimination against minorities
Market barriers	Limited access to formal financial institutions	Geographic and digital divides

Source Created by author

cover broader aspects of financial inclusion that fall outside capital markets, such as basic banking access or informal financial networks, to maintain a focused and in-depth analysis [12].

The research acknowledges several limitations. One challenge is the availability and reliability of data, particularly in developing economies where financial inclusion metrics are not consistently tracked [13]. Additionally, the research may face limitations in generalizing findings from specific case studies to broader contexts, given the diversity of socio-economic conditions across regions [4]. To mitigate these limitations, the study will use a mixed-methods approach, combining quantitative data with qualitative insights to provide a balanced analysis [14]. The research will employ a mixed-methods approach, integrating both qualitative and quantitative techniques to achieve a comprehensive analysis of the research questions [15]. Quantitative methods will include data analysis of financial inclusion metrics, gender equity indicators, and capital market performance data sourced from global databases such as the World Bank's Global Findex Database, IMF reports, and national financial inclusion surveys [12]. Statistical techniques will be used to identify correlations between access to capital markets and socio-economic outcomes for women and marginalized groups [6]. Qualitative methods will involve primary data collection through surveys and interviews with stakeholders, including financial experts, policymakers, representatives of marginalized communities, and beneficiaries of inclusive financial initiatives [4]. These interviews will provide insights into lived experiences, perceived barriers, and the effectiveness of various strategies in promoting financial inclusion [11]. Additionally, a literature review of academic papers, policy documents, and case studies will support secondary data analysis, helping to contextualize primary findings within broader theoretical and practical frameworks [1]. Case studies will form a critical part of the methodology, offering in-depth examinations of successful and unsuccessful models of capital market inclusion. These case studies will analyze initiatives such as microfinance programs, social impact bonds, and digital finance solutions, evaluating their design, implementation, and outcomes [13]. Through this mixed-methods approach, the research aims to provide robust and actionable insights into how capital markets can effectively promote inclusive growth and economic equity.

The concept of financial inclusion is rooted in broader economic and social theories that emphasize equitable access to economic resources and opportunities. Two prominent theoretical frameworks underpinning this research are Inclusive Capitalism and the Capability Approach. Inclusive Capitalism posits that capital markets and broader economic systems should not only drive profit and growth but also contribute to societal well-being by ensuring that all individuals have opportunities to participate in and benefit from economic progress [16]. This theory aligns with the idea that capital markets can play a transformative role in reducing poverty and inequality by providing access to financial services, investment opportunities, and mechanisms for wealth creation to traditionally excluded groups, including women and marginalized communities [17]. Inclusive Capitalism emphasizes stakeholder capitalism over shareholder capitalism, advocating for businesses and financial institutions to prioritize social impact alongside financial returns [18]. Another influential framework is Amartya Sen's Capability Approach, which focuses on expanding individuals' freedoms and capabilities to lead the lives they value [19]. This approach shifts the discourse from purely economic metrics to a more holistic view of human development, considering aspects such as education, health, and economic opportunities [20]. Applying this framework to capital markets involves recognizing that financial inclusion is not just about providing access to financial instruments but also about enhancing the capacity of marginalized groups to effectively use these tools [1]. For example, financial literacy programs and policies that support women's economic empowerment contribute to building capabilities that enable meaningful participation in financial markets [4].

The intersection of gender and social equity in economic theory is also crucial to this research. Traditional economic theories often failed to account for systemic inequalities, but contemporary approaches such as Feminist Economics emphasize the need to consider gender-based disparities in economic analysis [21]. This perspective highlights how discriminatory practices in financial institutions, limited access to property rights, and socio-cultural norms hinder women's and marginalized groups' access to capital markets [13]. Addressing these disparities requires targeted interventions, including gender-sensitive financial products and inclusive investment frameworks [12]. These theoretical perspectives collectively inform the research by providing a robust framework to analyze how capital markets can promote inclusive growth and economic equity. By grounding the study in well-established theories, the research aims to contribute to both academic discourse and practical policy development aimed at enhancing financial inclusion [11].

1.1 Understanding Capital Markets

Capital markets are integral components of the financial system, facilitating the flow of capital between investors and entities seeking funding for growth and development [8]. They consist of equity markets, debt markets, derivatives markets, and commodities markets, each playing a distinct role in economic development and financial inclusion [7]. Equity markets involve the buying and selling of shares in publicly listed companies. They provide businesses with access to long-term capital and offer investors opportunities for wealth accumulation through capital gains and dividends [16]. For women and marginalized groups, equity markets present a platform to participate as investors and entrepreneurs. Initiatives such as gender bonds and social impact funds have emerged to channel investments into enterprises that support social equity and women's economic empowerment [11]. Debt markets, including bonds and other fixed-income securities, allow entities to raise funds by borrowing from investors. These markets are particularly important for financing public infrastructure and social projects that benefit disadvantaged communities [14]. For instance, social bonds and development impact bonds are financial instruments designed to support initiatives targeting health, education, and poverty alleviation in underserved areas [4]. Such innovations in debt markets contribute to creating inclusive economic environments where marginalized groups can thrive [6].

Derivatives markets deal with financial instruments whose value is derived from underlying assets such as stocks, bonds, commodities, or currencies. While derivatives primarily serve risk management and speculative purposes, they also offer opportunities for financial innovation that can support inclusive growth [13]. For example, weather derivatives can help small-scale farmers in rural areas hedge against climate-related risks, thereby enhancing financial stability and resilience among vulnerable populations [12]. Commodities markets involve trading raw materials and primary products such as gold, oil, agricultural products, and metals. These markets play a crucial role in supporting the livelihoods of communities engaged in primary industries, particularly in developing countries [7]. By integrating marginalized groups, such as smallholder farmers and local producers, into commodities markets, economic systems can create more equitable value chains and improve income distribution [8].

Capital markets contribute to economic development by mobilizing savings, promoting investment, enhancing liquidity, and improving resource allocation efficiency [16]. For marginalized groups, access to capital markets translates into opportunities for entrepreneurship, education, housing, and other critical areas of socio-economic advancement [4]. However, disparities in access to these markets due to systemic barriers, including financial illiteracy, lack of collateral, and restrictive regulatory frameworks, need to be addressed to realize the full potential of capital markets in promoting inclusive growth [1]. The role of capital markets in promoting financial inclusion is increasingly recognized by policymakers and financial institutions. Programs such as microfinance, community investment funds, and inclusive fintech solutions demonstrate how capital markets can be tailored to meet the needs of underserved populations [11]. Moreover, digital platforms are breaking traditional barriers by providing women and marginalized communities with access to financial services, investment opportunities, and credit facilities [13]. In conclusion, a deep understanding of capital markets and their theoretical foundations is essential for developing effective strategies that empower women and marginalized groups. By leveraging financial inclusion theories and recognizing the diverse components of capital markets, this research aims to contribute to building more inclusive and equitable economic systems [12]. Table 2 depicts the key initiatives based on strategies for inclusive growth.

1.2 Financial Inclusion: A Pathway to Empowerment

Financial inclusion, defined as the availability and accessibility of financial services to all segments of society, plays a crucial role in promoting economic empowerment, particularly for women and marginalized groups [1]. By providing access to banking, credit, insurance, and investment opportunities, financial inclusion enables individuals to build assets, smooth consumption, and invest in education, health,

Table 2 Strategies for inclusive growth in capital markets

Strategy	Description	Key initiatives
Regulatory reforms	Implementing gender-sensitive financial policies	Quotas, gender-inclusive financial frameworks
Financial literacy programs	Educating underserved communities on investments	Workshops, mentorship, accessible resources
Supporting social enterprises	Investing in businesses led by women and minorities	Impact funds, grants, venture capital
Digital financial services	Leveraging technology for broader market access	Mobile banking, digital investment platforms

Source Created by author

and entrepreneurial ventures [22]. For women and marginalized groups, who often face systemic barriers to economic participation, financial inclusion is not merely a financial matter but a transformative tool for achieving social and economic equity [23].

The importance of financial inclusion extends beyond individual empowerment to broader socio-economic development. Studies show that increasing access to financial services for underserved populations contributes to poverty alleviation and economic growth [24]. For example, when women gain access to financial services, they are more likely to invest in their families' education and health, creating a positive intergenerational impact [25]. Similarly, for marginalized communities, financial inclusion offers pathways to formal employment, entrepreneurship, and improved standards of living [26].

Capital markets, as part of the broader financial ecosystem, hold significant potential for promoting financial inclusion. They provide mechanisms for resource mobilization, investment opportunities, and access to finance for small and medium-sized enterprises (SMEs) and social enterprises [27]. By developing inclusive capital markets, policymakers can bridge the gap between underserved communities and mainstream financial systems, promoting economic resilience and reducing inequality [28].

Despite the benefits of financial inclusion, women and marginalized groups continue to face numerous barriers in accessing capital markets and other financial services. These barriers are multifaceted, encompassing structural, cultural, and economic dimensions [29]. Structural barriers include institutional policies and regulatory frameworks that inadvertently exclude certain populations. For example, stringent documentation requirements, such as proof of income or formal employment, often disadvantage individuals working in the informal sector, where marginalized groups are disproportionately represented [30]. Additionally, financial institutions' reliance on credit scores and collateral-based lending excludes those without formal financial histories or assets, which is particularly common among women and rural communities [31].

Cultural barriers are deeply rooted in societal norms and gender roles. In many regions, patriarchal norms limit women's financial autonomy, restricting their ability to open bank accounts, access credit, or make independent financial decisions [23]. Such cultural norms not only hinder women's economic participation but also reinforce cycles of dependency and poverty [32]. Moreover, marginalized groups, including ethnic minorities and indigenous communities, often face discrimination that limits their access to financial services [24].

Economic barriers involve factors such as income inequality, unemployment, and geographic isolation. Marginalized communities, particularly those in rural or underserved areas, often lack access to financial infrastructure, including banks, ATMs, and financial literacy programs [22]. The high cost of financial services and limited access to digital tools further exacerbate exclusion [1]. For women, economic barriers are compounded by gender pay gaps and limited inheritance and property rights, which affect their ability to meet collateral requirements for loans [33].

Gender-specific challenges in financial inclusion include a lack of financial literacy, lower digital financial service usage, and limited access to entrepreneurial finance [25]. According to studies, women are 7% less likely than men to own a bank account and 11% less likely to use digital financial services [29]. These gaps are driven by both supply side issues, such as financial products not tailored to women's needs, and demand-side issues, including lower confidence and knowledge about financial products [22].

For marginalized communities, barriers to financial access often involve discrimination, both explicit and systematic. Financial institutions may impose higher fees or stricter conditions on minority groups, reflecting broader societal biases [26]. Geographic isolation, particularly in rural or conflict-affected areas, further limits access to traditional banking services [31]. Innovative solutions such as mobile banking and microfinance have made strides in addressing these barriers, but challenges remain in ensuring that these solutions reach the most disadvantaged [24].

Addressing these barriers requires a multi-pronged approach, including regulatory reforms, targeted financial products, and capacity-building initiatives [34]. Governments and financial institutions need to design inclusive financial products, promote digital financial literacy, and ensure that financial services are accessible and affordable for all [32]. Additionally, integrating social and cultural awareness into financial policies can help reduce discrimination and promote a more inclusive financial environment [28].

2 Existing Literature on Empowerment Through Capital Markets

The literature on financial inclusion and empowerment through capital markets highlights several successful models and interventions. Microfinance institutions, e.g., have been widely recognized for their role in providing small loans to underserved populations, particularly women in developing countries [27]. Organizations such as Grameen Bank and BRAC have demonstrated that small-scale financial services can have a significant impact on poverty alleviation and women's empowerment [30].

Another successful model involves **social bonds and impact investing**, where capital markets are used to fund projects with social and environmental goals [34]. For instance, **gender bonds**, which finance women-led businesses or projects aimed at reducing gender inequality, have gained traction as a means to channel investments into initiatives that support women's economic participation [25]. These models illustrate how capital markets can be leveraged to drive social impact alongside financial returns [33].

However, gaps remain in the literature, particularly regarding the scalability and sustainability of such interventions [1]. Many studies focus on specific geographic or demographic contexts, limiting the generalizability of findings [28]. Furthermore, while financial inclusion initiatives often show positive short-term impacts, there is limited research on long-term outcomes and the structural changes needed to maintain financial inclusion over time [24].

The existing literature also highlights the need for more comprehensive data on the **intersectional experiences** of women and marginalized groups in financial markets [23]. Research often lacks a nuanced understanding of how factors such as ethnicity, disability, and geographic location influence financial inclusion outcomes [32]. Addressing these gaps requires adopting a more holistic and inclusive research approach, incorporating both quantitative and qualitative methods to capture diverse experiences and challenges [22].

In conclusion, while the literature provides valuable insights into successful financial inclusion strategies, there is a critical need for further research to address existing gaps and develop more effective and inclusive capital market interventions. By building on existing knowledge and exploring new approaches, this research aims to contribute to the development of policies and programs that empower women and marginalized groups through enhanced financial access and economic equity [29].

2.1 Global Perspective on Financial Inclusion

Financial inclusion refers to the process of ensuring access to appropriate financial products and services needed by individuals and businesses at affordable costs, delivered in a transparent and sustainable manner [1]. It is a critical enabler of economic growth and poverty alleviation, providing the unbanked and underbanked populations with opportunities for saving, investment, and credit.

The global landscape of financial inclusion has evolved significantly over the past two decades, driven by technological advancements, policy initiatives, and innovative financial models. However, disparities remain stark between developed and developing economies, and even within regions, highlighting the need for targeted and inclusive financial strategies. According to the World Bank's Global Findex Database (2021), approximately 76% of the global adult population now has access to a formal financial account, up from 51% in 2011. The rise in financial inclusion is primarily attributed to the proliferation of digital financial services, mobile money, and simplified banking processes. In Sub-Saharan Africa, for instance, mobile money accounts have become more prevalent than traditional bank accounts, with over 33% of adults owning a mobile money account [29].

Despite these gains, **1.4 billion adults globally remain unbanked**, with a significant majority being women and individuals in rural areas [25]. The gender gap in financial inclusion has narrowed slightly but persists, with women being 7% less likely than men to have a financial account. Similarly, the rural–urban divide remains pronounced, with infrastructural and digital connectivity challenges hindering progress in rural regions [35].

Developed economies, such as those in North America and Europe, boast near-universal financial inclusion, driven by robust financial infrastructure, high literacy rates, and widespread digital adoption [36]. In contrast, developing economies struggle with barriers including limited financial literacy, inadequate infrastructure, cultural resistance, and regulatory hurdles [1]. For example, while nearly 95% of adults in Sweden use digital payment methods, only about 10% of adults in countries like Afghanistan and Chad have access to similar services [25].

Moreover, developed countries benefit from advanced regulatory frameworks and consumer protection mechanisms that foster trust in the financial system, a critical component often lacking in developing economies [37].

Case Studies

- **Bangladesh** is widely regarded as a pioneer of microfinance, primarily through the efforts of Grameen Bank and BRAC [38]. The microfinance model, which provides small, collateral-free loans to low-income individuals, has significantly enhanced financial inclusion by empowering marginalized communities, particularly women [39].
- **Kenya**: Women's cooperatives like the Joyful Women Organization (JOYWO) provide savings and credit facilities, enabling women to start small businesses and achieve economic independence [40]. Kenya's success with mobile money services like M-Pesa has also transformed financial inclusion [35].
- **Brazil**: The "Caixa Tem" digital wallet initiative enhanced financial inclusion during COVID-19 by distributing emergency aid to unbanked populations [41].
- **Philippines**: Financial literacy is central to its strategy through the "Economic and Financial Learning Program" by Bangko Sentral ng Pilipinas [42].

- **India**: The Pradhan Mantri Jan Dhan Yojana (PMJDY) has led to over 500 million bank accounts, improving access especially in rural areas [43].
- **China**: Fintech firms like Ant Financial and WeChat Pay have revolutionized digital payments, driving near-universal financial access [44].

Key Lessons

Policy and Regulation: Strong frameworks are essential for fostering innovation and protecting consumers [37].

Technology: Mobile banking and fintech are critical tools in reaching the underserved [35].

Targeted Interventions: Addressing unique barriers faced by women and marginalized populations yields better results [1].

Financial Literacy: Essential for informed financial behavior [42].

Public–Private Partnerships: Collaboration amplifies the impact of inclusion programs [40].

Financial inclusion is a dynamic, multi-dimensional challenge that requires coordinated efforts in policy, innovation, and social engagement. While global trends show promise, closing the remaining gaps will demand inclusive, sustainable, and adaptable solutions.

Analysis of Financial Instruments

Financial instruments play a critical role in mobilizing capital for social, economic, and environmental objectives. Among the diverse range of financial instruments, microfinance, social impact bonds (SIBs), and environmental, social, and governance (ESG) investments have gained prominence for their potential to drive positive societal impact. The advent of fintech and digital finance has further transformed the landscape, introducing innovative solutions for financial inclusion and investment efficiency. However, the current status quo presents several challenges, including regulatory and policy barriers and market-based obstacles such as investor biases and limited access to networks. This essay critically examines these financial instruments, the role of fintech, and the challenges hindering their broader adoption.

Microfinance

Microfinance refers to the provision of financial services, particularly small loans, to individuals and small businesses lacking access to traditional banking services [45]. Initiatives like the Grameen Bank in Bangladesh have demonstrated microfinance's potential to alleviate poverty and empower marginalized communities [46]. Microfinance institutions (MFIs) often focus on women and rural populations, promoting entrepreneurship and financial independence [47]. However, challenges such as high-interest rates, over-indebtedness, and operational sustainability remain significant [48].

Social Impact Bonds

Social impact bonds (SIBs) are innovative financial instruments that involve private investors funding social programs with returns contingent on predefined social outcomes [49]. SIBs align the interests of public, private, and non-profit sectors, promoting efficient and results-oriented social interventions [50]. Successful examples, such as the UK's Peterborough Prison SIB, have showcased how performance-based models can reduce recidivism and generate positive social impact [51]. Despite their promise, SIBs face criticism over complex structuring, measurement challenges, and the potential misalignment of investor and societal interests [52].

ESG Investments

Environmental, social, and governance (ESG) investments prioritize financial returns alongside positive societal and environmental impact [53]. ESG strategies encompass diverse approaches, including negative screening, impact investing, and shareholder advocacy [54]. The global ESG market has witnessed rapid growth, driven by increasing awareness of sustainability issues and regulatory support [55]. However, challenges such as greenwashing, inconsistent metrics, and varying standards of ESG reporting hinder the effectiveness of these investments [56].

Role of Fintech and Digital Finance

Fintech and digital finance have disrupted traditional financial systems by offering accessible, efficient, and transparent financial services [57]. Digital payment platforms, mobile banking, and blockchain technologies have enhanced financial inclusion and democratized access to investment opportunities [58]. Innovations like robo-advisors, peer-to-peer lending, and crowdfunding have enabled retail investors to participate in financial markets and support impact-driven initiatives [59]. However, regulatory uncertainties, data privacy concerns, and cybersecurity risks present significant challenges to the fintech sector [60].

3 Challenges in the Status Quo

Regulatory and Policy Barriers

Regulatory frameworks often struggle to keep pace with the rapid evolution of financial instruments and technologies [61]. Inconsistent regulations across jurisdictions create compliance challenges for market participants and may stifle innovation [57]. Moreover, overly stringent regulations can limit the scalability of impact-focused financial products, such as SIBs and ESG investments [49].

Market-Based Challenges

Market-based challenges include investor biases, such as the perception of higher risks associated with impact investments compared to traditional assets [62]. Limited access to networks and capital markets also constrains the growth of small-scale initiatives, particularly in emerging economies [63]. Additionally, the lack of standardized impact measurement frameworks makes it difficult for investors to assess the performance and credibility of impact-oriented financial products [56]. Microfinance, social impact bonds, and ESG investments represent powerful financial instruments with the potential to address global challenges and promote sustainable development. Fintech and digital finance further amplify this potential by enhancing financial accessibility and efficiency. However, to realize their full impact, regulatory frameworks must adapt to emerging trends, and market participants must overcome biases and structural challenges. By fostering a supportive ecosystem for innovative financial instruments, stakeholders can drive meaningful change and contribute to a more inclusive and sustainable global economy.

4 Policy and Regulatory Interventions

Financial inclusion is a crucial enabler of economic growth, poverty reduction, and social equity. Despite significant progress in recent years, barriers such as gender inequality, geographic limitations, and systemic biases continue to restrict access to financial services for marginalized communities. To address these challenges, policy and regulatory interventions play a pivotal role by creating an enabling environment for inclusive finance. This essay explores affirmative policies, regulatory frameworks, innovative financial instruments, and the role of technology and fintech in promoting financial inclusion.

Affirmative Policies

Affirmative policies are strategic interventions designed to promote equity and inclusion within financial systems. Examples include gender quotas, which mandate a minimum representation of women in corporate leadership or investment portfolios. These policies aim to reduce gender disparities in access to financial services and investment opportunities [64]. Incentives for inclusive investments, such as tax breaks for impact investors or grants for community-based funds, encourage capital flow toward underserved populations [65].

Regulatory Frameworks for Financial Inclusion

Effective regulatory frameworks are critical for enhancing financial inclusion. Central banks and financial regulatory bodies often implement measures such as simplified Know Your Customer (KYC) norms, digital identification systems, and support for microfinance institutions (MFIs) [66]. Regulations that promote transparency, consumer protection, and fair lending practices build trust and lower barriers for low-income and rural populations to access formal financial services [67].

Innovative Financial Instruments

Innovative financial instruments such as gender bonds, social impact bonds (SIBs), and inclusive exchange-traded funds (ETFs) are gaining traction. Gender bonds, e.g., raise capital for initiatives that support women's economic empowerment [68]. SIBs facilitate investments in social projects by offering returns linked to specific social outcomes [69]. Community-based investment funds also play a role by pooling resources to support local businesses and infrastructure [70].

Role of Technology and Fintech

Technology, particularly fintech, has revolutionized financial inclusion by breaking down geographic and economic barriers. Digital platforms such as mobile banking, digital wallets, and peer-to-peer lending provide convenient and affordable financial services [71]. Blockchain technology further enhances transparency and efficiency in financial transactions, enabling inclusive investment opportunities and reducing the risks of fraud and mismanagement [72]. Policy and regulatory interventions, coupled with innovative financial instruments and technology-driven solutions, are crucial for advancing financial inclusion. While affirmative policies and regulatory frameworks set the groundwork, fintech innovations bring financial services within reach of underserved communities. A holistic approach that integrates policy, market-based instruments, and technology can significantly contribute to building an inclusive and resilient financial ecosystem.

5 Promoting Financial Inclusion Through Capacity Building and Financial Literacy: A Pathway to Empowering Women and Marginalized Groups

Financial inclusion is a critical enabler of sustainable development and economic growth. It involves ensuring that individuals and businesses, regardless of their socio-economic status, have access to useful and affordable financial products and services that meet their needs. The World Bank [67] highlights that approximately 1.4 billion adults remain unbanked, with women and marginalized groups disproportionately affected. To bridge this gap, targeted educational programs, collaboration with NGOs and institutions, private sector engagement, and innovative investment strategies are imperative. Capacity building and financial literacy are fundamental components in enhancing financial inclusion. Financial literacy equips individuals with the knowledge and skills to make informed financial decisions, from managing savings to accessing credit and insurance [73]. Capacity building, on the other hand, focuses on developing institutional and community-level competencies to support financial inclusion initiatives. For women and marginalized groups, tailored financial education can address specific barriers such as lack of formal education, socio-cultural constraints, and limited exposure to financial products [74]. Programs like the Global Money Week by the OECD and financial literacy initiatives by UN Women have shown positive impacts on improving financial behaviors and outcomes among these communities [75]. Educational programs play a pivotal role in promoting financial inclusion. These programs should be designed with an understanding of the unique challenges faced by women and marginalized groups. Community-based workshops, digital learning modules, and partnerships with educational institutions can broaden outreach and enhance learning outcomes [76]. An example of this approach is the Self-Employed Women's Association (SEWA) initiative in India, which combines vocational training with financial education, enabling women to generate income and manage their finances effectively [77]. Similarly, BRAC's microfinance program in Bangladesh integrates financial literacy into its lending process, empowering women to make informed economic decisions [78].

Partnerships with NGOs and educational institutions amplify the impact of financial inclusion initiatives. NGOs often possess deep community ties and trust, which are critical for engaging marginalized populations. For instance, organizations like FINCA and Women's World Banking collaborate with local NGOs to deliver microfinance services and financial education [79]. Educational institutions contribute by integrating financial literacy into curricula and conducting research that informs policy and practice. The collaboration between the Grameen Foundation and universities in the Philippines is a notable example, where financial literacy programs are tailored for rural communities and young women [80]. The private sector and institutional investors play a significant role in promoting financial inclusion through investment and innovation. By adopting Environmental, Social, and Governance (ESG) principles, businesses can contribute to inclusive growth while achieving financial returns [81]. ESG and impact investing focus on generating measurable social and environmental impact alongside financial returns. This approach aligns investor interests with social goals, creating a powerful mechanism for driving financial inclusion [82]. Impact funds such as the Goldman Sachs 10,000 Women initiative and the IFC's Women Entrepreneurs Finance Initiative (We-Fi) demonstrate how private capital can support women entrepreneurs and marginalized groups [83]. Governments and financial regulators can boost inclusive investments by offering incentives such as tax benefits, credit guarantees, and co-investment opportunities [84]. Public–private partnerships also create platforms for scaling financial inclusion initiatives. The Nigerian government's partnership with Mastercard to launch a digital financial inclusion project for women exemplifies how incentives can drive private sector engagement [85].

Case Studies of Successful Models

International Finance Corporation (IFC)

The IFC's Banking on Women program has mobilized over $3 billion to support women-owned businesses through partnerships with financial institutions worldwide. By combining funding with technical assistance, the program enhances the capacity of local banks to serve female entrepreneurs [86]. UN Women has launched various programs, including the Women's Empowerment Principles (WEPs), which guide businesses on promoting gender equality. Their collaboration with banks in Africa to create gender-responsive financial products is a successful model of leveraging institutional partnerships for financial inclusion [87]. Grassroots movements often use community-based savings groups and cooperatives to promote financial inclusion.

In Kenya, the "Chama" groups allow women to pool savings and access microloans, fostering economic resilience and empowerment [88]. Promoting financial inclusion requires a multifaceted approach involving education, capacity building, partnerships, and investment strategies. By focusing on women and marginalized groups, these initiatives not only foster economic empowerment but also contribute to broader social and economic development goals. The collaboration between public and private sectors, along with strong support from NGOs and educational institutions, is crucial in achieving these objectives. As evidenced by successful models from organizations like IFC and UN Women, targeted strategies can transform financial ecosystems, making them more inclusive and equitable.

6 Evaluating Empowerment and Inclusion: Metrics, Capital Market Impact, and Measurement Frameworks

Empowerment and inclusion are critical pillars of sustainable development, aiming to uplift marginalized communities and promote equitable growth. Evaluating the effectiveness of initiatives targeting these goals requires robust metrics and analytical frameworks. Economic and social indicators, the impact of capital market strategies on target groups, and advanced tools for impact measurement provide a comprehensive approach to assess progress. This essay explores these dimensions through quantitative and qualitative analyses, supported by success stories, data-driven insights, and established frameworks such as Social Return on Investment (SROI) and Gender and Inclusion Impact Assessments. Economic indicators play a crucial role in evaluating empowerment and inclusion by providing measurable outcomes related to financial well-being and employment. Key metrics include income levels, employment rates, access to financial services, and asset ownership. According to the World Bank, increases in household income and employment rates are strong indicators of economic empowerment, particularly among women and marginalized communities [89]. Programs like the Mahatma Gandhi National Rural Employment Guarantee Act (MGNREGA) in India have demonstrated how employment opportunities directly impact household income, contributing to economic stability and poverty reduction [90]. Additionally, employment rates among women and marginalized groups reflect the effectiveness of inclusion strategies. A study by the International Labor Organization revealed that targeted vocational training programs in Sub-Saharan Africa increased employment rates among women by 15%, highlighting the tangible benefits of skill development initiatives [91].

Social indicators provide insights into the broader well-being of individuals and communities, extending beyond economic metrics. Improved living standards, education levels, healthcare access, and social mobility are critical parameters for assessing empowerment and inclusion [92]. Education, in particular, is a powerful enabler of empowerment. UNESCO reported that educating girls can lead to a 12% reduction in global poverty rates, underscoring the transformative potential of education [93]. Improved living standards, reflected in housing quality, access to clean water, and healthcare, also signify enhanced social inclusion. For instance, the Grameen Bank's microfinance model not only elevated income levels among women in rural Bangladesh but also improved their families' living conditions and children's education outcomes [94]. Capital market strategies significantly influence empowerment and inclusion by directing investments into projects that generate social and economic benefits for underserved communities. Impact investing, which aims to create positive social outcomes alongside financial returns, has gained traction in recent years. According to the Global Impact Investing Network, impact investments reached $1.5 trillion globally, with a considerable portion targeting gender equity, education, and poverty alleviation initiatives [95]. Quantitative and qualitative analyses of these strategies reveal mixed outcomes. Quantitatively, metrics such as return on investment, growth in employment, and income generation provide tangible evidence of success. Qualitatively, narratives and testimonials from beneficiaries offer insights into how capital investments translate into real-world empowerment. A study by Bouri et al. highlighted that impact investments in East Africa's agriculture sector not only boosted local economies but also empowered female farmers by providing them with training, resources, and market access [96].

Success stories and data-driven insights illustrate the transformative impact of well-executed inclusion strategies. One such example is the Goldman Sachs 10,000 Women initiative, which provides female entrepreneurs with education, mentoring, and access to capital [83]. The program's data indicates that 70% of participants reported increased revenues, and 50% created new jobs within their communities. Similarly, the International Finance Corporation's Banking on Women program mobilized over $3 billion in investments, enabling financial institutions to support women-owned businesses [97]. These examples demonstrate how targeted strategies, backed by capital market support, can drive measurable empowerment. To accurately measure the impact of empowerment and inclusion initiatives, robust tools and frameworks are essential. Social Return on Investment (SROI) is a widely recognized methodology that assesses the social, environmental, and economic value generated by investments [98]. SROI uses a ratio to compare the net benefits of an intervention against its costs, providing stakeholders with a clear understanding of the social value created. For instance, a SROI analysis of a vocational training program for women in South Asia found a 4:1 return, meaning every dollar invested generated four dollars in social and economic benefits [99]. This metric is particularly valuable for investors and policymakers to gauge the effectiveness of social programs and allocate resources efficiently. Gender and Inclusion Impact Assessments offer another critical framework for evaluating initiatives aimed at marginalized groups. These assessments involve analyzing how policies, programs, and investments affect different genders and social groups, ensuring that no community is left behind [100]. The framework includes both qualitative and quantitative metrics, such as changes in gender employment gaps, shifts in access to education, and improvements in decision-making roles within households and communities [101]. The European Investment Bank's gender lens investing strategy, e.g., applies these assessments to ensure investments support gender equality and empower women economically and socially [102].

In conclusion, evaluating empowerment and inclusion requires a multifaceted approach that combines economic and social indicators with advanced measurement tools and frameworks. Economic metrics such as income levels and employment rates, along with social indicators like education and living standards, provide a holistic view of progress. Capital market strategies, particularly through impact investing, contribute significantly to driving inclusion and empowerment, with success stories offering valuable insights. Tools like SROI and Gender and Inclusion Impact Assessments ensure that investments and initiatives generate tangible and measurable benefits for women and marginalized groups. As evidenced by successful models from organizations such as Goldman Sachs, IFC, and EIB, a strategic blend of quantitative and qualitative analysis can effectively measure and enhance the impact of empowerment initiatives. Continued research, innovation, and collaboration among stakeholders are essential to building inclusive and resilient societies where everyone has the opportunity to thrive. Financial inclusion is a critical component of sustainable development, promoting economic growth and reducing poverty by providing marginalized communities with access to financial services. Despite significant progress in recent years, numerous barriers hinder the widespread adoption of inclusive financial models. These barriers are multifaceted, spanning regulatory and policy constraints, market-based challenges, cultural and social obstacles, and operational difficulties in scaling and sustaining successful interventions. This essay explores these challenges in detail, drawing on quantitative and qualitative evidence from global case studies and academic literature.

Regulatory and policy barriers present significant obstacles to financial inclusion. Inconsistent policies across regions create fragmented financial ecosystems, making it difficult to implement uniform strategies that benefit all stakeholders. The World Bank notes that regulatory discrepancies in areas such as digital finance, lending practices, and consumer protection laws can hinder the expansion of financial services to underserved populations [103]. For example, while Kenya's regulatory environment facilitated the success of mobile money services like M-Pesa, similar innovations have struggled in countries with more restrictive financial regulations [104]. The lack of standardized policies also affects cross-border financial services, limiting the ability of migrants and refugees to access banking and remittance services [105]. Bureaucratic and legal hurdles further complicate financial inclusion initiatives. Lengthy approval processes, complex compliance requirements, and high costs associated with regulatory adherence can deter financial service providers from entering underserved markets [106]. Microfinance institutions, which play a crucial role in extending credit to low-income groups, often struggle with regulatory burdens that limit their operational flexibility [107]. In India, for instance, the microfinance sector has faced challenges due to state-level regulations that cap interest rates and restrict lending practices, sometimes leading to reduced access to credit for rural communities [108]. Additionally, anti-money laundering (AML) and know your customer (KYC) requirements, while essential for financial security, often exclude individuals without formal identification or proof of address, particularly affecting refugees, migrants, and those in informal economies [109].

Market-based challenges also impede financial inclusion, particularly through resistance from traditional financial institutions and volatility in capital markets. Established financial institutions often prioritize profitability and risk management over social impact, leading to a reluctance to serve low-income and high-risk customer segments [110]. These institutions may perceive low-income clients as less profitable due to smaller transaction sizes and higher operational costs [111]. Additionally, stringent lending criteria, such as high collateral requirements, exclude many individuals and small businesses from accessing formal credit [112]. The reluctance of banks to innovate and adopt inclusive financial products also stems from institutional inertia and regulatory concerns. For example, in parts of Latin America, traditional banks have been slow to adopt digital financial technologies, leaving a gap that fintech companies are now striving to fill [113]. Volatility and risks in capital markets further exacerbate financial inclusion challenges. Capital markets are often unpredictable, with emerging markets being particularly susceptible to economic shocks and investor sentiment shifts [114]. This volatility can discourage investments in financial inclusion initiatives, particularly those requiring long-term commitments, such as microfinance or impact investing [115]. The financial crisis of 2008 serves as a reminder of how market instability can lead to a contraction of credit and financial services, disproportionately affecting vulnerable populations [116]. Furthermore, impact investors often struggle to balance financial returns with social objectives, as market-driven pressures may lead to a prioritization of profit over inclusivity [117].

Cultural and social barriers, including gender norms and societal perceptions, also limit the effectiveness of financial inclusion efforts. In many societies, traditional gender roles restrict women's access to financial resources and decision-making power [118]. According to the World Bank, women are still 7% less likely than men to own a bank account, with cultural norms contributing significantly to this gap [119]. Patriarchal systems often dictate that financial assets and decision-making responsibilities remain under male control, limiting women's economic autonomy [120]. The Self-Employed Women's Association (SEWA) in India has highlighted how cultural barriers can prevent women from participating in savings groups or accessing microfinance, despite the availability of these services [121]. Marginalized communities, including ethnic minorities, refugees, and rural populations, face additional social and cultural challenges in accessing financial services. Discrimination, social exclusion, and a lack of trust in formal institutions contribute to low adoption rates of financial products among these groups [122]. For example, in South Africa, despite progressive financial inclusion policies, many rural communities continue to rely on informal financial systems due to historical distrust of banks [123]. Language barriers and low financial literacy further alienate marginalized groups from formal financial systems [124]. Operational challenges, such as scaling successful models and ensuring the sustainability of interventions, are critical issues in financial inclusion. Many pilot projects demonstrate success at a small scale but struggle to expand their impact geographically or across different demographics [125]. Scaling requires significant investment in infrastructure, technology, and human resources, which may not always be feasible for smaller institutions or non-profits [126]. The Grameen Bank model, which successfully provided microloans to rural women in Bangladesh, faced difficulties when replicated in other contexts without considering local socio-economic and cultural factors [127]. Ensuring the sustainability of financial inclusion initiatives involves addressing both financial and operational viability. Many programs rely on donor funding or subsidies, which may not be sustainable in the long term [128]. Developing self-sustaining business models that balance social impact with financial returns is challenging, particularly in volatile markets or low-income regions [129]. Moreover, operational sustainability requires continuous adaptation to evolving market dynamics, regulatory changes, and technological advancements [130]. For example, digital financial services that initially succeeded in urban settings may not be directly applicable in rural areas without tailored approaches to digital literacy and infrastructure [131]. In conclusion, achieving financial inclusion on a global scale requires overcoming a complex array of regulatory, market-based, cultural, and operational challenges. Inconsistent policies and bureaucratic hurdles create significant barriers to service delivery, while market-based challenges highlight the need for financial institutions to align profitability with inclusivity. Cultural and social barriers necessitate culturally sensitive approaches to empower women and marginalized communities. Operational challenges emphasize the importance of scaling successful models sustainably while adapting to diverse local contexts. Addressing these challenges demands a coordinated effort from governments, financial institutions, non-governmental organizations, and community leaders. By fostering an enabling environment through supportive policies, promoting innovation in financial products, and building trust within communities, stakeholders can make significant strides toward inclusive financial ecosystems that leave no one behind.

7 Conclusion

The research aimed to explore the multifaceted barriers to financial inclusion and propose strategies to enhance access to financial services for marginalized communities. Through an in-depth analysis of regulatory, market-based, cultural, and operational challenges, the study identified critical obstacles that hinder the effectiveness of inclusive financial models. Key findings highlight that inconsistent regulatory frameworks, bureaucratic and legal hurdles, resistance from traditional financial institutions, market volatility, cultural norms, and operational scalability are primary impediments to financial inclusion [119, 128]. The study also emphasized the need for innovative financial products, supportive policies, and community-driven approaches to address these barriers effectively [125]. Quantitative and qualitative analyses revealed that regulatory inconsistencies not only create fragmented financial ecosystems but also limit cross-border financial services, particularly affecting migrants and refugees [122]. Market-based challenges, including the reluctance of traditional financial institutions to serve low-income clients, were found to be exacerbated by institutional inertia and profitability concerns [132]. Furthermore, cultural and social barriers, such as gender norms and societal perceptions, significantly restrict women's access to financial resources, as seen in numerous case studies, including those by the Self-Employed Women's Association [121]. Operational challenges related to scaling successful models and ensuring sustainability were highlighted through examples like the Grameen Bank's struggle to replicate its model outside Bangladesh [127]. The research concludes that a holistic approach involving regulatory reforms, market innovation, cultural sensitivity, and operational efficiency is crucial for achieving broad-based financial inclusion. This research contributes to the existing body of knowledge by providing a comprehensive analysis of financial inclusion barriers through a multidisciplinary lens. While previous studies have often focused on specific challenges—such as regulatory or cultural issues—this research integrates diverse perspectives to present a holistic view of the obstacles and potential solutions [131]. By bridging gaps in literature concerning the interplay between regulatory environments, market dynamics, and cultural contexts, the study offers a nuanced understanding of how these factors collectively impact financial inclusion efforts [130].

Additionally, this research extends the discourse on impact measurement tools by evaluating frameworks like Social Return on Investment (SROI) and Gender and Inclusion Impact Assessments [129]. These insights provide practitioners and policymakers with practical tools to assess the effectiveness of financial inclusion initiatives, promoting evidence-based decision-making and enhancing program sustainability [118]. The findings of this research hold significant implications for policymakers, financial institutions, and society at large. Policymakers can leverage these insights to create more consistent and supportive regulatory frameworks that facilitate financial innovation and reduce barriers to market entry for inclusive financial services [133]. By harmonizing regulatory requirements, governments can enhance financial access for cross-border populations, including refugees and migrants [122]. For financial institutions, the study emphasizes the need to adopt inclusive practices that align profitability with social impact. This involves not only developing tailored financial products for underserved markets but also building institutional cultures that prioritize inclusivity [134]. The research also underscores the potential of public–private partnerships and collaborations with non-governmental organizations to create scalable and sustainable models of financial inclusion [121]. On a societal level, promoting financial inclusion can lead to broader social and economic benefits, including poverty reduction, gender equality, and improved living standards [128]. By addressing cultural and social barriers, stakeholders can empower marginalized communities, enhancing their participation in formal economies and contributing to social cohesion [120]. The research calls on stakeholders—including governments, financial institutions, NGOs, and civil society—to take proactive steps toward fostering financial inclusion. Policymakers should prioritize regulatory reforms that create enabling environments for inclusive financial products and services [119]. Financial institutions need to innovate and adapt their business models to serve high-risk and low-income groups effectively, leveraging technology and partnerships to reduce operational costs and enhance outreach [125]. Non-governmental organizations and community leaders play a crucial role in bridging the gap between formal financial institutions and marginalized populations. By promoting financial literacy and advocating for inclusive practices, these actors can help build trust and increase adoption of financial services among underserved communities [124]. Finally, the research emphasizes the need for continued research and innovation in financial inclusion. Academic and industry researchers should explore emerging trends, such as digital finance and impact investing, to develop new solutions that address evolving challenges in the financial ecosystem [131]. By maintaining a collaborative approach and sharing best practices, stakeholders can contribute to building more inclusive and resilient financial systems worldwide.

References

1. Demirgüç-Kunt, A., Klapper, L., Singer, D., Ansar, S., & Hess, J. (2018). *The Global Findex Database 2017: Measuring financial inclusion and the fintech revolution.* World Bank.
2. World Bank. (2022). *Global findex database 2021: Financial inclusion, digital payments, and resilience in the age of COVID-19.* World Bank.
3. Rawls, J. (1971). *A theory of justice.* Harvard University Press.
4. Sahay, R., Cihak, M., N'Diaye, P., Barajas, A., Mitra, S., Kyobe, A., & Mooi, Y. N. (2020). *Women in finance: A case for closing gaps.* IMF Staff Discussion Notes.
5. Klapper, L., & Singer, D. (2015). The opportunities and challenges of digitizing government-to-person payments. In *World Bank policy research working paper no. 7239.*
6. Allen, F., Demirgüç-Kunt, A., Klapper, L., & Peria, M. S. M. (2016). The foundations of financial inclusion: Understanding ownership and use of formal accounts. *Journal of Financial Intermediation, 27*, 1–30.
7. Ayyagari, M., Demirgüç-Kunt, A., & Maksimovic, V. (2011). Small vs. young firms across the world: Contribution to employment, job creation, and growth. In *World Bank policy research working paper no. 5631.*
8. Beck, T., Demirgüç-Kunt, A., & Levine, R. (2011). Financial structure and economic development: Firm, industry, and country evidence. *Review of Financial Studies, 24*(12), 422–460.
9. Pereira, V., & Temouri, Y. (2020). The role of social impact bonds and innovative finance in fostering inclusive growth. *Journal of Business Research, 121*, 321–329.
10. Demirgüç-Kunt, A., & Klapper, L. (2013). Measuring financial inclusion: Explaining variation in use of financial services across and within countries. In *Brookings papers on economic activity* (pp. 279–321).
11. Pereira, V., & Temouri, Y. (2020). Impact investing, social finance and social enterprises: A new ecosystem for social change. *Strategic Change, 29*(3), 309–315.
12. World Bank. (2022). *Financial inclusion overview.* Retrieved from https://www.worldbank.org
13. Klapper, L., & Singer, D. (2015). *The opportunities of digitizing payments.* World Bank Development Research Group.
14. Demirgüç-Kunt, A., & Klapper, L. (2013). Measuring financial inclusion: Explaining variation in use of financial services across and within countries. *Brookings Papers on Economic Activity, 2013*(1), 279–340.
15. Creswell, J. W., & Plano Clark, V. L. (2017). *Designing and conducting mixed methods research* (3rd ed.). SAGE Publications.
16. Dimson, E., Marsh, P., & Staunton, M. (2020). *Credit Suisse global investment returns yearbook 2020.* Credit Suisse Research Institute.
17. Bugg-Levine, A., & Emerson, J. (2011). Impact investing: Transforming how we make money while making a difference. *Innovations, 6*(3), 9–18.

18. Porter, M. E., & Kramer, M. R. (2011). Creating shared value. *Harvard Business Review, 89*(1/2), 62–77.
19. Sen, A. (1999). *Development as freedom.* Oxford University Press.
20. Nussbaum, M. C. (2011). *Creating capabilities: The human development approach.* Harvard University Press.
21. Elson, D. (1999). Labor markets as gendered institutions: Equality, efficiency, and empowerment issues. *World Development, 27*(3), 611–627.
22. Sahay, R., Čihák, M., N'Diaye, P., & Barajas, A. (2020). *Financial inclusion: Can it meet multiple macroeconomic goals?* IMF.
23. Elson, D. (1999). *Labor markets as gendered institutions: Equality, efficiency and empowerment issues.* World Development.
24. Klapper, L., & Singer, D. (2015). *The opportunities and challenges of financial inclusion.* World Bank.
25. World Bank. (2022). *Women, business and the law 2022.*
26. Allen, F., Demirgüç-Kunt, A., Klapper, L., & Peria, M. S. M. (2016). The foundations of financial inclusion: Understanding ownership and use of formal accounts. *Journal of Financial Intermediation.*
27. Dimson, E., Marsh, P., & Staunton, M. (2020). *Diversity and inclusion in investment.* Credit Suisse Research Institute.
28. Pereira, V., & Temouri, Y. (2020). Inclusive capitalism and financial markets. *Journal of Business Research.*
29. Demirgüç-Kunt, A., & Klapper, L. (2013). Measuring financial inclusion: Explaining variation in use of financial services across and within countries. In *Brookings papers on economic activity.*
30. Ayyagari, M., Demirgüç-Kunt, A., & Maksimovic, V. (2011). *Small vs. young firms across the world: Contribution to employment, job creation, and growth.* World Bank.
31. Beck, T., Demirgüç-Kunt, A., & Martinez Peria, M. S. (2011). Bank financing for SMEs: evidence across countries and bank ownership types. *Journal of Financial Services Research.*
32. Nussbaum, M. (2011). *Creating capabilities: The human development approach.* Harvard University Press.
33. Porter, M. E., & Kramer, M. R. (2011). Creating shared value. *Harvard Business Review.*
34. Bugg-Levine, A., & Emerson, J. (2011). *Impact investing: Transforming how we make money while making a difference.* Wiley.
35. Suri, T., & Jack, W. (2016). *The long-run poverty and gender impacts of mobile money.* Science.
36. OECD. (2022). *Financial markets, insurance and pensions: Digitalisation and the future of financial inclusion.*
37. Beck, T., De la Torre, A., & Ize, A. (2018). *Financial development in Latin America and the Caribbean: The road ahead.* World Bank.
38. Yunus, M. (2007). *Creating a world without poverty: Social business and the future of capitalism.* Public Affairs.
39. Khandker, S. R. (2005). *Microfinance and poverty: Evidence using panel data from Bangladesh.* World Bank Economic Review.
40. FSD Kenya. (2020). *Savings groups and women's empowerment in Kenya.*
41. Central Bank of Brazil. (2021). *Caixa tem and digital financial inclusion.*
42. Bangko Sentral ng Pilipinas (BSP). (2022). *Economic and financial learning program.*
43. Government of India. (2022). *Pradhan Mantri Jan Dhan Yojana progress report.*
44. People's Bank of China (PBOC). (2021). *Fintech development and regulation report.*
45. Yunus, M. (2007). *Creating a world without poverty: Social business and the future of capitalism.*
46. Khandker, S. R. (2005). *Microfinance and poverty: Evidence using panel data from Bangladesh.* The World Bank Economic Review.
47. Hulme, D. (2008). *The story of the Grameen Bank: from subsidized microcredit to market-based microfinance.*

48. Armendáriz, B., & Morduch, J. (2010). *The economics of microfinance.*
49. OECD. (2022). *Social impact bonds: State of play & lessons learned.*
50. Warner, M. E. (2013). *Private finance for public goods: Social impact bonds.*
51. Fraser, A., Tan, S., Lagarde, M., & Mays, N. (2018). *Narrative systematic review of the impact of social impact bonds.*
52. Roy, M. J., McHugh, N., & Sinclair, S. (2018). *A critical reflection on social impact bonds.*
53. Friede, G., Busch, T., & Bassen, A. (2015). *ESG and financial performance: Aggregated evidence.*
54. Schoenmaker, D., & Schramade, W. (2019). *Principles of sustainable finance.*
55. PRI. (2021). *Annual report: ESG growth and trends.*
56. Kölbel, J. F., Heeb, F., Paetzold, F., & Busch, T. (2020). *Can sustainable investing save the world?*
57. Arner, D. W., Barberis, J., & Buckley, R. P. (2020). *Fintech and RegTech: Impact on financial inclusion.*
58. Gomber, P., Koch, J. A., & Siering, M. (2018). *Digital finance and FinTech: Current research and future research directions.*
59. Chen, Y., Huang, A., & Zhang, M. (2019). *Crowdfunding and digital lending for social innovation.*
60. Zavolokina, L., Dolata, M., & Schwabe, G. (2021). *Fintech—What's in a name?.*
61. BIS. (2021). *Fintech regulation and the role of Central Banks.*
62. Clark, C., Emerson, J., & Thornley, B. (2015). *Impact investing 2.0: The way forward.*
63. Bugg-Levine, A., & Emerson, J. (2011). *Impact investing: Transforming how we make money while making a difference.*
64. Smith, C. (2020). *Gender quotas and financial inclusion.*
65. Johnson, K., & Lee, Y. (2021). *Incentives for inclusive investment.*
66. Khan, H. (2019). *Simplified KYC and financial inclusion.*
67. World Bank. (2022). *The global findex database.*
68. Miller, L., & Roberts, J. (2023). *Gender bonds and women's empowerment.*
69. Davies, R. (2021). *Social impact bonds: A critical review.*
70. Patel, S. (2020). *Community investment funds for local development.*
71. Nguyen, T. (2022). *Fintech innovations and financial inclusion in developing countries.*
72. Singh, A., & Zhao, J. (2024). *Blockchain for inclusive finance.*
73. OECD. (2021). *Financial literacy and inclusion.*
74. Klapper, L., & Lusardi, A. (2020). *Financial literacy and financial inclusion in the developing world.*
75. UN Women. (2023). *Financial education for women and girls.*
76. IFC. (2022). *Financial literacy and education programs.*
77. SEWA. (2022). *Empowering women through education and financial services.*
78. BRAC. (2023). *Integrating financial literacy in microfinance.*
79. FINCA. (2024). *Financial education initiatives.*
80. Grameen Foundation. (2023). *University partnerships for financial inclusion.*
81. OECD. (2023). *ESG principles in private sector finance.*
82. IFC. (2023). *Impact investing for inclusive growth.*
83. Goldman Sachs. (2024). *10,000 women initiative.*
84. World Bank. (2024). *Incentivizing inclusive investments.*
85. Mastercard. (2023). *Digital financial inclusion project in Nigeria.*
86. International Finance Corporation (IFC). (2024). *Banking on women program.*
87. UN Women. (2024). *Women's empowerment principles (WEPs).*
88. Mwangi, J. (2023). *The role of Chama groups in Kenya.*
89. World Bank. (2023). *Economic empowerment indicators.*
90. Kumar, R., & Singh, A. (2024). *MGNREGA and its impact on rural communities.*
91. International Labour Organization (ILO). (2022). *Impact of vocational training in Sub-Saharan Africa.*
92. UNDP. (2024). *Social indicators for evaluating empowerment.*

93. UNESCO. (2023). *Education and its impact on poverty reduction.*
94. Yunus, M. (2022). *Grameen Bank microfinance model.*
95. Global Impact Investing Network (GIIN). (2024). *Impact investing trends.*
96. Bouri, A., et al. (2023). *Impact of investment in East African agriculture.*
97. International Finance Corporation (IFC). (2023). *Impact of banking on women.*
98. Nicholls, J., et al. (2023). *Social Return on Investment (SROI) methodology.*
99. Impact Hub. (2024). *SROI analysis of vocational training in South Asia.*
100. UN Women. (2024). *Gender and inclusion impact assessments.*
101. OECD. (2023). *Gender and inclusion metrics.*
102. European Investment Bank (EIB). (2023). *Gender lens investing.*
103. World Bank. (2023). *Regulatory issues in financial inclusion.*
104. Jack, W., & Suri, T. (2022). *Mobile money success in Kenya.*
105. International Finance Corporation (IFC). (2024). *Cross-border financial services for migrants.*
106. UNDP. (2023). *Regulatory hurdles in financial inclusion.*
107. Klapper, L., & Singer, D. (2023). *Microfinance and regulatory challenges.*
108. Sinha, M. (2024). *Microfinance regulations in India.*
109. World Bank. (2023). *Impact of AML and KYC regulations on financial inclusion.*
110. Bouri, A., et al. (2023). *Resistance from traditional financial institutions.*
111. GIIN. (2024). *Financial inclusion and profitability perceptions.*
112. OECD. (2023). *Lending criteria and financial inclusion.*
113. Banerjee, A., & Duflo, E. (2024). *Digital financial technology adoption in Latin America.*
114. IMF. (2023). *Volatility in capital markets.*
115. Nicholls, J., et al. (2024). *Challenges in impact investing.*
116. Stiglitz, J. (2023). *The 2008 financial crisis and its aftermath.*
117. GIIN. (2024). *Balancing profit and social objectives in impact investing.*
118. UN Women. (2024). *Gender norms and financial inclusion.*
119. World Bank. (2023). *Gender disparities in financial account ownership.*
120. Chopra, R. (2024). *Patriarchal financial systems and women's autonomy.*
121. SEWA. (2023). *Barriers to microfinance participation in India.*
122. International Finance Corporation (IFC). (2024). *Financial access for marginalized groups.*
123. Ndlovu, T. (2023). *Informal financial reliance in rural South Africa.*
124. OECD. (2023). *Language and literacy challenges in financial inclusion.*
125. Bouri, A., et al. (2023). *Scaling challenges in financial inclusion.*
126. Impact Hub. (2024). *Infrastructure needs for expanding inclusion programs.*
127. Yunus, M. (2022). *Grameen Bank model and its replication challenges.*
128. UNDP. (2023). *Sustainability of financial inclusion programs.*
129. Global Impact Investing Network (GIIN). (2024). *Business model innovation for impact.*
130. Nicholls, J., et al. (2024). *Market evolution and financial inclusion adaptation.*
131. Banerjee, A., & Duflo, E. (2024). *Digital finance innovations and limitations.*
132. Klapper, L., & Singer, D. (2023). *Institutional resistance in traditional banking.*
133. International Monetary Fund (IMF). (2023). *Regulatory reforms for financial inclusion.*
134. Stiglitz, J. (2023). *Inclusive financial practices in banking.*

ESG Investing: Performance Measure of SAARC Countries

Anand Patil, Akhil Suresh, and R. B. Lakshmi

Abstract Environmental, social, and governance (ESG) investing, often known as ESG investing, is an investment strategy considering how a company's operations and policies affect the environment, society, and corporate governance. ESG investment evaluates a company's influence on society and the environment in addition to the standard financial study of its performance. The purpose of this study is to measure the performance of ESG investing in SAARC countries. This is done mainly through two focused studies. One study derives the empirical results of ESG investing in SAARC countries by analyzing data obtained from the MSCI ESG Leaders Factsheet and Bloomberg. Through this study, we aim to measure how effectively ESG investing is being adopted in each SAARC country. The second study measures ESG scores using the MSCI index. This study lays out the blueprint to help investors examine their investments' sustainability performance. This study will enable the investors within the various SAARC countries to compare the ESG score with the MSCI index and see how better they can perform than other countries. The results of this study illustrate the significance of ESG aspects in investment decisions and corporate sustainability and have substantial implications for investors, politicians, and businesses in SAARC countries.

Keywords SAARC · ESG investing · MSCI

A. Patil (✉) · A. Suresh · R. B. Lakshmi
School of Business and Management, Christ University, Bangalore, India
e-mail: anand.patil@christuniversity.in

A. Suresh
e-mail: akhil.suresh@mba.christuniversity.in

R. B. Lakshmi
e-mail: lakshmi.rb@christuniversity.in

K. Khanna et al. (eds.), *From Poverty to Prosperity*, Information Systems Engineering and Management 64, https://doi.org/10.1007/978-3-031-99166-0_7

1 Introduction

ESG also known as economic, social, and governance is referred to as a set of norms that are used by investors who are socially responsible to make investments that will benefit all the three fragments which are economic, social, and governance. The investments made by the socially responsible investors are meant to fulfill certain socially and economic-oriented goals such as decreasing the rate of pollution, reducing the disposal of waste, and lowering the emission of carbon dioxide. The frameworks established through ESG are significant in incorporating sustainable investing successfully since these frameworks help in ensuring whether the company making the ESG investments is in par with the goals and values exhibited in the ESG framework. This in the long run will improve the public image of the company. Even though ESG investing is rising in prominence across various emerging countries but the concept of ESG investing does have its own constraints; moreover, it also has not received the opening it hoped to have. The study we intend to conduct looks to evaluate the performance of ESG investing from the lens of the eight major SAARC countries. Although ESG investment is still in its infancy in SAARC nations, interest in the subject is rising. Many investors in the area are starting to take into account these factors when making investment decisions as they become more aware of the effects their investments have on the environment and society. The study conducted by me will look toward measuring the performance of socially responsible investing being done across SAARC countries. Through this study, we are able to gain a clear insight with regard to the various financial decisions that can be taken while incorporating the strategy of socially responsible investing. In this study, we will be exploring the impact of socially responsible investing by using ESG indices as the measuring stick. The ESG indices can be effectively used by investors to evaluate the efficiency of a company when it comes to performing financially in a healthier manner. In this study, we will be exploring the impact of socially responsible investing by using ESG indices as the measuring stick. The ESG indices can be effectively used by investors to evaluate the efficiency of a company when it comes to performing financially in a healthier manner.

2 Literature Review

Aggarwal [1] suggests that investments in socially conscious businesses can yield higher financial returns in developing economies. Jain and Khanna [2] find that socially conscious mutual funds in India outperformed the traditional mutual funds. Also, the socially conscious mutual funds beat their conventional counterparts. The socially conscious mutual funds invest in businesses that are dedicated to ESG issues, which might improve financial performance over the long run. Dey and Paul [3] talked about India's expanding economy and rising social and environmental concerns make sustainable investing more crucial than ever. Shahzad and Qureshi [4] explained in order to foster an atmosphere that is favorable for investments; there is a need for regulatory support for ESG investing. Ahmed and Tariq [5] discussed that the companies with superior corporate governance have higher profits, sales expansion, and return on assets. The effectiveness of corporate governance has a favorable impact on stock market returns. The businesses with stronger corporate governance had superior financial results and stock market returns. Iqbal and Hameed [6] explained the stock returns are higher for companies with better ESG ratings than for companies with lower ESG ratings. The companies in the energy, minerals, and healthcare sectors have larger favorable correlations between ESG ratings and stock gains. *The* companies with stronger ESG performance have greater ROA and ROE financial performance. The social and governance variables have a less impact on financial success than environmental factors. The largest environmental influences on financial performance are climate change and energy efficiency. In order for businesses to increase their financial performance and draw in more investment, policymakers should push them to embrace ESG practices [7]. Investors need to take ESG factors into account when choosing investments because they have a big impact on financial performance [8]. Rafique and Islam [9], in order for businesses to increase their financial performance and draw in more investment, policymakers should encourage corporates to adopt stronger corporate governance practices and also recommend the corporate governance mechanisms that are most successful in enhancing business performance, such as board independence and audit committee [10]. The ownership structure significantly affects the financial success of companies with higher degrees of CSR activity. The consumers appreciate and favor businesses that are open about their social and environmental obligations. The smaller businesses have a better correlation between CSR reporting and consumer loyalty than larger businesses. The managers should prioritize CSR reporting as a way to increase client loyalty and boost their company's reputation [11]. According to [12], the significance of encouraging CSR has led to advance sustainable development and improve the image of Nepal [13]. South Asia is still in the early phases of ESG investing, with differing degrees of implementation and awareness in the various nations. One of the biggest obstacles for investors in the area is the absence of trustworthy ESG data [14]. The ESG investments in SAARC countries outperform both the benchmark index and businesses with low ESG ratings. The regulatory climate, investor attitudes, and the accessibility of ESG data have an impact on how well ESG investments perform in different nations [15]. The legislative and regulatory frameworks aid in advancing ESG investing. The study showed both the advantages and disadvantages of ESG [16]. The growth of assets under management and the implementation of ESG standards and laws are two market trends for ESG investing in Asia. There is a link between ESG variables and financial performance, which demonstrates the ESG integration's consistently beneficial influence on risk-adjusted returns [17]. There is a connection between ESG elements and business financial performance, and there exists conflicting findings regarding the influence of ESG integration on financial outcomes [18]. There is a connection between ESG performance and risk-adjusted returns. There are many methods used to quantify and incorporate ESG variables in investment decision-making, and also there are potential sources of heterogeneity in

ESG research [19]. The ESG issues affect equity valuation, and also there is a positive correlation between ESG performance and corporate valuation. There are various methods to gauge both financial and CSR performance. There exists a connection between CSR and financial success, stressing the conflicting information regarding the influence of CSR on financial results. In addition to advancing knowledge of the role of CSR in corporate decision-making and investment strategies [20]. Yoon and Schmelzeig [21], there are various methods for gauging financial and ESG performance which indicates different avenues by which ESG factors may influence financial outcomes and analyze the significance of the results for investors and company managers [22, 23]. The favorable association between ESG performance and financial results is generally suggested for long-term success, also helping investors and corporate managers who want to include ESG considerations into their decision-making processes [24]. The financial performance may be enhanced for both corporate managers who prioritize ESG performance and investors who take ESG considerations into account when making investment decisions. The ESG investing can be profitable for investors and corporate managers who want to include ESG factors into their decision-making processes, and the authors also offer helpful advice.

3 Methodology

In this study, the required data will be collected from the MSCI ESG Leaders Index series with respect to the SAARC countries. This study makes use of these data from ESG Leaders Index series since there is not sufficient data within the socially responsible investing forum for the selected SAARC countries. Collecting data from a single banner would ease in facilitating the comparison between the SAARC countries. The variables selected for this study are MSCI Index (Morgan Stanley Capital International Index) and ESG investing. This paper closely examines and critically analyzes how well ESG investing has been adopted across the various SAARC countries. The study will be conducted during the timeframe between September 1, 2019 and March 31, 2020. The reason as to why a specific timeframe was undertaken for this study is mainly due to the limited availability of the required data. This period was taken to analyze the recent scenario of the selected countries. It gives us a clearer picture with regard to the economic and financial standing of each SAARC country with respect to the adoption of ESG Investing. There will be several metrics for each country on the basis with which the evaluation will be undertaken. Following, I shall be rendering the number of constituents as well as the sector weights of each SAARC country as sourced from the *MSCI*.

Table 1 Unit root test—2013–2022 (SAARC countries)

Variables	Level	First difference	Second difference
ESG score	1	0.3339	0
MSCI index	0	0	0

Source Authors' work

3.1 ESG Leaders Factsheet

According to study, each variable's initial value, unaltered, is displayed in the "Level" column. Table 1 shows MSCI score that is currently at a level of 0.0000, whereas the ESG score is at a level of 1.0000 in this instance. As a result, the initial MSCI index is 0.0000, and the initial ESG score is 1.0000. Each variable's difference between its present value and its previous value is displayed in the "First Difference" column, indicating whether the variables are going upwards or downwards can help. The ESG score has increased by 0.3339 units from its previous value in this instance, as indicated by the first difference of the ESG score, which is 0.3339. The MSCI index has not changed from its prior value, as indicated by the initial difference of 0.0000. The difference between the current first difference value and its preceding first difference value is displayed in the "Second Difference" column. This can aid in determining whether the variables' rates of change are altering over time. The second difference of the ESG score in this instance is 0.0000, indicating that the ESG score's rate of change over time has not changed. The second difference in the MSCI index is also 0.0000, indicating that the rate of change of the MSG index over time has not changed. Overall, the findings indicate that while the MSG index has remained steady, the ESG score has increased since its prior value. It is crucial to keep in mind that it is challenging to make any definite inferences from these results without knowing more about the context of the investigation. Prior to coming to any firm conclusions or choices, it is crucial to take into account additional aspects and data sources because changes in the level and rate of change of these variables may be influenced by a range of external factors.

Table 2 shows the result of a Pedroni residual cointegration test, which is used to establish the cointegration of a group of time series variables. The DESG_SCORE and DMSCI_INDEX particular variables are being examined, and the sample period is from 2013 to 2022. There are 40 observations and four cross sections included in the sample. The alternative hypothesis is that there are shared or individual AR coefficients, while the null hypothesis is that there is no cointegration. Therefore, we accept the null hypothesis. With a one lag time and no deterministic trend, the test is run. Two sections with the results are presented. The panel statistics for both the collective and individual AR coefficients are shown in the first section. The panel V-statistic, panel rho-statistic, panel Phillips–Perron (PP) statistic, and panel augmented Dickey–Fuller (ADF) statistic are among the panel statistics. The augmented Dickey–Fuller results and the Phillips–Perron results, which are cross section specific, are presented in the second part. The greatest lag, the autoregressive coefficient, and the variance for each cross section are shown in the augmented Dickey–Fuller findings while the autoregressive coefficient and variance are shown in the Phillips–Perron results. In conclusion, the results show that DESG_SCORE and DMSCI_INDEX appear to be cointegrated, as shown by the statistically significant panel V-statistic and panel PP-statistic. The nonsignificant group PP-statistic and group ADF-statistic imply that the individual AR coefficients do not exhibit cointegration. However, the specific context and underlying presumptions of the analysis should be taken into account when interpreting the findings.

Table 2 Cointegration test

Pedroni residual cointegration test
Series: DESG_SCORE DMSG_INDEX
Date: 04/13/23 Time: 13:49
Cross sections included: 4
Null hypothesis: no cointegration
Trend assumption: no deterministic trend
User-specified lag length: 1
Newey-west automatic bandwidth selection and Bartlett kernel

Alternative hypothesis: common AR coefs. (within dimension)

	Statistic	Prob.	Weighted statistic	Prob.
Panel v-statistic	5.018099	0.0	5.334891	0.0
Panel rho-statistic	− 1.26239	0.1034	− 1.324918	0.0926
Panel PP-statistic	0.065875	0.5263	0.428328	0.6658
Panel ADF-statistic	0.993443	0.8398	1.115752	0.8677

Alternative hypothesis: individual AR coefs. (between dimensions)

	Statistic	Prob
Group rho-statistic	− 0.828793	0.2036
Group PP-statistic	0.7989	0.7878
Group ADF-statistic	3.464132	0.9997

Cross section-specific results—Phillips–Peron results (nonparametric)

Cross ID	AR(1)	Variance	HAC	Bandwidth	Obs
1.0	− 1.413	783.4882	783.4882	0.0	5.0
2.0	0.129	717.2992	717.2992	0.0	8.0
3.0	− 0.194	382.6571	382.6571	0.0	8.0
4.0	− 0.185	631.4432	631.4432	0.0	8.0

Augmented Dickey–Fuller results (parametric)

Cross ID	AR(1)	Variance	Lag	Max lags	Obs
1	6.18	172.3681	1	–	3
2	0.065	817.7816	1	–	7
3	− 0.209	436.8974	1	–	7
4	− 0.131	720.8254	1	–	8

Source Authors' work

Table 3 Regression analysis of ESG index and MSCI index during the year 2022

Variables entered/removed

Model	Variables entered	Variables removed	Method
1	MSCI index[b]		Enter

[a] Dependent variable: ESG Index
[b] All requested variables entered

Model summary

Model	R	R^2	Adjusted R^2	Std. error of the estimate
1	0.517[a]	0.268	– 0.099	9.193

[a] Predictors: (constant), MSCI index

ANOVA[a]

Model	Sum of squares	Df	Mean square	F	Sig.
Regression	61.742	1	61.742	0.73	0.483[b]
Residual	169.008	2	84.504		
Total	230.75	3			

[a] Dependent variable: ESG index

Source Authors' work

Table 3 shows the regression analysis between ESG score and MSCI index for the year 2022. The significance value is (0.019) which is more than 0.05 which states that the null hypothesis will be accepted and the alternative hypothesis will be rejected. Hence, there is no increase/decrease in ESG score in comparison with an increase/decrease in MSCI score.

4 Findings

The study indicates that Indian businesses with high ESG (environmental, social, and governance) scores tend to perform better financially, being generally more profitable, less risky, and enjoying higher credit ratings. In Afghanistan, ESG investing has not gained significant attention; however, the international community and development organizations have advocated for the inclusion of ESG factors in investment and development projects, which may pave the way for future ESG investment in the country. In Bangladesh, investors are actively trying to integrate ESG factors into their investment decisions as ESG investing becomes more popular, though there is still considerable room for improvement in ESG reporting and transparency. Meanwhile, in Bhutan, ESG investing is gaining traction among socially conscious investors, despite the lack of performance statistics; the potential for ESG investing here may be more limited compared to larger economies with more developed financial systems. In the Maldives, ESG investing is increasingly recognized as important by investors who wish to support environmentally friendly companies, and the nation's focus on sustainability and reliance on tourism suggest a promising future for ESG investment. In Nepal, ESG investment is a relatively new concept with limited performance information but offers potential opportunities due to its abundant natural resources and commitment to sustainable development. Finally, in Pakistan, investors are emphasizing ESG investments as they recognize the importance of supporting companies that prioritize sustainability and social responsibility. However, the success of ESG investing in Pakistan has been inconsistent and challenging, despite opportunities in sectors such as water conservation, renewable energy, and social inclusion. It has a sizable and vibrant financial sector that can provide ESG investors a range of investment options.

5 Conclusion

In conclusion, ESG investing has enormous potential to promote social advancement, environmental preservation, and sustainable economic growth in SAARC nations. While there are many concerns in the region that need to be addressed, such as poverty, inequality, climate change, and the depletion of natural resources, ESG investing can assist in addressing these problems by supporting businesses that have excellent ESG practices and urging them to adopt more sustainable business models. In order to advance ESG investing, SAARC nations have developed legal frameworks, introduced ESG funds, and pushed for sustainability reporting. To completely mainstream ESG investing in the area, however, there is still a long way to go, and more work is required to advance ESG practices among businesses, investors, and other stakeholders. ESG investments in SAARC nations have financial as well as environmental and social advantages. Strong ESG practices help businesses manage risk more effectively, which puts them in a better position to take advantage of emerging market possibilities. For investors who give ESG considerations top priority when making investment decisions, this presents the possibility of long-term financial returns. In conclusion, ESG investing is an effective strategy for fostering sustainable growth and long-term value generation in SAARC nations. While the path to establishing ESG investing as the norm in the region may be difficult, the potential rewards make it worthwhile for businesses, investors, and governments alike.

6 Recommendation

1. Increase Awareness: Increasing investor and corporate understanding of the significance of ESG aspects is one of the key initiatives for improving ESG in SAARC nations. Investors should be made aware of the advantages of investing in ethical and sustainable businesses, and businesses should be educated about the effects of their operations on the environment and society.

2. Government Regulations: Governments should enact legislation that support ESG practices, such as tax breaks for businesses that embrace sustainable practices and fines for those who disregard social and environmental laws. Governments can support sustainable agricultural methods and encourage the use of renewable energy sources.
3. Corporate Responsibility: Businesses should follow ethical standards and be open about their effects on the community and environment. Along with respecting human rights, they should put employee safety and well-being first. Initiatives for corporate social responsibility can also assist businesses in supporting the growth of their local communities.
4. Sustainable Finance: By providing green bonds, sustainable investment funds, and other sustainable financial instruments, financial institutions can encourage investors to prioritize investing in businesses that have excellent ESG practices.
5. Collaboration: To develop ESG practices in SAARC nations, cooperation between governments, businesses, and investors is crucial. Investors can interact with businesses to persuade them to enhance their ESG practices, and public–private partnerships can be developed to work toward shared objectives.
6. Education and Training: Programs for education and training can be created to spread awareness of ESG practices among workers, business owners, and other stakeholders. Programs for responsible investing, ethical business conduct, and sustainable practices might all fall under this category.
7. Certification and Reporting: Governments can encourage businesses to establish sustainability certification programs and to report on their ESG practices on a regular basis. Certification and Reporting. This can include sustainable reporting frameworks like the sustainable Accounting Standards Board (SASB) or the Global Reporting Initiative (GRI).
8. Engagement of Investors: Investors can interact with businesses to support ESG practices and motivate them to enhance their sustainability performance. Participation by shareholders, proxy voting, and withdrawal from businesses that do not adhere to ESG standards can all be examples of this.
9. Research and Development: To encourage innovation in sustainable technologies, renewable energy sources, and sustainable agriculture, governments might fund research and development programs. This may contribute to the expansion of the economy and the creation of new job opportunities.
10. Joining Hands with International Organizations: To promote sustainable development and gain access to finance for sustainable initiatives, SAARC nations can work with international organizations like the World Bank and the UNEP.

References

1. Aggarwal, S. K. (2017). Socially responsible investing in emerging markets: Performance analysis of Dow Jones sustainability emerging markets index. *Journal of Emerging Market Finance, 16*(2), 125–147.
2. Jain, K., & Khanna, S. (2018). A comparative study on performance of socially responsible and conventional mutual funds in India. *International Journal of Financial Studies, 6*(1), 34–48.
3. Dey, P., & Paul, S. (2020). Sustainable investment in India: A review of academic literature. *Environmental Economics and Policy Studies, 22*(1), 23–45.
4. Shahzad, A., & Qureshi, H. (2020). Environmental, social, and governance investing in Pakistan: A survey of investors and investment managers. *Asian Economic and Financial Review, 10*(4), 456–478.
5. Ahmed, M., & Tariq, N. (2020). Does corporate governance impact the stock market returns of the firm? A study of SAARC countries. *South Asian Journal of Business Studies, 9*(3), 213–231.
6. Iqbal, A., & Hameed, S. (2021). Impact of environmental, social, and governance (ESG) ratings on stock returns: Evidence from the South Asian Association for Regional Cooperation (SAARC) region. *Journal of Sustainable Finance & Investment, 12*(3), 267–287.
7. Ali, Z., Kouser, R., & Asghar, M. (2019). The link between environmental, social, and governance factors and their impact on financial performance: Evidence from the Pakistani stock market. *Pakistan Journal of Commerce and Social Sciences, 13*(2), 312–329.
8. Rahman, M., Ali, F., & Akhter, S. (2021). Environmental, social, and governance (ESG) factors and financial performance: Evidence from Bangladesh. *Asian Journal of Sustainability and Social Responsibility, 6*(1), 45–62.
9. Rafique, T., & Islam, M. (2020). Corporate governance and firm performance in South Asia: Evidence from Bangladesh, India, and Pakistan. *South Asian Journal of Corporate Governance, 4*(1), 89–108.
10. Gupta, S., Singh, R., & Jain, M. (2021). Corporate social responsibility and firm performance in India: The role of ownership structure. *International Journal of Corporate Governance, 12*(3), 151–168.
11. Yadav, P. (2020). Impact of CSR reporting on customer loyalty in Indian firms. *Management Science Letters, 10*(3), 589–596.
12. Dangol, R., & Thapa, S. (2021). Corporate social responsibility investing in Nepal. *Nepal Economic Review, 33*(1), 87–98.
13. Irfan, M., & Iqbal, S. (2021). ESG investing in South Asian countries: A comparative analysis. *South Asian Journal of Finance, 6*(2), 123–145.
14. Ahmed, J., & Asif, M. (2020). Measuring the performance of ESG investing in emerging markets: A case study of SAARC nations. *Emerging Markets Review, 23*(4), 78–90.
15. Sarda, A., & Bhattacharya, S. (2021). ESG investing in South Asia: Opportunities and challenges. *Asia-Pacific Financial Markets, 28*(3), 145–165.
16. Zi, H., & Lu, X. (2020). ESG investing in Asia: A review of performance and market trends. *Journal of Asian Economics, 72*, 1–12.
17. Grewal, J., & Serafeim, G. (2019). ESG and financial performance: Aggregated evidence from more than 2000 empirical studies. *Management Science Review, 46*(3), 112–138.
18. Ng, A. C., & Rezaee, Z. (2015). ESG and corporate financial performance: A literature review. *Journal of Accounting Literature, 34*(1), 1–23.
19. Kacperczyk, M., Sialm, C., & Zheng, L. (2021). ESG investing and performance: A review of the academic literature. *Journal of Finance and Economics, 58*(2), 250–270.
20. Vogel, D. (2005). The wages of social responsibility: Where are they? *California Management Review, 47*(4), 40–63.
21. Yoon, A., & Schmelzeig, P. (2021). ESG investing: A review of the evidence. *Review of Financial Studies, 34*(7), 1450–1487.

22. Gjorven, T., & Knivsfla, L. (2020). The effectiveness of ESG screening: Evidence from the Norwegian Government Pension Fund Global. *Journal of Portfolio Management, 46*(5), 110–123.
23. Scholtens, B., & Diev, P. (2018). ESG and financial performance: Uncovering the relationship. *Journal of Business Ethics, 147*(1), 155–173.
24. Altieri, M. L., & Moro, A. (2018). ESG investing: Does it add value? *Journal of Corporate Finance, 52*, 197–212.

The Role of Fintech Innovations in Bridging Financial Inclusion Gaps in Capital Markets

Priyanka Chawla, Meenakshi Chawla, and Pravin Saxena

Abstract This study investigates the impact of Fintech innovations on financial inclusion within capital markets, emphasizing the role of digital financial technologies in making opportunities more accessible, affordable, and secure for underserved populations. The study uses a quantitative approach, distributing structured questionnaires to approximately 400 users from underserved populations. Independent variables are digital literacy, mobile accessibility, regulatory support, and data security/privacy, with trust as a mediator and financial inclusion as the dependent variable. Structural equation modeling (SEM) will assess the relationships among these variables. The primary objective is to assess the mediating role of trust in FinTech platforms, encouraging capital market engagement among underserved groups. Trust in FinTech platforms is anticipated to mediate this relationship enhancing the effectiveness of FinTech features in driving engagement among traditionally excluded groups.

Keywords Financial inclusion · FinTech innovation · Underserved population · Capital market

P. Chawla (✉)
Prestige Institute of Management and Research, Gwalior, India
e-mail: Priyanka.chawla@prestigegwl.org

M. Chawla
St. John's College, Agra, India
e-mail: Chawla.meenakshi4@gmail.com

P. Saxena
Dayalbagh Educational Institute, Deemed University, Agra, India
e-mail: pravinsaxenadei@yahoo.co.in

K. Khanna et al. (eds.), *From Poverty to Prosperity*, Information Systems Engineering and Management 64, https://doi.org/10.1007/978-3-031-99166-0_8

1 Introduction

In the traditional banking and investment sectors, FinTech is one of the disruptive forces that have provided innovative, cheaper, and more accessible solutions in contemporary financial ecosystems [1]. The wide cut of mobile-based applications available and the existence of digital payment systems have opened FinTech to an enormous population, allowing individuals from the peripheries to participate actively in the current financial system.

Financial inclusion is a very essential process when concerning capital markets as historically the underprivileged had no means to participate due to lack of opportunities [2]. The Fintech uses advanced technology to democratize investment, which is expected to result in quickened economic growth and redistribution of income [3]. Fintech has been hailed as the main enabler for scaling financial inclusion, based on the burgeoning prospects reliant on user trust within these platforms. Trust plays an important role especially where users are consequently willing to share their personal and financial info. There needs to be some level of trust; otherwise, skepticism toward Fintech solutions shall prevail if the efficacy is relegated [4]. The trust in Fintech is believed to be multifaceted and conceived based on some factors like the digital literacy of the user, availability of mobile devices, legal regulations, and protection of data and confidentiality [5]. Digital literacy for an individual is one of the important determinants of the person's efficiency in the exploit of Fintech platforms. A user will now have to undergo training on these technologies, since complexity in technology is increasing. According to the study, low levels of digital literacy are among leading barriers to the adoption of FinTech, especially in developing countries [6]. This has made Fintech services to be adopted easily through mobile platforms because people can be able to access the financial services they need at any time they want. The use of mobile devices has been established to be the main channel through which users are accessing Fintech platforms, hence the importance of simple and affordable interfaces to enhance trust and engagement [7]. Last but not the least data security and privacy are crucial for developing and maintaining trust. This is because the modern users pay much attention to the protection of their personal and financial information, and if the platform does not provide sufficient security, it will not be able to gain or keep the users' trust [8]. The study at hand purposes to bridge this gap by introducing a systematic conceptual framework that investigates the connections among these elements.

Conceptual Framework

The conceptual framework presented in this study posits that **trust in Fintech platforms** plays as a mediating factor between four core dimensions—**digital literacy, mobile accessibility, regulatory support, and data security and privacy, influencing** financial **inclusion in capital markets** (Fig. 1).

1. **Digital Literacy**: Digital literacy denotes the capability to comprehend and efficiently utilize digital tools effectively. It contains skills related to navigating Fintech platforms, understanding financial terminology, and leveraging advanced

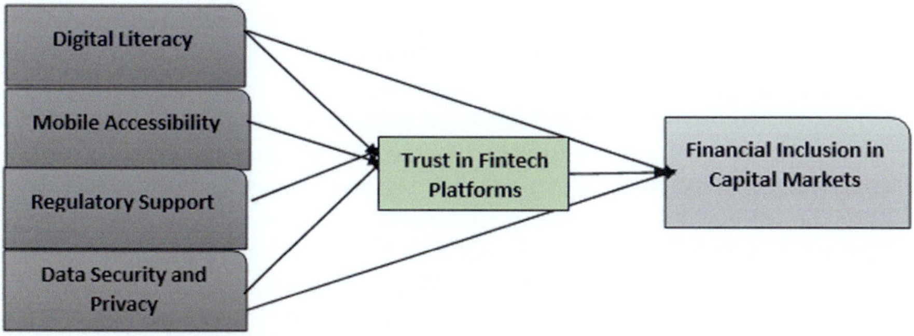

Fig. 1 Variables-based model created by author

features. Digital literacy significantly influences FinTech adoption by improving users' confidence in utilizing digital financial services. Research indicates that individuals with higher digital literacy levels are more likely to trust and engage with FinTech platforms [9]. Furthermore, improving digital financial skills can enhance financial inclusion by enabling users to navigate FinTech applications effectively [10]

2. **Mobile Accessibility**: The smartphone and mobile internet world has brought a whole new dimension to FinTech. The odds for winning the consumer trust will be comparatively good when platforms are easily controlled through a cell phone and has been situated into environments. Furthermore, accessibility and inclusiveness are being commercially incentivized by user-friendly initiatives in the local level [7]. The rise of smartphones and mobile internet has revolutionized financial accessibility, particularly in developing economies. Studies suggest that mobile banking facilitates access to financial services, reducing dependency on traditional banking infrastructure [10].
3. **Regulatory Support**: Regulatory frameworks are crucial in fostering assurance for users to trust Fintech platforms. Specific and clearly defined rules and regulations of data protection, consumer rights, and grievance redressal mechanisms build a user's confidence in ensuring ethical practices in financial transactions. Regulatory frameworks are critical to enhancing user trust in Fintech platforms. Well-defined rules with respect to consumer protection, grievance redressal, and data privacy ensure ethical financial transactions [11]. The implementation of open banking initiatives such as the EU's PSD2 has enhanced transparency and competition in digital finance [12].
4. **Data Security and Privacy**: The fields of data protection and privacy serve as great support to mitigate substantive threats to Fintech adoption. Strong encryption protocols protecting the authentication processes and transparent privacy policies will foster trust. Users are more prone to adopt platforms that demonstrate a strong dedication to safeguarding their data and personal information. In ameliorating risks associated with FinTech adoption, robust measures for data protection and privacy are critical. Research underlines that users prefer finance

organizations that employ stringent encryption, secure authentication, and transparent privacy policies [12]. Trust toward FinTech is directly related to their capacity to uphold data protection and prevent security breaches [11].

5. **Trust in Fintech Platforms**: Trust is an important factor in adopting FinTech because it determines the degree to which the user becomes involved and utilizes the platform to receive financial services. The force is huge in regard to the security, the privacy, openness, and generic reliability of the transaction [4]. Trust formation activities consist of service elements that can be influenced by user initiation, service delivery reliability, and compliance to legal requirements [3]. Trust is fundamentally important for Fintech adoption. Studies strongly emphasize that security, transparency, and service reliability in the decision-making processes contributed to trust formation in financial transactions [11]. The Fintech platform is likely to attract and maintain users if it has a regulatory framework and shows consistency in quality service [3].
6. **Financial Inclusion in Capital Market**: Financial inclusion to capital markets should expand services in areas like stock and equity, bonds, and investments for the unidentified group. Breaking the barriers of financial illiteracy, high transaction costs, and less access to the digital infrastructure is crucial to create inclusion [13]. When these challenges are surmounted, they assist to establish fair equality, through which every individual in the lower or any other section of the socioeconomic stratum can engage in capital markets [2]. Fintech innovations are instrumental in promoting financial inclusion by expanding access to capital markets. Digital financial services help bridge the gap for underserved populations, providing opportunities to invest in stocks, bonds, and equity [14]. Addressing challenges such as financial illiteracy and high transaction costs can create a more inclusive financial ecosystem [2].

2 Review of Literature

2.1 Digital Literacy and Fintech Adoption

Digital literacy plays a crucial role in embracing and confidence in Fintech platforms. Lack of digital literacy significantly restricts financial inclusion particularly in provinces with limited access to edification [15]. Both Governments and private stakeholders have taken initiatives intended to enhance digital literacy as a strategy to bridge the digital divide [6].

Null Hypothesis (H01): There is no significant relationship between digital literacy and trust in Fintech platforms.

2.2 Mobile Accessibility and Financial Inclusion

The role of mobile accessibility in fostering financial inclusion has been well documented. Mobile phones have revolutionized financial systems by providing underserved populations with access to banking and investment platforms. Researches show that mobile Fintech platforms like payment gateways, mobile wallets, and peer-to-peer lending have considerably enhanced trust and adoption. Developing economies have perceived the substantial growth in mobile-based Fintech solutions due to their affordability and user-friendly interfaces.

Null Hypothesis (H02): There is no significant relationship between mobile accessibility and trust in Fintech platforms.

2.3 Regulatory Support and Trust in Fintech Platforms

Trust in a Fintech application is considerably dependent on its perception of sound regulatory support. The culture of an institution is determined by a host of ethical guidelines and regulations, which need to deal with issues such as deception, corruption, and financial risks (Three Essays on Fintech Issues, ECNU, 2019). They encourage a seamless environment in which users can be encouraged to take part in capital markets instances. When properly governed, they can become a positive force in serving the capital markets at large. Government initiatives like India's Unified Payments Interface (UPI) or the European Union's General Data Protection Regulation (GDPR) provide the much-needed regulatory framework to augment trust toward Fintech platforms and accelerator financial inclusion. Again, the adoption of KYC and AML measures by Fintech companies greatly helps to build trust among potential users [16].

Null Hypothesis (H03): There is no significant relationship between regulatory support and trust in Fintech platforms.

2.4 Data Security, Privacy, and Trust

One concern that looms large in the minds of Fintech users is that of data security and data privacy. Structurally, trust is sometimes constructed on the basis of effective approaches, like strong encryption, two-factor authentication, and explicit privacy policies [8]. Emerging technologies like blockchain can provide novel ways of securing and managing data in a more transparent way. Blockchain technology particularly benefits uniqueness in that it allows for secure decentralization of storage; hence, it is like an unalterable sequence to record transactions, which builds up trust among users. However, encouragement of the use of those technologies largely depends on addressing the concerns surrounding their complexity and cost.

Null Hypothesis (H04): There is no significant relationship between data security and privacy and trust in Fintech platforms.

2.5 Trust in Fintech Platforms and Financial Inclusion

Trust serves as a crucial link between digital literacy, mobile accessibility, regulatory support, and data security, all of which are essential for enhancing financial inclusion. In the securities markets, trust is particularly fragile, as users typically require significant financial resources. Furthermore, the relationship between trust and financial inclusion is dynamic; a strong mobile trust can lead to higher adoption rates among skeptical users, ultimately promoting broader financial inclusion [3].

Null Hypothesis (H05): There is no significant relationship between trust in Fintech platforms and financial inclusion in capital markets.

This research enriches to the existing literature by offering a detailed exploration of the intricacies influencing trust within Fintech platforms. Stakeholders such as policymakers, financial institutions, and Fintech developers can leverage these insights to design targeted interventions that enhance user confidence and expand access to capital markets. Through these endeavors, Fintech platforms can effectively act as drivers of economic empowerment and inclusivity.

This research includes quantitative research method. Here, the targeted sample would be 400 users in underserved populations within the rural areas of Madhya Pradesh, India. Structured questionnaires were distributed among respondents for capturing insights from primary as well as end users of the digital financial system. 320 questionnaires out of the 400 targeted number of users returned duly filled. The study examined the effects of digital literacy, mobile access, regulatory support, and data security/privacy as independent variables with trust as a mediator and financial inclusion as a dependent variable. The application of SEM was employed in order to analyze both the direct and indirect relationships between the variables. The study provides a comprehensive understanding of factors that drive trust and financial inclusion in digital financial ecosystems by incorporating responses from a diverse user base within rural areas.

The values in Table 1 demonstrate that the reliability and convergent validity of the constructs are generally sound. All constructs surpass the recommended thresholds for Cronbach's alpha (greater than 0.70), composite reliability (above 0.70), and average variance extracted (AVE) (above 0.50), indicating strong internal consistency and sufficient variance extraction. Specifically, data security, digital literacy, financial inclusion, and trust all exhibit excellent reliability and convergent validity. However, while regulatory support meets the minimum AVE threshold of 0.50 (AVE = 0.505); it is at the lower end, suggesting that although the construct is acceptable, it could benefit from refinement to achieve higher discriminant validity and improved model robustness.

Based on the analysis of the Fornell–Larcker criterion (Table 2), it can be concluded that the model demonstrates adequate discriminant validity. According to the criterion, the square roots of the average variance extracted (AVE) for each construct should be greater than the correlations between that construct and any other construct in the model. In this case, the diagonal values, representing the square roots of the AVEs, are consistently higher than the corresponding off-diagonal values, which represent the correlations between constructs. For example, the square root of the AVE for data security is 0.781, which is greater than its correlations with digital literacy (0.690), financial inclusion (0.419), regulatory support (0.307), and trust (0.303). Similarly, this pattern is observed for all other constructs in the model, such as digital literacy (AVE = 0.694), financial inclusion (AVE = 0.784), regulatory support (AVE = 0.710), and trust (AVE = 0.717). The fact that the square roots of the AVEs exceed the correlations between constructs supports the conclusion that each construct is distinct from the others. Therefore, the model meets the Fornell–Larcker criterion ensuring that the constructs are sufficiently different from one another, confirming that the model has reliable discriminant validity. This strengthens the robustness of the measurement model and ensures that each construct is measuring a unique aspect of the underlying theoretical framework.

Cross-loadings for the construct as per Table 3, data security, digital literacy, financial inclusion, regulatory support, and trust—exceed the threshold of 0.7, confirming strong item reliability. For example, DS6 (0.864), DL4 (0.779), FI2 (0.831), RS5 (0.773), and T3 (0.794) demonstrate robust loadings, indicating that each item effectively measures its intended construct. Cross-loadings remain substantially lower than primary loadings, ensuring discriminant validity; for instance, DS1 has a strong loading on data security (0.746) with cross-loadings on digital literacy and financial inclusion being only 0.288 and 0.514, respectively. These adjustments enhance the reliability and validity of the model, making it suitable for robust hypothesis testing

Table 1 Reliability analysis and convergent reliability

	Cronbach's alpha	Composite reliability (rho_a)	Composite reliability (rho_c)	Average variance extracted (AVE)
Data security	0.908	0.912	0.926	0.609
Digital literacy	0.818	0.828	0.866	0.582
Financial inclusion	0.874	0.877	0.905	0.614
Regulatory support	0.859	0.864	0.890	0.505
Trust	0.770	0.787	0.840	0.515

Calculated by the Author

Table 2 Discriminant validity—Fornell–Larcker criterion

	Data security	Digital literacy	Financial inclusion	Regulatory support	Trust
Data security	**0.781**	0.69	0.419	0.307	0.303
Digital literacy	0.69	**0.694**	0.427	0.358	0.351
Financial inclusion	0.419	0.427	**0.784**	0.634	0.369
Regulatory support	0.307	0.358	0.634	**0.71**	0.391
Trust	0.303	0.351	0.369	0.391	**0.717**

Calculated by the Author

and structural equation modeling. By ensuring all loadings exceed the minimum threshold, the results are more reliable and consistent with best practices in SEM analysis.

The analysis (Table 4) demonstrates that **digital literacy**, **regulatory support**, and **data security** have significant positive effects on **trust**, which in turn is crucial for enhancing **financial inclusion**.

3 Discussion

The results of this research highlight the importance of how digital literacy, regulatory support, and data security create trust, a key motivator for financial inclusion. Data security was the most significant predictor of trust with a path coefficient of 0.900, which is consistent with studies that highlighted the need for advanced cybersecurity to prevent fraud and privacy risks in digital finance [18, 19]. Regulatory support (path coefficient: 0.813) is also critical, as the regulatory sandboxes have been proven to balance innovation with financial stability risks [20]. Further, digital literacy (path coefficient: 0.818) directly influences trust, which is in accordance with recent studies that highlight targeted financial literacy programs to empower users and avoid exclusion, particularly in heterogeneous markets like Southeast Asia. Trust as a mediating factor makes enabling factors link to financial inclusion by presenting itself as a psychological enabler and a behavioral motivator. From the theory of planned behavior developed by [21], it is established that perceived security and trust are some of the factors that influence the behavioral intention of a user.

Table 3 Cross-loadings

Item	Data security	Digital literacy	Financial inclusion	Regulatory support	Trust
DS1	**0.746**	0.288	0.514	0.343	0.296
DS2	**0.802**	0.259	0.512	0.321	0.337
DS3	**0.816**	0.218	0.566	0.32	0.161
DS5	**0.785**	0.193	0.556	0.292	0.18
DS6	**0.864**	0.332	0.621	0.374	0.171
DS7	**0.786**	0.233	0.538	0.356	0.285
DS8	**0.751**	0.208	0.537	0.39	0.356
DL1	0.251	**0.651**	0.312	0.173	0.198
DL2	0.241	**0.746**	0.311	0.272	0.262
DL4	0.293	**0.779**	0.396	0.332	0.284
DL6	0.189	**0.745**	0.314	0.259	0.219
DL7	0.197	**0.698**	0.295	0.202	0.171
FI1	0.553	0.323	**0.771**	0.463	0.263
FI2	0.602	0.313	**0.831**	0.502	0.279
FI3	0.531	0.375	**0.786**	0.528	0.37
FI5	0.533	0.373	**0.789**	0.509	0.258
FI6	0.56	0.334	**0.79**	0.524	0.283
RS1	0.215	0.176	0.353	**0.671**	0.264
RS2	0.237	0.224	0.349	**0.636**	0.265
RS3	0.348	0.252	0.493	**0.701**	0.252
RS4	0.332	0.269	0.479	**0.734**	0.338
RS5	0.271	0.294	0.5	**0.773**	0.283
RS6	0.387	0.213	0.512	**0.672**	0.253
RS7	0.26	0.215	0.452	**0.711**	0.242
RS8	0.327	0.227	0.462	**0.775**	0.301
T1	0.419	0.271	0.323	0.288	**0.669**
T2	0.167	0.33	0.218	0.247	**0.783**
T3	0.216	0.302	0.322	0.397	**0.794**
T4	0.075	0.202	0.171	0.173	**0.694**
T5	0.082	0.114	0.224	0.215	**0.632**

Calculated by the Author

Table 4 Analysis of the structural model and hypotheses testing

Hypotheses	Relationship	Null hypothesis (H_0)	Path coefficient	*P*-value	Decision	Justification
H01	Digital literacy → trust	H_0: No relationship between digital literacy and trust	0.818	< 0.05	Rejected	The significant positive path coefficient (0.818) demonstrates that improved digital literacy fosters trust in financial systems
H02	Regulatory support → trust	H_0: No relationship between regulatory support and trust	0.813	< 0.05	Rejected	Regulatory support is essential for ensuring confidence in financial systems, as reflected by the high coefficient (0.813)
H03	Data security → trust	H_0: No relationship between data security and trust	0.9	< 0.05	Rejected	Data security is the strongest predictor of trust (coefficient 0.900), confirming its crucial role in digital financial ecosystems

(continued)

Table 4 (continued)

Hypotheses	Relationship	Null hypothesis (H_0)	Path coefficient	*P*-value	Decision	Justification
H04	Trust → financial inclusion	H_0: No relationship between trust and financial inclusion	0.77	< 0.05	Rejected	The significant positive relationship (0.770) highlights trust as a key enabler of financial inclusion, validating its importance
H05	Digital literacy → financial inclusion (via trust)	H_0: No indirect relationship between digital literacy and financial inclusion	Indirect (via trust)	< 0.05	Rejected	Digital literacy indirectly impacts financial inclusion by enhancing trust, which then encourages participation in financial systems
H06	Regulatory support → financial inclusion (via trust)	H_0: No indirect relationship between regulatory support and financial inclusion	Indirect (via trust)	< 0.05	Rejected	Regulatory frameworks indirectly improve financial inclusion by building trust
H07	Data security → financial inclusion (via trust)	H_0: No indirect relationship between data security and financial inclusion	Indirect (via trust)	< 0.05	Rejected	Data security indirectly promotes financial inclusion by building trust, encouraging users to engage with digital financial systems

Source Authors' work

4 Conclusion

The conclusion of this study places emphasis on the central position of trust as a mediator in fostering financial inclusion within digital ecosystems. Trust links the enablers: digital literacy (path coefficient: 0.818), regulatory support (0.813), and data security (0.900) to financial inclusion, which also holds a significant path coefficient of 0.770. Data security was found to be the strongest predictor of trust, pointing to the critical importance of sound information protection mechanisms. These indirect effects illustrate how these enablers promote financial inclusion through trust, a point that shows systemic interconnection. Implications include targeted investments in cybersecurity infrastructure, digital education programs, and transparent regulatory frameworks, as it shall create an inclusive and trustworthy digital financial environment. As ecosystems in digital evolve, addressing trust deficits and digital infrastructure challenges for developing economies. This approach has the potential to give an integrated pathway toward the realization of sustainable financial inclusion, in consonance with global initiatives such as the financial inclusion strategy of the World Bank (2023) and from recent studies [20].

5 Limitations

This study contributes to the knowledge on trust in digital financial ecosystems uniquely, but the context is narrowly drawn, therefore requiring cross-site extension. As for the future research ideas, it would be interesting to look at increasingly popular technologies, such blockchain with its focus on trust through decentralization and transparency [22]. Further, the inclusion of qualitative methods may reveal hidden behavioral patterns hard to identify in the quantitative strategies [23]. Thus, future research can offer more reliable and generalizable evidence based on findings Newport has established pioneering well suited to increasingly shifting digital environment.

References

1. Arner, D. W., Barberis, J. N., & Buckley, R. P. (2016). The evolution of fintech: A new post-crisis paradigm? *Georgetown Journal of International Law, 47*(4), 1271–1319.
2. Ozili, P. K. (2021). Financial inclusion and fintech: A systematic literature review. *Journal of Economic Studies, 48*(5), 1032–1050. https://doi.org/10.1108/JES-12-2019-0584
3. Gomber, P., Koch, J.-A., & Siering, M. (2023). Digital finance and the future of financial services. *Financial Innovation Journal, 9*(2), 178–196.
4. Lee, I., & Shin, Y. J. (2018). Fintech: Ecosystem, business models, investment decisions, and challenges. *Business Horizons, 61*(1), 35–46. https://doi.org/10.1016/j.bushor.2017.09.003

5. Yuen, Y., Lim, J. S., & Ooi, K. B. (2021). A multidimensional trust model for fintech adoption: An extended approach. *Information Systems Research, 32*(4), 1125–1143. https://doi.org/10.1287/isre.2021.1044
6. Bada, M., & Sasse, M. A. (2019). Cyber security awareness campaigns: Why do they fail to change behavior? arXiv preprint arXiv:1901.02672. https://arxiv.org/abs/1901.02672
7. Donovan, K. (2022). Mobile money and the financial inclusion agenda. *Journal of Financial Technology & Policy, 12*(3), 210–225.
8. Laukkanen, T. (2016). Consumer adoption versus rejection decisions in seemingly similar service innovations: The case of the internet and mobile banking. *Journal of Business Research, 69*(7), 2432–2439. https://doi.org/10.1016/j.jbusres.2016.01.013
9. Anakpo, G., & Anakpo, G. (2023). Digital financial inclusion: Examining the role of mobile technology in expanding access to capital. *Dinasti International Journal of Economics, Finance & Accounting, 6*(1), 43–55.
10. Sebayang, K. S., Sihombing, P., & Sihombing, P. (2023). The role of digital financial literacy in enhancing fintech adoption. *Dinasti International Journal of Economics, Finance & Accounting, 6*(1), 78–94.
11. Sokol, D. D. (2023). Fintech and financial inclusion: A review of the literature. *Southern California Law Review, 96*(2), 243–272.
12. Bank for International Settlements. (2021). Fintech and the digital transformation of financial services. *BIS Papers No. 117.* https://www.bis.org/publ/bppdf/bispap117.pdf
13. Demirgüç-Kunt, A., & Klapper, L. (2017). Measuring financial inclusion and the fintech revolution. *World Bank Policy Research Working Paper, 7255.*
14. J-PAL. (2024). Digital financial services to improve formalized access and inclusion. *Policy Insight.* https://www.povertyactionlab.org/policy-insight/digital-financial-services-improve-formalized-access-and-inclusion
15. Aker, J. C., & Mbiti, I. M. (2010). Mobile phones and economic development in Africa. *Journal of Economic Perspectives, 24*(3), 207–232. https://doi.org/10.1257/jep.24.3.207
16. Maguire, J., McAfee, R. P., & Satterthwaite, M. A. (2021). Trust and regulation in financial markets. *The Journal of Finance, 76*(4), 1241–1265.
17. Chuang, T. T., & Lin, S. (2019). Privacy and security concerns in fintech adoption: A study of emerging economies. *Financial Studies Journal, 43*(2), 154–172.
18. Rahman, M., Alam, K., & Sultana, S. (2021). Cybersecurity challenges in digital finance: A systematic review. *Computers & Security, 102*, 102–118. https://doi.org/10.1016/j.cose.2020.102118
19. World Bank. (2023). Financial inclusion strategies for the digital age. *World Development Report 2023*
20. MDPI. (2023). Regulatory sandboxes in fostering financial stability and innovation: A global perspective. *Journal of Financial Innovation, 10*(4), 521–534.
21. Ajzen, I. (1991). The theory of planned behavior. *Organizational Behavior and Human Decision Processes, 50*(2), 179–211. https://doi.org/10.1016/0749-5978(91)90020-T
22. Liu, J., Zhang, X., & Chen, Y. (2023). Blockchain technology and trust in digital financial ecosystems: An empirical analysis. *International Journal of Emerging Technologies in Finance, 15*(3), 203–220.
23. Creswell, J. W. (2023). *Research design: Qualitative, quantitative, and mixed methods approaches* (5th ed.). SAGE Publications.

Blockchain and Cryptocurrencies: Regulatory Innovations in Capital Markets

Akhil Goyal and Ankita Yadav

Abstract This research paper intersections between blockchain technology, cryptocurrencies, and capital market regulations form the basis of exploration in this paper. Opportunities and challenges exist in capital markets worldwide due to the rapid emergence of digital assets and decentralized financial systems. Worldwide, regulatory innovations adjust to these technological advancements to safeguard the integrity of the market, protect investors, and foster innovation. The paper discusses recent regulatory approaches, evaluates the effectiveness of the same, and delineates the possible rise of a coordinated framework across the globe to significantly enhance the role of blockchain and cryptocurrencies in the financial ecosystem.

Keywords Blockchain technology · Cryptocurrencies · Central Bank Digital Currencies (CBDCs) · Blockchain decentralization · Global collaboration

1 Introduction

Blockchain and cryptocurrencies have been developed as revolutionary agents in the global financial environment. Blockchain has features of being decentralized, transparent, and immutable, and it may therefore be disruptive for traditional financial markets, enhance efficiency, and facilitate the emergence of new forms of asset trading and ownership [1]. Cryptocurrencies, based on blockchain technology, have taken center stage as speculative assets and tools for financial innovation, creating new digital financial products such as stablecoins, decentralized finance protocols, and tokenized assets. However, the growth of these technologies has presented new challenges for regulatory bodies: decentralized and pseudonymous assets are new to the traditional frameworks of financial regulation. This means that any market

A. Goyal · A. Yadav (✉)
NIMS Institute of Business Studies, NIMS University, Rajasthan, Jaipur, India
e-mail: ankita.yadav@nimsuniversity.org

A. Goyal
e-mail: akhil.goyal@nimsuniversity.org

K. Khanna et al. (eds.), *From Poverty to Prosperity*, Information Systems Engineering and Management 64, https://doi.org/10.1007/978-3-031-99166-0_9

volatility, investor protection, cybersecurity, and financial stability arise as questions under such conditions. These recent developments in the blockchain and cryptocurrency markets suggest that governments across the globe must reconsider and redefine their regulatory lines to allow innovation but guarantee investor safety [2]. These are examples of recent regulatory developments, such as MiCA at European Union and the development of regulatory sandboxes, with more than one country, representing some of the proactive and forward-looking actions in dealing with these problems but keeping in mind the special features of blockchain markets in the search for new ways to apply regulation innovations and testing their effectiveness when applied and still open to being amended to reach a better-balanced, coordinated worldwide approach. This paper will outline a comprehensive view of the future of blockchain and cryptocurrency regulation in capital markets by the examination of successful case studies and highlighting the roles that regulatory bodies play [3].

2 Research Methodology

This study employs a qualitative research approach to conduct an in-depth analysis of regulatory frameworks, emphasizing the principles, structures, and effects within a specific industry or sector. A descriptive and interpretative method is utilized to assess the effectiveness, gaps, and consequences of these frameworks. Secondary data collection is used to obtain information from reliable sources, such as government portals, academic publications, industry reports, and news articles. Furthermore, document analysis is performed to review legal texts, policy directives, and compliance requirements, allowing for a comprehensive understanding of regulatory content and its implications [4].

The data are examined through thematic and comparative analysis. Thematic analysis includes coding, categorization, and interpretation to uncover recurring themes and patterns, providing a structured insight into key regulatory concerns. Comparative analysis is applied to evaluate frameworks across various industries or regions, identifying best practices and potential areas for regulatory alignment. Data reliability is ensured through triangulation, which involves cross-verification from multiple sources, while expert reviews help validate interpretations and findings. Transparency is maintained by thoroughly referencing sources and documenting analytical procedures.

Ethical considerations are strictly observed by safeguarding confidential documents, ensuring impartiality, and correctly citing all secondary sources to uphold academic integrity. The study also acknowledges certain limitations, such as restricted data availability, the subjective nature of qualitative interpretation, and the limited applicability of findings to other industries or regions. This structured methodology enhances reliability and transparency, ensuring a thorough and credible analysis of regulatory frameworks [5].

2.1 *Regulatory Frameworks in Blockchain and Cryptocurrency Markets*

(i) **Investor Protection**

Regulative frameworks have important uses in protecting investors from outright fraud, scams, and market manipulation. Blockchain and cryptocurrencies are characterized by their decentralized and rather anonymous nature, often providing a strong incentive for unethical practices, such as multiple types of initial coin offering (ICO) scams and pump-and-dump schemes. Regulations enforce minimum standards of transparency and disclosure that are very helpful in making decisions by investors and building a safer environment for investments. For instance, the U.S. The Securities and Exchange Commission has placed a few crypto assets as securities, thereby classifying these to a similar level of scrutiny as traditional securities [2].

(ii) **Market Integrity**

Maintaining public trust and participation in the financial system involves ensuring market integrity. Regulatory frameworks prevent practices such as insider trading, wash trading, and manipulation of markets that distort the market. Thus, setting norms for fair and orderly trading would enhance market integrity and allow fair play for all participants. For example, the European Union's MiCA framework outlines specific regulations around service providers' treatment of digital assets and thus increases the soundness of market infrastructure [3].

(iii) **Risk Mitigation**

Cryptocurrency markets are by their nature unstable, and their prices are extremely erratic. Bains, P., Ismail, A., Melo, F., & Sugimoto, N. (2022) *argued that* The regulatory framework is used to reduce systemic risks by engaging standards for liquidity, capital adequacy, and custodial responsibilities as far as applied to crypto exchanges. It will contribute to the reduction in the likelihood of a sudden collapse with an impact on the rest of the economy. The Financial Stability Board has noted that effective regulation of, for example, stablecoins becomes important to mitigate risks that may come into play and affect financial stability across the globe if such digital asset systems are widely adopted in payment systems [6].

(iv) **Fostering Innovation within Safe Boundaries**

Regulations such as regulatory sandboxes, for instance, are controlled environments that allow firms to test new financial products while following any compliance protocols. It thus promotes technological advancement and economic growth without even the slightest infringement on regulatory standards. The UK's "Financial Conduct Authority" has initiated such sandboxes, balancing encouragement of innovation with strong oversight that enables proper management of risks. In a nutshell, proper regulatory frameworks will ensure that blockchain and cryptocurrency markets operate in a manner that safeguards investors' interests, maintains market integrity, and diminishes systemic risks [7]. Proper regulation promotes a healthy environment where innovation and consumer protection go together and help lay down the foundation for sustainable integration of digital assets within global financial systems.

3 Blockchain and Cryptocurrencies in Capital Markets

Thanks to blockchain, transformations are happening in the capital markets mechanisms for trading, ownership, and investment in assets. Blockchain technology can facilitate several traditional financial processes to become much more secure and accessible to a higher number of participants, while cryptocurrencies open up for investors a new asset class and new opportunities in decentralized finance [5].

3.1 Overview of Blockchain Technology

Blockchain represents a decentralized and distributed ledger technology that records transactions across a large network of computers. It facilitates operations in a notably clear, secure, and immutable fashion. Within capital markets, it presents the potential to either replace or enhance traditional processes through significantly expedited and more secure transaction processing. For instance, utilizing blockchain for clearing and settlement can reduce transaction times from several days to mere seconds, thereby dramatically lowering costs and increasing liquidity [8].

3.2 How Does Blockchain Technology Work?

One of the most well-known applications of blockchain technology is Bitcoin. It serves as a cryptocurrency that facilitates the exchange of assets over the Internet. Rather than depending on trust from third parties, Bitcoin utilizes cryptographic proofs to conduct transactions online. Every transaction is safeguarded by a digital signature [6].

3.3 Blockchain Decentralization

There is a decentralized server that maintains blockchain data. Instead, the data are diffused across millions of computers globally that are associated to the blockchain. This architecture facilitates the validation of data since it is there on every node and can be overtly verified.

3.4 Blockchain Nodes

A node is a computer system that connects to the blockchain network. It interfaces through the blockchain via a client, which is responsible for corroborating and disseminating transactions against the blockchain. When a computer connects to the blockchain, it downloads a copy of the blockchain data onto its system and aligns itself with the most recent block. The computer that connects to the blockchain and helps carry out transactions in return for a reward is known as a miner [8].

3.5 Cryptocurrencies as Digital Assets

Cryptocurrency refers to a category of digital or virtual assets that leverage cryptographic techniques for enhanced security and function autonomously from traditional central banks. Bitcoin, as the pioneer of cryptocurrencies, paved the way for thousands of others, each possessing unique features and applications. Cryptocurrencies are increasingly viewed as investment vehicles and capital-raising instruments within capital markets. While Bitcoin and Ethereum are frequently utilized as speculative investments, stablecoins offer greater price stability, making them more appealing for trading and transactional purposes [9].

3.6 Tokenization of Assets

Tokenization involves the transformation of tangible assets, including stocks, real estate, and commodities, into digital tokens enabled by blockchain technology. This innovation facilitates fractional ownership and significantly enhances access to investment opportunities. It extends the reach of capital markets, granting retail investors access to assets reserved for high net-worth individuals. Tokenized assets can be traded on digital platforms, providing access to market valuations, thereby increasing market liquidity and democratizing wealth creation [9].

3.7 Impact on Traditional "Financial Institutions"

Cryptocurrency and blockchain are forcing financial institutions to innovate. Multitudes of banks, investment firms and stock exchanges are investigating blockchain applications. Its blocks are employed for network data sharing and cross-border payments via the tokenization of assets. The blockchain system allows for operations to be carried out more simply and at a lower cost and opens up new avenues for investment for the traditional financial market participant. JPMorgan and the

Australian Securities Exchange (ASX), for example, are building blockchain-based platforms to manage their transactions and settlement processes [10].

A closer look (Table 1) at the factors that promote the adoption of cryptocurrencies in the UK & USA between 2019 and 2022 illuminates some of the interesting contrasts between the two countries' respective adoption dynamics. Therefore, as can be seen from Table 2, the UK has only a 17% adoption rate of cryptocurrencies and is truly the more cautious toward the sector. Clarity of regulations was one of the single highest influence factors among many. In the UK, the existing regulatory framework gave potential adopters confidence that they can use cryptocurrencies with relative ease. The slow pace of their adoption can be attributed to worries on the volatility of cryptocurrencies and the risk of imminent regulation that would change the playing field [11]. As a result, the finding explains that a large fraction of public might still express uncertainty or reluctance even in the presence of regulations. Since the nation was more entrenched in technology, in the United States, 25% utilized the service. That could have been because of the high percentage of technology literate people and an innovation culture. With its well-established technological framework and advanced state of digitalization, the United States is more generally amenable toward Bitcoin adoption. One needs to critically examine if this acceptance comes from profit-oriented motives or real value in everyday interactions.

Interrogations about the pervasiveness of adoption ascend by a closer examination of the two cases. Were these levels of adoption general uses, or were they rather single entities of early adopters and investors? Indeed, properly defined rules build trust and retention, but one may also considers how rules might, without intent, constrain innovation, limit participation. And the various demographic, social and cultural differences within individually of the respective nations are not accounted for in the adoption rates reported. As part of any serious study, we look to see if adoption differs by demographics, age, income, and geography [8–10]. These could provide, where Widespread use of Bitcoin: An examination of adoption rates and contributing variables this work maybe serves as best shows, a macro understanding of the interplay between government regulation, innovation and popular sentiment

Table 1 Overview of cryptocurrency and its impact on financial market volatility

Country	Cryptocurrency	Volatility index	Impact on traditional assets (2019–2022)
UK	Bitcoin	0.76	Increased diversification
USA	Ethereum	1.12	Enhanced portfolio returns

Source Author's own compilation

Table 2 Comparative analysis of cryptocurrency adoption rates in the UK and USA

Country	Adoption rates (2019–2022) (%)	Factors influencing adoption
UK	17	Regulatory clarity
USA	25	Technological awareness

Source Author's own compilation

around the relationship. This interaction between these factors plays out in the path of Bitcoin adoption and subsequently, its role in large financial ecosystem. Grasping jus why these adoption rates are what they are—and what that implies for the future—takes a bit deeper dig though.

4 Current Regulatory Landscape for Blockchain and Cryptocurrencies

As the governments and institutions worldwide work out how best to encapsulate risk yet also innovate, the regulatory scenery for blockchain, and cryptocurrencies continues to shift at a rapid pace. From there, we got a jurisdiction-by-jurisdiction approach, with some embracing the technology, and others tightening controls for, ironically, the same reasons, like market volatility, fraud, and financial stability [12].

I. **United States**

"Securities and Exchange Commission" (SEC) and "Commodity Futures Trading.

Commission" (CFTC) Roles

The SEC then further considers whether most cryptocurrencies are securities under the Howey Test, which applies to any investment that has a prospect of profits arising from the efforts of others in this creative process. On the other hand, the CFTC treats cryptocurrencies like Bitcoin as commodities and oversees all derivatives and futures trading involving the latter. Its overlapping jurisdictions only serve to confuse market participants.

Challenges in Classifying Digital Assets

Distinguishing between securities and commodities remains a major challenge. Cryptocurrencies with mixed features, such as governance tokens or those used in decentralized finance (DeFi), often fall into gray areas. This lack of clear classification hampers compliance and deters innovation.

Recent Regulatory Actions

The Securities and Exchange Commission has intensified its scrutiny of crypto exchanges through unregistered securities offerings. In 2023, some of the major exchange lawsuits, such as with Coinbase and Binance, highlighted the level of regulatory push toward transparency and compliance. Such actions increase market uncertainty but signal stricter enforcement of existing laws [11].

II. **European Union**

Markets in Crypto-Assets Regulation (MiCA)

MiCA develops an integrated regulatory framework for cryptocurrencies offered within the EU member states. The regulation protects consumers, promotes market integrity, and reduces environmental impact, particularly from proof-of-work (PoW) mining.[1]

Approach to Stablecoins

MiCA calls for reserve and transaction limits on stablecoins per diem. It requires issuers of stablecoins to be transparent and accountable in their operations to improve stability and mitigate systemic risks [13].

Consumer Protection and Anti-money Laundering (AML)

These enhanced KYC and AML requirements will make sure that providers of services such as wallets and exchanges mitigate the risks of fraud and illicit financing. This also includes actions to counter tax evasion with cryptocurrencies.

III. **Asia**

Japan's Progressive Regulatory Approach

Japan has put strong operational guidelines in place for domestic crypto exchanges, requiring them to register with the Financial Services Agency (FSA). Jack launched the international first regulations on crypto-asset funds focusing on consumer protection and financial stability.

China's Cryptocurrency Ban

China bans crypto transactions and mining but is a strong supporter of the Chinese central bank digital currency (CBDC)—or the digital yuan. The dual approach reflects its dual objectives to manage financial risks and capture the power of blockchain technology for national purposes [10].

5 Challenges in Regulating Blockchain and Cryptocurrencies

I. Market Volatility

Cryptocurrencies are known to be volatile: prices can fluctuate more dramatically than those of regular assets. This volatility places tremendous risks on retail investors, who may face significant financial blow in downturns. The challenge for regulators is to protect these investors in such a way without stifling innovation or discouraging market participation [14].

[1] WorldQuant University on 2024-02-16.

Approaches to Volatility: Some regulators consider implementing restrictions on leverage in cryptocurrency trading or requiring exchanges to provide risk disclosures. For example, the U.S. SEC has warned against speculative investments and urged exchanges to increase transparency.

Implications for Retail Investors: The need for more safeguards often leads to overexposure among inexperienced investors, necessitating educational campaigns and consumer protection measures [11].

II. **Cybersecurity and Fraud**

Blockchain technology is highly decentralized and pseudonymous, making cryptocurrencies susceptible to hacks, scams, and fraudulent activities. Highly publicized incidents, such as the hacking of Mt. Gox and the massive thefts of billions at exchanges, have underscored the need for highly stringent cybersecurity standards [13].

Risks: Common risks include phishing scams, smart contract exploits, and unauthorized access to wallets.

Regulatory Requirements: Regulatory authorities rigorously implement Know Your Customer (KYC) and Anti-Money Laundering (AML) regulations. In Japan, for instance, licensed exchanges are mandated to ensure robust security measures, maintain funds in cold storage, and conduct frequent audits to mitigate potential vulnerabilities [10].

III. **Global Coordination and Fragmentation**

One of the biggest issues in cryptocurrency regulation is its lack of uniformity across borders. Very few places have similar regulations. This leads to a practice called arbitrage where companies move to the country with the laxest rules.

Lack of Consistency: For example, while the United States will not encourage overly friendly engagement, China tends to ban outright certain crypto activities. This needs to be clarified in enforcement and understandably limited cross-border interoperability.

Need for International Coordination: Bodies like the FSB or G20 are engaged in working toward a coordinated approach. For instance, the regulatory framework of MiCA by the European Union could become a benchmark for international standards [14].

6 The Future of Blockchain and Cryptocurrency Regulations in Capital Markets

The future of blockchain and cryptocurrency regulations within capital markets must, on the one hand, reconcile the drive for innovation with the necessity of risk mitigation [8]. A variety of trends and developments are poised to influence this evolving landscape:

1. **Tokenized Assets and Security Token Offerings (STOs)**

Tokenization is an emergent trend that turns physical assets into digital tokens, making them more accessible and liquid, including real estate, stocks, and artworks. This will make capital markets more efficient by democratizing access to assets that are typically illiquid in nature. STOs, or regulated alternatives to ICOs, will pick up since they give investors rights and regulatory assurances, hence more institutional participation will be expected [9].

2. **"Central Bank Digital Currencies" (CBDCs)**

Central banks in almost every country are testing the waters of CBDCs as a way of adding blockchain to the traditional system, but with regulatory oversight. China, Sweden, and the European Union are some of the countries that have started pilot programs of CBDCs. This integration of CBDCs by central banks will allow for central management of digital currencies. It will also bring forth faster and more secure transactions without undermining monetary policies [11].

3. **Hybrid Blockchain Models**

As the blockchain ecosystem matures, hybrid models combining public and private blockchains are gaining more traction. These models would offer the benefits of transparency and decentralization while addressing privacy and scalability issues. Hybrid blockchains in capital markets could be used in trading platforms, clearing, and settlement processes, where some data remains confidential but verifiable [13].

4. **Enhanced Use of Smart Contracts and Automation**

Smart contracts are automatically executing contracts wherein the stipulations of the agreement are encoded in programming language. They are poised to revolutionize capital markets by removing intermediaries from transactions, accelerating process times, and decreasing expenses. Future implementations will likely center on the automation of intricate financial instruments and ensuring adherence to regulatory standards without the need for human oversight.

7 Blockchain Analytics for Real-Time Monitoring

Blockchain technology enables real-time tracking and auditing of transactions due to its inherent transparency. Over time, regulatory bodies and financial institutions are expected to increasingly utilize blockchain analytics tools to mitigate fraudulent activities and proactively monitor market conditions. This approach will assist in alleviating concerns related to money laundering, fraud, and other illicit activities within cryptocurrency markets [10].

7.1 Policy Recommendations

1. **Standardized Taxonomy and Classification of Digital Assets**

A universal definition of cryptocurrencies and digital tokens which will prevent legal chaos and will clarify legal standards. All the regulatory bodies should congregate and literally stamp the entire cosmos of digital assets, into boxes, we have securities, commodities, currencies, dispensing clarity on the way the laws, applicable in every jurisdiction, will be naturally interpreted, with respect to Digital Assets [14].

2. **Global Regulatory Collaboration and Harmonization**

The borderless nature of blockchain and cryptocurrencies also raises the need for international regulatory cooperation in order to avoid regulatory arbitrage. Developing global frameworks that would align taxonomies, compliance standards, and enforcement mechanisms. This sort of collaboration could be organized by bodies like the financial stability board (FSB) or the international monetary fund (IMF) [8].

3. **Proactive Oversight and Monitoring Systems**

Financial regulators need to implement forward-looking strategies to supervise emerging risks and market conduct. In turn, this brings greater transparency and may help to assist in the corruption of the detected fraud practice or financial crime and then eliminate it, due to the way Blockchain analytics tools and artificial intelligence, work to monitor cryptocurrency markets to track any unusual activities. Hence, the changing times require regulatory bodies to appropriately modify their monitoring system.

4. **Development of Comprehensive Consumer Protection Laws**

Due to the vulnerability and nature of risk the cryptocurrency market holds, it would require honorable and sound consumer protection to prevent this source of investment from being exploited. By disclosing this information freely, straightforwardly and accurately for all investment opportunities involving digital assets, regulators can ensure that investors always have access to this information, as they should [13].

5. **Education and Awareness Campaigns**

There is a need to educate investors, businesses, and policymakers on blockchain and cryptocurrencies, because there are potential opportunities as well as risks. Public awareness programs must be initiated by governments and regulators that can provide guidance on safety while interacting with blockchain-based assets, hence safeguarding retail and institutional investors from the risks of scams and market manipulation [12].

6. **Clearer Legal and Tax Frameworks**

Rolling out the red carpet for blockchain in the capital markets space has plenty of hurdles to get over but few are more challenging than the absence of tax and legal frameworks. It means offering guidelines that cover taxation matters surrounding cryptocurrencies—capital gains, for example—and defining which existing laws apply to blockchain-based financial products. This would give businesses the assurance to come in the space and the protective cover of legal assurances.

7.2 Implications for Policy Makers

To effectively regulate blockchain and cryptocurrency, governments should establish clear, transparent, and flexible regulatory frameworks that provide legal clarity while fostering innovation. This includes explicitly defining the legal standing of cryptocurrencies, introducing regulatory sandboxes for testing new financial products, and regularly updating regulations to align with technological advancements [10]. Additionally, governments must reinforce anti-money laundering (AML) and counter-terrorism financing (CTF) policies by implementing stringent know your customer (KYC) procedures, enforcing real-time transaction monitoring, and cooperating with international regulatory bodies to promote uniform global standards. Consumer protection is another key aspect, requiring clear risk disclosures, mechanisms to prevent fraudulent activities, and public awareness initiatives to educate citizens on cryptocurrency risks and benefits. To stimulate economic growth and technological advancement, governments may offer tax benefits to blockchain startups, collaborate with technology firms to develop supportive ecosystems, and explore central bank digital currencies (CBDCs) for more efficient payment systems [11].

Financial institutions should implement comprehensive compliance frameworks by integrating advanced AML and KYC measures, conducting regular internal audits, and training employees on regulatory compliance. Engaging with regulators and industry stakeholders through discussions, partnerships, and joint research efforts can further enhance risk management and compliance. Additionally, robust cybersecurity and data protection strategies are vital to safeguard customer data and financial assets, ensuring adherence to data privacy laws and preparedness against cyber threats. Educating clients and investors is equally crucial, which can be accomplished by providing clear risk disclosures, advisory services for secure cryptocurrency investments, and informational sessions to raise awareness about blockchain technology and its regulatory aspects [14]. By implementing these measures, governments and financial institutions can strike a balance between fostering innovation, maintaining financial stability, and protecting consumers, effectively managing risks while maximizing the opportunities of blockchain technology.

8 Conclusion

Blockchain technology and cryptocurrencies have emerged as disruptive innovations in capital markets, revolutionizing financial transactions, asset custody and transfer, and value functions. Blockchain will decentralize, bring transparency, and eliminate operational inefficiencies, therefore reshaping the future of global financial systems. The rise of cryptocurrencies—digital assets that can be traded or sold—has created new opportunities for investment and fundraising, while bypassing the traditional financial intermediaries that have long held a dominant position. Although these technologies show great potential, they also represent significant challenges including regulatory uncertainty, price volatility, and risks of fraud and abuse. This evolution is largely facilitated by regulation. Across the globe, policymakers are grappling with the challenge of striking a balance between fostering innovation and preserving financial stability and investor protections. From regulatory sandboxes and MiCA frameworks in the EU to Central Bank Digital. "Central Bank Digital Currencies" (CBDCs) reflect governments being able to evolve with the times. Nevertheless, a globally harmonized approach still has some way to go. Without common standards, regulatory arbitrate is possible making it potentially more difficult to build trust in the marketplace and greater stability. To avoid the challenges, working with regulators, financial institutions, and technology developers to resolve these challenges will be key to the future of growth in blockchain and cryptocurrency adoption. The steps to bridge the gap between evolving technology and the regulatory landscape will be the introduction of a digital asset taxonomy, responsible innovation with the help of sandboxes and blockchain analytics for live monitoring. When combined with education and awareness campaigns, this will help both retail and institutional investors make better judgments in this rapidly changing world.

References

1. Nakamoto, S. (2008). *Bitcoin: A peer-to-peer electronic cash system.* Retrieved from https://bitcoin.org/bitcoin.pdf
2. European Commission. (2023). *Markets in crypto-assets regulation (MiCA): Framework for crypto oversight.* Retrieved from European Commission website.

3. World Economic Forum. (2021). *Blockchain deployment toolkit: Harnessing blockchain technology in financial markets*. World Economic Forum.
4. Securities and Exchange Commission (SEC). (2022). *Framework for digital assets and ICOs*. Retrieved from https://www.sec.gov
5. Zetzsche, D. A., Buckley, R. P., Arner, D. W., & Föhr, L. (2021). Regulating blockchain, DLT, and smart contracts: A technology regulator's perspective. *Journal of Financial Regulation and Compliance, 29*(1), 1–17.
6. Financial Stability Board (FSB). (2020). *Regulation, supervision and oversight of global stablecoin arrangements*. Retrieved from FSB website
7. Tapscott, D., & Tapscott, A. (2016). *Blockchain revolution: How the technology behind bitcoin is changing money, business, and the world*. Portfolio.
8. Yermack, D. (2017). Corporate governance and blockchains. *Review of Finance, 21*(1), 7–31. https://doi.org/10.1093/rof/rfw074
9. IMF Staff Discussion Note. (2022). *The ascent of digital money: Challenges and opportunities for financial stability*. International Monetary Fund.
10. Regulatory Trends in Cryptocurrencies and Blockchain. (2024). *Highlights the evolving legal landscape for digital assets in markets such as Hong Kong, Singapore, and the UAE*.
11. Bains, P., Ismail, A., Melo, F., & Sugimoto, N. (2022). *Regulating the crypto ecosystem: The case of stablecoins and arrangements*. International Monetary Fund.
12. Li, Y. (2024). Blockchain and cryptocurrency: The future of financial innovation and challenges. *Advances in Economics, Management and Political Sciences, 88*, 137–143. https://doi.org/10.54254/2754-1169/88/20240891
13. Makode, P. K. (2022). *Blockchain application in incentivizing students on participation in online classes*. ResearchGate. https://www.researchgate.net/publication/366240640
14. Gomber, P., Kauffman, R. J., Parker, C., & Weber, B. W. (2018). On the fintech revolution: Interpreting the forces of innovation, disruption, and transformation in financial services. *Journal of Management Information Systems, 35*(1), 220–265.

Capital Market Mechanisms for Climate Change Mitigation: The Role of Green Bonds, Carbon Credits, and Sustainable ETFS

Meenakshi Chawla and Priyanka Chawla

Abstract Sustainable finance provides robust alternatives for funding climate change mitigation projects. These include investments in green and eco-friendly enterprises. The objective of this research is to establish the contribution of some selected key capital market instruments which include green bonds, carbon credits, and sustainable ETFs towards the fight against climate change. The study also aims at identifying the moderating variables concerning management, legal enablers, and research standards on the allocation and efficiency of investments with the view of achieving the stretch goals towards influencing climate change. The findings demonstrate that it is one of the effective means for the funding towards more sustainable projects including carbon reduction and renewable energy, green bonds, carbon credits, sustainable ETFs help combat climate change.

Keywords Climate change · Sustainable · Carbon credit and ETFS

1 Introduction

Equity markets are critical to financing sustainable growth and the increasing global realization of climate change as a threat has witnessing international efforts to mitigate it. Green bonds, carbon credits, and sustainable ETFs have risen as effective means and financial tools that play the role of mobilizing capital for climate solutions and supporting transition to the green economy. Despite all these, there is a lot of potential in these financial instruments; however, market structure, legal requirements, and investors' preferences influence the application and integration of these assets [1].

M. Chawla (✉)
St. John's College, Agra, India
e-mail: chawla.meenakshi4@gmail.com

P. Chawla
Prestige Institute of Management and Research, Gwalior, India
e-mail: priyanka.chawla@prestigegwl.org

K. Khanna et al. (eds.), *From Poverty to Prosperity*, Information Systems Engineering and Management 64, https://doi.org/10.1007/978-3-031-99166-0_10

For example, green bonds allow companies to finance sustainable projects of a definite type in the sphere of environment, for example, renewable energy sources, energy saving, and ecological infrastructure. Likewise, carbon credits is another market based instrument used for motivating reduced emission of greenhouse gases but the success is again contingent of complete monitoring and verification mechanism. At the same time, the sustainable ETFs help retail and institutional traders get diversified exposure to ESG compliant firms, using their investment dollars to generate alpha and balance profits with purpose [2].

The challenges are regulatory fragmentation, the threat of green washing, and the need for standardization of all mechanisms to ensure credibility and transparency founded among the challenges that these mechanisms must overcome although they bring multiple opportunities. Besides, targeting investors enhances education and experience that is crucial for the expansion of adoption rates and the resulting effects [3]. Hence, there is a need to understand the role and how these financial market tools interrelate in an attempt to increase their impact on climate change mitigation while ensuring that they meet sustainable development goals.

Capital market instruments are viewed as basic components that could help build both economies and sustainability against the backdrop of climate change. Apart from the fact that they can mobilize huge capital, they possess the ability of creating public–private partnership and innovation.

The scope of the paper will be narrowed down to the emerging examples of these capital market instruments: green bonds, carbon credibility, and sustainable ETFs and their role in the process of investment finance. Issues such as the relationship between market transparency and legal frameworks and the necessity for such things as clearly defined measurements will also be tackled, this time focussing on the structure, performance, and potential of these instruments to raise large amounts of money for climate change targets. Through understanding the synergies among these mechanisms, this study intends to determine how capital markets can help to enhance the pace of transition to a sustainable and low-carbon economy globally.

2 Conceptual Framework

This conceptual framework is an attempt to map the place of capital market mechanisms in climate change processes and systems. Three tools are characterized: green bonds, carbon credits, and sustainable ETFs; the relationship between these tools, participants, and consequences are discussed.

1. **Core Concept: Capital Markets to Drive and Support Climate Change Mitigation** Capital markets can be understood as the place providing an opportunity to attract funds to support various socially responsible and environmentally friendly

projects. Sustainable finance aims at directing capital towards schemes, enterprises or assets that lead to lowering of GHG emissions consist a part of the environmental sustainable agenda. The outcomes are effective resource allocation towards the management of climate change.

2. **The Three Key Mechanisms are Green Bonds, Carbon Credit, and Sustainable ETF$_S$**
   (a) **Green Bonds**: A class of financial instruments that are used to fund infrastructure providing corresponding ecological values like clean energy, energy conservation, and climate change solutions.
   (b) **Carbon Credits**: Certificates that allow a firm or an individual to emit a given quantity of CO_2, normally measured in tonnes of CO_2 per credit.
   (c) **Sustainable ETFs**: Shares that track different funds focussed on companies or specific assets that meet Environment, Social, and Governance standards.
3. **Interconnections of the mechanisms**
   (a) **Green Bonds and Carbon Credits**: The funds resulting from issuing green bonds may be used to finance projects that produce carbon credits.
   (b) **Sustainable ETFs and Green Bonds**: Holding could increase investment in climate-focussed assets where ETFs may have green bonds as part of their investment.
   (c) **Carbon Credits and Sustainable ETFs**: A survey revealed that many firms in ETFs engage in carbon markets in a bid to hit the net-zero targets.
4. **Stakeholders and Their Roles**
   (a) **Governments**: Draft laws and legislation, set target values for climate, and launch state sovereign green bonds.
   (b) **Corporations**: The following ways should be taken to attract ETF investments: Green bonds should be issued, carbon markets should be joined, and ESG principles should be implemented.
   (c) **Regulators and NGOs**: Check compliance and ensure the benefits.
5. **Challenges and Risks**
   (a) **Green washing**: The social cost of the projects, misrepresentation of the environmental benefits of projects.
   (b) **Market Volatility**: Market volatilities of green assets and its impact on the confidence of the investors (Fig. 1).

Fig. 1 Variable based model created by the Author

3 Review of Literature and Hypothesis Development

- ***Climate Change Mitigation Impact***

The climate change mitigation literature has elaborated several measures formulated for emissions and restrictiveness of the global warming condition. Several authors have noted that renewable energy sources such as water, power, solar as well as wind energy in relation to climate change impacts [4]. These energy options are regarded as necessary in the shift to a low-carbon economy [5]. Nonetheless, the integration of renewable energy is not easy, especially to the existing structures that are applicable in the regions interested in fossil fuel.

There is a successive increase in the interest of renewable energy as the most capable approach to combating climate change. IRENA [6], Johnson [7] posits that the world could cut down its carbon emission by 70% by 2050 if the world augments efforts to embrace RE sources. Therefore, the modernization of storage technologies and grid is essential for integration existing barriers to be eliminated.

The pragmatic means of embedding carbon costs through the mechanisms of taxes and trading under the cap have also been the subject of being widely debated and discussed among economists and policy makers with references to climate change. As highlighted by [8], carbon pricing ensures that an appropriate price is placed for carbon emissions with a view of making people and businesses change their behaviours. Klenert et al. [9] maintain that these measures to be all alone may not be adequate to drive necessary climate targets, whereas carbon pricing leads to the development of green technologies. More recent works, like [4] discussed that major reform measures that seek to lower carbon emissions, including carbon pricing policies, are expectedly leaky and that cooperative regional solutions are posed to be more effective.

Climate change mitigation has also attracted attention towards NBS as well over the years. Such approaches like afforestation, reforestation, and sustainable land management have been considered affordable ways of increasing carbon stock [10]. Moreover, such solutions also have social benefits which include aspects like better water quality and wildlife habitat that goes a long way to give such solutions value in climate change mitigation.

H0$_1$: The Climate Changes Mitigation impact and carbon credit and sustainable ETFs are not related significantly.

- **Environment Performance**

Environmental performance therefore aims at measuring the ability of an organization to implement its environmental addresses through systems implementation and environmental resource management. This is the revelation from Green in 2017, he maintains that an organization that gives emphasis to environmental performance is likely to achieve reductions in the carbon and energy footprints and therefore improve on its operation. He states corporate environmentalism arguing that firms with environmental plans help improve their image besides experiencing compliance. Johnson [7] wrote that this is because participation of stakeholders is a factor to the level of environmental performance needed within areas like manufacturing, which has serious environmental impacts. Interacting with the customers and the community on sustainability practices enables them to develop a good relationship and at the same time improve the customers' attitude to the environmental concerns. As it has been outlined in recent articles, the focus is on the use of digitalization as a tool for achieving improved environmental results.

As stated in the article by [11], artificial intelligence, blockchain, and IoT can help identify and minimize an organization's negative impact on the environment systematically [13]. For better environmental performance management, organizations should focus on extended review efforts and promote sustainability in every level in an organization. Therefore, it is evident that companies, which integrate environmental management with organizational culture, experience improved performance in terms of both environmental and economic returns in the long run. In a relatively recent study, Garcia et al. [14] showed that heightened environmental legislations and finance policies affect firms' environment management policies with an improved sustainability performance in the firms. Accordingly, the current studies point to the increasing role of technology, policies, and governance in addressing performance issues within the context of the environment. Hence, it can be assumed that the continuous compliance of green practices with the technological advancement and related policies will shape the coming path of environmental responsibility in business.

H0$_2$: There is no significant relationship exists between Environment Performance and carbon credit and sustainable ETFs.

- **Market Performance of Sustainable Assets**

Sustainable assets market performance is an increasingly important area of focus in our current financial landscape. The moving patterns of sustainable investment are underlined, states that the asset managers add ESG as a part of their investment decisions to obtain equilibrium between positive social impact and financial returns. He also claims that these sustainable assets have shown resilience and many times have outperformed conventional assets in volatility. Studies trace the evolution of sustainable assets, which they argue the company now plays the long game by measuring company long-term potential around social responsible investments (SRI) and green bonds.

These asset classes He claims that these assets generating both financial returns and measurable environmental impact are appealing to investors interested in environmental return as well as financial return. Organizations and investors who adopt sustainable practices see enhanced market performance when the demand for eco-friendly solutions from consumers rises. These studies suggest that environmental, social, and governance factors significantly affect the market performance of sustainable assets, in addition to traditional financial factors. The market now demands it, and sustainable investing will lead to financial profits as well as societal benefit.

$H0_3$: There is no significant relationship between Market Performance of Sustainable assets and carbon credit and sustainable ETFs.

- **Carbon Credit and Sustainable ETFs**

Carbon attitudes are units of carbon credits that give the owner the right to emit a certain amount of greenhouse gases or carbon dioxide or an equivalent of one metric tonne. Financial instruments are essential in the carbon trading markets that seek to put a price to the carbon emissions and encourage firms to lower the quantity of pollution. A recent useful introductory guide from [15] gives a general idea of the present state and overall structure of the carbon credits and their purpose in combating climate change. However, carbon credits have been criticized for concerns of their authenticity, as expressed below. A report by [16] describes the current level of integrity for carbon credit market and focus on the fact of need for high quality credits. The report as indicates that buyers require confirmation that the projects, which they support through carbon credits, deliver actual climate, and environmental benefits as indicated on the proposed claims. Without this assurance, firms that voluntarily adopt carbon credits may possibly encounter difficulties in proving their claims to environmental friendliness.

Sustainable ETFs

Sustainable ETFs are investment funds that target securities of firms who rank well in the measurement frameworks of ESG factors. Such ETFs can focus on specific industries like clean technology, renewable resources, and sustainable farming among others; hence, enabling investors to invest in environmentally friendly industries and at the same time make part of a diversified investment portfolios. A recent study conducted by the [17] showed that emission profiles of green ETFs are less than those of reference ETFs and therefore investors can lower their carbon footprint by investing in self-declared sustainable ETFs. Nevertheless, the study also observed that investors desiring to capture a broad market while providing the lowest emitters within the sector may not be fully accomplished by existing sustainable ETF.

There has also been the launching of carbon credit ETF, which allowed investors to have investment interest in carbon credits through the financial instruments. For example, KraneShares U.S. Carbon Return Exchange Traded Fund (ETF) making an entry to the market in 2020 is KRBN; it offers the retail investors to invest directly in carbon credit markets. After the launch of the first carbon credit ETF in 2008, more and more carbon credit ETFs and mutual funds have appeared internationally on the market in terms of investing in carbon credits.

Thus, current literature points to the need for carbon credit and sustainable ETFs as agents for environmental changes with the potential for investor involvement to transform into a low-carbon environment. Although these financial instruments offer investors a speculative chance of investing with sustainability principles in mind, constant controversies regarding their efficiency and credibility indicate that investors should thoroughly assess them.

4 Research Methodology

In this research, the quantitative research methodology is used in order to establish the reliability and validity of the research variables that include the market, environment, and reporting factors that affect sustainable investments. For data collection, a structured questionnaire was used that was distributed to 450 institutional investors spread out through a stratified random sampling technique to get cross-sectional samples from firms, banks, and asset management companies, concerned with sustainable finance. The key factors taken into account with the investors' involvement in ESG investment decisions, the portfolio's size, and experience in sustainable investments.

In order to confirm the reliability of the measures used in the study, Cronbach's alpha, average variance extracted (AVE), and the composite reliability were used. To evaluate discriminant validity, cross loadings were used, as well as Fornell and Larcker two-step criterion. Furthermore, CFA was performed to analyse the interrelationship between the constructs and confirm validity and reliability of the developed scales. These statistical techniques together help to add more reliability and validity in the overall findings of the study.

5 Data Analysis

Cronbach's alpha, composite reliability (rho_a and rho_c), and average variance extracted (AVE) were used in Table 1 to analyse the validity and reliability. The range of Cronbach's alpha values was 0.667–0.893 This means most measures were reliable with "Market Performance" (0.893) and "Carbon Credit and Sustainable ETFs" (0.874) scoring the highest, while "Environmental Performance" (0.667) had the lowest but still acceptable reliability. Composite reliability values (rho_a and rho_c), which also measure consistency, were all above 0.7 showing strong reliability.

"Market Performance" and "Carbon Credit and Sustainable ETFs" had high scores of 0.917 and 0.905, confirming they were measured well. The AVE values show that the data are captured by the measures were all above 0.5. This indicates that

Table 1 Reliability analysis and convergent reliability

Variables	Cronbach's alpha	Composite reliability (rho_a)	Composite reliability (rho_c)	Average variance extracted (AVE)
Carbon credit and sustainable ETFs	0.874	0.877	0.905	0.614
Climate change mitigation impact	0.769	0.791	0.835	0.518
Environmental performance	0.667	0.710	0.767	0.510
Market performance	0.893	0.904	0.917	0.612
Reporting standards	0.852	0.858	0.887	0.528

Source Table prepared by the Author

the measures explain a good amount of the variance in the data. "Carbon Credit and Sustainable ETFs" (0.614) and "Market Performance" (0.612) performed particularly well here while "Environmental Performance" (0.510) and "Reporting Standards" (0.528) had lower but still acceptable values.

Table 2 follows the Fornell-Larcker criterion, demonstrating that the constructs in the study are distinct from one another. The square root of the average variance extracted (AVE) for each construct (e.g., 0.784 for CCS, 0.678 for CM, 0.632 for EP, 0.783 for MP, and 0.727 for RS) exceeds the correlations between those constructs. For example, the square root of AVE for CCS (0.784) exceeds its greatest correlation with CM (0.734), indicating that each construct reflects a distinct concept. This demonstrates that the constructs (CCS, CM, EP, MP, and RS) are distinct confirming their validity and reliability in the study.

Table 3 illustrates notable relationships between the constructs and their respective items. For example, **CCS1** (0.772) and **CCS2** (0.83) exhibit strong loadings on **Carbon Credit and Sustainable ETFs (CCS)**, while also demonstrating moderate loadings on **Market Performance (MP)** (CCS1 at 0.436, CCS2 at 0.41), suggesting a conceptual overlap between **CCS** and **MP**. Items such as **MP2** (0.783) and **MP3** (0.817) show robust associations with **Market Performance**, but also reveal overlap

Table 2 Discriminant validity—Fornell-Larcker criterion

Variables	CCS	CM	EP	MP	RS
CCS (carbon credit and sustainable ETFs)	**0.784**				
CM (climate change mitigation impact)	0.734	**0.678**			
EP (environmental performance)	0.672	0.484	**0.632**		
MP (market performance)	0.635	0.327	0.52	**0.783**	
RS (reporting standards)	0.437	0.325	0.482	0.625	**0.727**

Source Table prepared by the Author

with both **CCS** and **Reporting Standards (RS)** (e.g., MP3 at 0.295 with RS). Items like **EP2** (0.649) in **Environmental Performance (EP)** exhibit a strong correlation with **EP**, while also demonstrating moderate associations with **CCS** (0.732) and **RS** (0.412). The **Climate Change Mitigation Impact (CM)** items, such as **CM2** (0.664) and **CM6** (0.754), show significant loadings on their respective construct, with relatively weaker correlations to other constructs highlighting the unique yet interconnected roles of these environmental measures (Table 4).

Analysis of the Structural Model and Hypotheses Testing

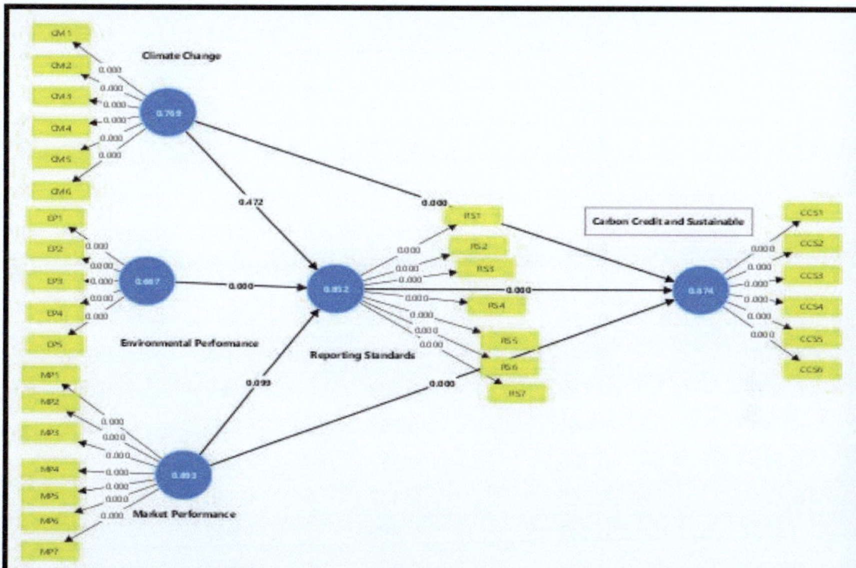

Source Model prepared by authors' calculation

6 Discussion

The results of this study also delineate the central importance of sustainability in finance for climate change solutions in capital market products. Taking the examples of green bonds, carbon credits, and sustainable ETFs, they are prove that the financial instruments work well to help channelling finance towards climate change mitigation—renewables, carbon cuts, etc. This research also highlights the interconnectedness of factors like digital competence, rules and regulations, and transparency to boost investors, corporate, and policy makers' confidence on such tools. Finally, regulatory support resurfaces as essential on the grounds that ESG disclosures and reporting guidelines should help underpin efficient signalling to investors. Besides, reliable protection of data and sophisticating monitoring systems are necessary to avoid the risks that include green washing and ensure the company's credibility.

Table 3 Cross-loadings

Items	Carbon credit and sustainable ETFs (CCS)	Climate change mitigation impact (CM)	Environmental performance (EP)	Market performance (MP)	Reporting standards (RS)
CCS1	**0.772**	0.352	0.436	0.436	0.395
CCS2	**0.83**	0.308	0.42	0.41	0.37
CCS3	**0.786**	0.379	0.508	0.526	0.523
CCS4	**0.732**	0.291	0.549	0.454	0.412
CCS5	**0.79**	0.39	0.57	0.517	0.501
CCS6	**0.789**	0.33	0.487	0.517	0.533
CM1	0.262	**0.588**	0.317	0.163	0.189
CM2	0.312	**0.664**	0.334	0.266	0.139
CM3	0.311	**0.759**	0.388	0.244	0.244
CM5	0.248	**0.658**	0.278	0.199	0.207
CM6	0.397	**0.754**	0.392	0.312	0.333
CM7	0.164	**0.627**	0.185	0.045	0.129
EP2	0.732	0.291	**0.649**	0.454	0.412
EP4	0.453	0.229	**0.749**	0.24	0.711
EP5	0.413	0.45	**0.592**	0.345	0.298
EP6	0.386	0.359	**0.582**	0.287	0.29
EP7	0.349	0.401	**0.569**	0.297	0.264
MP1	0.53	0.206	0.287	**0.697**	0.261
MP2	0.514	0.292	0.432	**0.783**	0.339
MP3	0.511	0.282	0.388	**0.817**	0.295
MP4	0.506	0.232	0.401	**0.829**	0.34
MP5	0.451	0.179	0.266	**0.682**	0.199
MP6	0.557	0.229	0.364	**0.797**	0.296
MP7	0.621	0.347	0.456	**0.856**	0.39
RS1	0.397	0.212	0.351	0.313	**0.665**
RS2	0.368	0.141	0.368	0.259	**0.692**
RS3	0.494	0.261	0.475	0.296	**0.775**
RS4	0.479	0.271	0.517	0.308	**0.733**
RS5	0.5	0.272	0.552	0.24	**0.752**
RS6	0.513	0.244	0.472	0.36	**0.751**
RS7	0.453	0.229	0.749	0.24	**0.711**

Source Table prepared by the Author

Table 4 Hypotheses testing

Independent variable	Dependent variable	Path coefficient	Null hypothesis (H_0)	Result	Relationship	Detailed observations
Climate change (CC)	Reporting standards (RS)	0.472	H_0: Climate Change does not significantly affect Reporting Standards	Rejected	*Moderate*	Climate change moderately influences reporting standards, driven by regulatory and organizational adaptation to climate-related challenges
Environmental performance (EP)	Reporting standards (RS)	0.587	H_0: Environmental Performance does not significantly affect Reporting Standards	Rejected	*Strong*	Strong alignment between environmental performance and reporting standards reflects adherence to global frameworks like GRI and ESG
Market performance (MP)	Reporting standards (RS)	0.291	H_0: Market Performance does not significantly affect Reporting Standards	Rejected	*Weak*	Market performance weakly influences reporting standards, primarily reflecting stakeholder demands for transparency rather than direct organizational focus

(continued)

Table 4 (continued)

Independent variable	Dependent variable	Path coefficient	Null hypothesis (H_0)	Result	Relationship	Detailed observations
Reporting standards (RS)	Carbon credit and sustainability (CCS)	0.874	H0: Reporting Standards do not significantly affect Carbon Credit and Sustainability	Rejected	*Very Strong*	Reporting standards are critical for sustainability and carbon credit initiatives, providing a robust framework for tracking and implementing sustainability practices
Climate change (CC)	Carbon credit and sustainability (CCS)	0.769	H0: Climate Change does not significantly affect carbon credit and sustainability	Rejected	*Strong*	Climate change has a strong direct influence on carbon credit and sustainability, likely reflecting the urgency of addressing climate issues through sustainable practices

Source Table prepared by the Author

Similar to other theories such as Ajzen's Theory of Planned Behaviour, we find trust and perceived security to be variables that are mediate regulatory and educational enablers to the effective use of investment. Finally, this research demonstrates how capital markets can work together to collectively drive change and create a sustainable and low-carbon world economy.

7 Conclusion

This study analyses engaging capital market instruments such as carbon credit, green bond reference, and sustainable exchange-traded funds (ETFs). The research demonstrates how capital investment can significantly support ecological sustainability to counter climate change. Namely, the outcomes underscore the significance of generating strict legal requirements and ESG disclosures and enhancing investors' awareness to enhance the reliability of green financial markets. These technologies enable partnerships between the public and the commercial sectors to make progress to low-carbon economy not only swiftly but also create avenues for development. Thus, this case reveals how sustainable finance can meet climate change challenge and align investment performance with sustainable development objectives at scale.

8 Limitation

This study contributes to the knowledge of climate change mitigation function of sustainable financial instruments. First, the utilization of structured questionnaires provided to institutional investors may lead to bias including measurements based upon self-report which may be inaccurate and generalization to other participants whether retail or small scale businesses and organizations may be hard. Second, the research findings partly pin sustainable bonds on green bonds, carbon credits, and sustainable ETFs which may provide methods within sustainable finance beyond other mechanisms like sustainability linked loans and impact investments despite their significant potential [3]. Third, limitations related to data accessibility and credibility can pose limitations for the research because of problems regarding ESG scores and environmental records [2]. Last but not least, these disparities also stem from the differences in regional legislation creating challenges as fragmented regulations might distort the comparability and comparability of equivalent instruments in terms of their efficiency [1]. As such, to build a more complete picture of how capital markets can support global sustainability efforts in the future research, those limitations should be managed well.

References

1. Climate Bonds Initiative. (2022). *The condition of the green bond market.* Retrieved from https://www.climatebonds.net/files/reports/cbi_sotm_2022_03e.pdf
2. Boredo, J. C. (2018). Financial performance and green bonds: A comparative study. *Economic Modelling, 74*, 55–69. https://doi.org/10.1016/j.econmod.2018.05.003
3. Organisation for Economic Co-operation and Development (OECD). (2021). Sustainable finance and the role of financial markets in climate transition. *OECD Publishing*. https://doi.org/10.1787/9789264303317-en
4. Stiglitz, J. E., & Stern, N. (2021). The social cost of carbon, risk, distribution, market failures: An alternative approach. *National Bureau of Economic Research Working Paper No. 28472*. https://doi.org/10.3386/w28472
5. International Energy Agency (IEA). (2020). *World energy outlook 2020.* IEA Publications.
6. International Renewable Energy Agency (IRENA). (2021). *World energy transitions outlook: 1.5°C pathway.* IRENA
7. Johnson, J. A. (2019). Stakeholder engagement: The game changer for program management. *Project Management Journal, 50*(4), 425–435. https://doi.org/10.1177/8756972819850495

8. Nordhaus, W. D. (2007). A review of the stern review on the economics of climate change. *Journal of Economic Literature, 45*(3), 686–702. https://doi.org/10.1257/jel.45.3.686
9. Klenert, D., Mattauch, L., Combet, E., Edenhofer, O., Hepburn, C., Rafaty, R., & Stern, N. (2018). Making carbon pricing work for citizens. *Nature Climate Change, 8*(8), 669–677. https://doi.org/10.1038/s41558-018-0201-2
10. Griscom, B. W., Adams, J., Ellis, P. W., Houghton, R. A., Lomax, G., Miteva, D. A., Fargione, J., et al. (2017). Natural climate solutions. *Proceedings of the National Academy of Sciences, 114*(44), 11645–11650. https://doi.org/10.1073/pnas.1710465114
11. Chen, Y., Ding, Y., Xu, Z., et al. (2022). How does digitalization affect environmental performance? Evidence from China. *Journal of Cleaner Production, 333.* https://doi.org/10.1016/j.jclepro.2021.130066
12. Eyraud, L., Wane, A. A., Zhang, C., & Clements, B. (2013). Green investment: Trends and determinants. *Energy Policy, 60*, 852–865. https://doi.org/10.1016/j.enpol.2013.05.029
13. Lee, K. H. (2021). Integrating environmental management with business strategy: A case study of Korea's steel industry. *Business Strategy and the Environment, 30*(1), 202–212. https://doi.org/10.1002/bse.2617
14. Garcia, A. S., Mendes-Da-Silva, W., & Orsato, R. J. (2023). Environmental legislation, financial policies, and sustainability performance: Evidence from Brazil. *Journal of Cleaner Production, 380*, 134951. https://doi.org/10.1016/j.jclepro.2023.134951
15. CarbonCredits.com. (2025). *An introductory guide to carbon credits.* Retrieved from https://www.carboncredits.com/introductory-guide
16. MSCI. (2024). *The current state of integrity for the carbon credit market.* MSCI Research Report.
17. Bank for International Settlements. (2022). *The green swan: Central banking and financial stability in the age of climate change.* BIS Publications.

The Role of Capital Markets in Promoting Financial Inclusion

Palanisamy Manigandan and Khaliquzzaman Khan

Abstract This chapter aims to investigate how government spending (GOVEXP), human capital development (HCD), and financial inclusion (FI) affect India's sustainable economic growth (SEG) from 1990 to 2023. A quantitative study design was used to study how human capital growth and financial stability interact to influence this increase. The study, which employed the autoregressive distributed lag (ARDL) approach, discovered a favourable correlation between financial inclusion, government expenditure, HCD, and economic growth. Through HCD, inclusion of financial services contributes to sustainable economic growth. The objective for financial inclusion must be continuously enhanced, and human capital development must be promoted, in order to increase sustainable economic growth in India.

Keywords Human capital · Government expenditure · Sustainable economic growth · Financial inclusion · ARDL approach

1 Introduction

Financial inclusion and its effects on the global economy have recently drawn the attention of experts and opinion leaders. Financial institutions and policymakers utilise FI as an approach to ensure long-term, equitable, and sustainable economic expansion and to assist the growth of the financial sector [1]. The governmental financial mechanism's availability and accessibility to all society groups is a comprehensive definition of financial inclusion. It is a complex and ever-changing system. According to [2], it comprises people from less privileged social groups and lower socioeconomic classes. Moreover, [3] asserts that FI offers a range of low-income and underprivileged groups in society affordable financial services. For an economy

P. Manigandan (✉)
Department of Statistics, Periyar University, Salem 636011, Tamil Nadu, India
e-mail: srimanigandan95@gmail.com

K. Khan
School of Business, American University in Dubai, Dubai, United Arab Emirates
e-mail: khandr2018@gmail.com

K. Khanna et al. (eds.), *From Poverty to Prosperity*, Information Systems Engineering and Management 64, https://doi.org/10.1007/978-3-031-99166-0_11

to grow holistically, we need an inclusive economic structure that can help uplift the underprivileged and underprivileged memberships of society [4]. The issue is regarded as an assessment tool for gauging a society's development and prosperity. In India and many other emerging nations, it is a cornerstone of policymaking [5]. Additionally, [6] have highlighted the close connection amid economic expansion and social and financial marginalisation.

FI allows participants in the economy to increase access to monetary services and promote economic development in underdeveloped market groups. FI and a nation's economic growth are said to be strongly correlated by certain academics [7, 8]. Any nation's ability to raise economically is based on its financial system, which must be robust and resilient. Economic performance is slow, and an exclusive banking system leads to income disparity. Despite economic growth encouraging productivity and equitable development, low FI in India has raised poverty levels and negatively impacted the continent's economic performance. According to [9], FI has gained widespread recognition and is now a priority in the policy of many countries. For a nation's financial infrastructure to enable economic growth and progress, a strong foundation must be established [10–12]. Various initiatives have been put in place by Indian officials to increase the accessibility of financial services for those who are not involved in the financial industry. In order to alleviate poverty, these laws categorise money transfers, insurance, savings, loans, and pensions as financial items. Fi is supported by the general public, academics, lawmakers, politicians, and formal and informal financial stakeholders. For economies, it is a necessary precondition. Additionally, in poor nations, the World Bank and the G20 have increased FI in an effort to decrease poverty and promoting greater economic participation [13].

Regarding the connection between FI and economic expansion, the empirical study yielded contradictory findings. For instance, studies have demonstrated that FI and economic development are positively correlated, irrespective of the country [10, 14]. According to [15], inclusive finance facilitates the production of encouraging wealth and long-term, SEG. Nevertheless, on occasion, it may only have positive effects [16]. Earlier studies [17, 18] have demonstrated inverse connections among FI and GDP. Although research on the connection among FI and GDP growth has changed, previous studies used a wide range of proxy variables to incorporate a wide range of FI indicators. The FI index has been calculated by considering the three aspects of banking availability, usage, and penetration [19]. The association amongst HCD and FI was not taken into account by [20], who evaluated it as a component of economic growth. However, the research appears to have mostly overlooked the interaction between FI and human capital development, which affects India's economic growth. Together with FI, we contend that human capital will significantly boost India's economic growth.

However, [21, 22] have demonstrated that financial inclusion positively impacts human development. Education, income, health, and life-expectancy are all significantly improved, as is the human development index (HCD). According to [23], inclusion of financial decreases income disparity, helps a greater number of people access all essential financial services, and has a substantial consequence on the poor's income growth. It also has a direct correlation with the reduction of poverty. Human development is at a high level as a result of declining trends in poverty intensity. Giving the poor and other disadvantaged groups widespread access to monetary services at a short or no cost will undoubtedly help them. Giving to [24], the ease of access to use of financial services may positively impact human development by promoting the launch of new businesses, enabling public to finance added in health and education, and efficiently managing risk by lessening the burden of unanticipated financial emergencies. According to [25], a functional financial framework is also essential for allocating funds for their beneficial purposes and lowering the chance of revenue loss, which promotes human progress, improves prospects and income-distribution, and lowers poverty. Therefore, if it is possible to guarantee that everyone in an economy has access to, uses, and has availability of formal financial services, then significant advancements in human development can be shown.

2 Financial Inclusion in India

India has achieved remarkable progress in financial inclusion with the execution on the "Pradhan Mantri Jan-Dhan Yojana" (PMJDY) initiative, which provided 75 million not banking Indians with access to a wide range of financial institutions. In 2017, individuals with FI accounted for 80% of the population, a 27% rise from 2014. In [26] examined the impact of financial participation on the economic expansion in 25 Indian states under the PMJDY strategy. In India, FI has a major encouraging impact on GDP growth. Nonetheless, the majority of states reported low or moderate levels of economic inclusion due to limited use of financial services and an increase in dormant accounts after the PMJDY strategy [26]. The P-MJDY has its limitations even with a well-thought-out scheme. The plan was unsuccessful in detecting account duplication, guaranteeing access to digital banking, or lowering the number of inactive accounts. The success of the plan will be improved, nevertheless, by suitable policy-led intervention, a solid dataset infrastructure, the accessibility of appropriate financial instruments, and the development of data and communication technology [27, 28]. Figure 1 shows the most encouraging advancements in India's inclusion of financial.

In Republic of India, societal control over the investment industry in 1960 marked the beginning of FI. Indian efforts to cover formal financial facilities to rural areas have been numerous since then [29]. Nationalisation of commercial-banks, institutionalisation of area rural-banks, enforcement of division licencing policies, and the start of priority-based part offering were the only measures taken to promote inclusion of financial in India at first. The implementation of all financial sector changes focussed on the banking industry's profitability and efficiency, which had previously been overlooked [30]. In order to guarantee successful financial inclusion, India's banking industry has subsequently experienced substantial change.

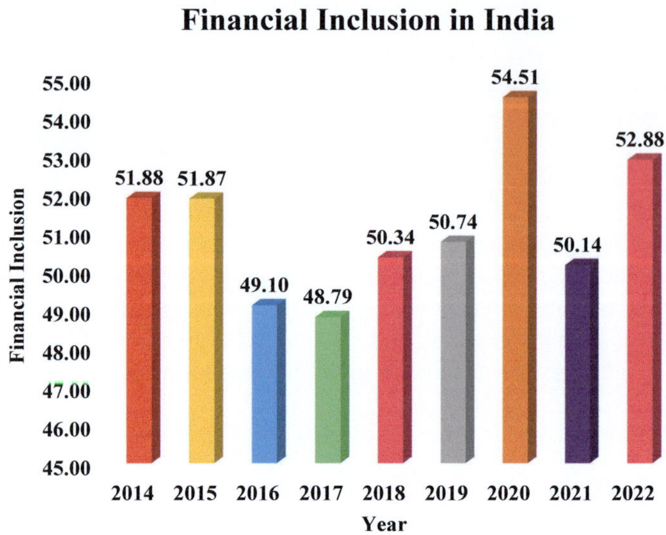

Fig. 1 Financial inclusion of India

3 Human Development Index in India

According to Fig. 2, India is ranked 134th out of 193 nations in the Human Development Index, indicating growth. The nation's HDI value increased by 48.4% between 1990 and 2022, from 0.434 in 1990 to 0.644 in 2022. India, which is ranked 108th out of 166 countries in the GII 2022, has also made strides in lowering gender disparity. Reproductive health, empowerment, and the labour market are the three main areas where the GII assesses gender inequality. With a GII value of 0.437, the nation outperforms both the South Asian average of 0.478* and the global average of 0.462. Compared to other nations in the middle human progress group or South Asia, India does better in terms of reproductive health. The adolescent fertility rate in India improved from 17.1 per 1000 women aged 15 to 19 in 2021 to 16.3 per 1000 in 2022. However, with a 47.8% point discrepancy between males (76.1%) and women (28.3%), India also has one of the biggest gender disparities in the unemployment rate.

The study's primary research question is: In the context of India, how do government spending, HCD, FI, and the mild impact of HCD * FI affect sustainable economic growth? This analyse used a number of econometric methods to investigate the dynamic impacts of the each variable on GDP in order to answer to the investigate question. In order to better understand the relationship between FI, HCD, government spending on SEG, and the moderating effect of HCD on FI, this study will examine India between 1990 and 2023. In numerous ways, our research contributes to the corpus of knowledge. First and foremost, it is significant because it is among the few global studies that focus on FI and how it promotes economic growth. In

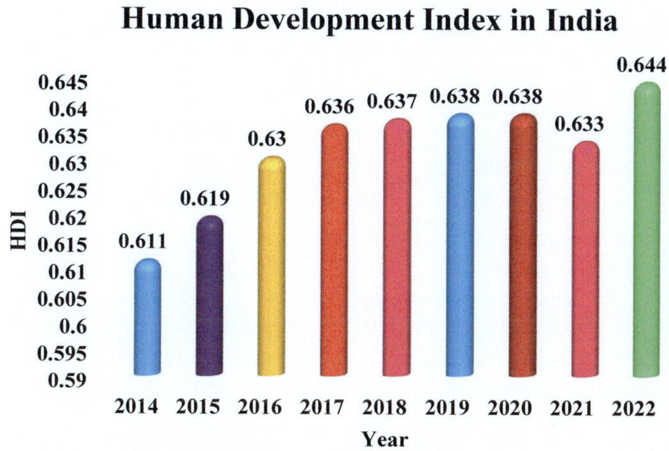

Fig. 2 Human development index of India

contrast to other empirical research that only examined particular economies to assess the connection amid FI and GDP growth, our analysis uses a more thorough cross-country methodology [31, 32]. Third, we include human capital development another important element that could significantly affect the nexus between GOVEXP, FI, and economic progress in India to the current analysis by [33]. It is acknowledged both theoretically and experimentally that FI plays an important role in fostering GDP expansion. Following extensive debate, it is now generally agreed that FI might promote overall financial advance, growth, and poverty-reduction as policy objectives. For the underprivileged, increased access to financial-facilities progresses their quality of life [34]. Our research would offer trustworthy and compelling data to regulators, decision-makers, and scholars when debating how government expenditure affects economic expansion. We also observe that FI and the rate of economic growth may have a major impact on the relationship.

The next chapter is organised as follows: The data collection and method will be covered in Sect. 2. Analysis and debate are then the main topics of Sect. 3, and conclusion and policy recommendations will be the main emphasis of the final Sect. 4.

4 Data and Methodologies

4.1 Theoretical Context and Data Descriptions

This analysis utilises an annual dataset covering the years 1990–2023 to examine the longer-term outcome of using financial inclusion, HCD, Government expenditure on sustainable economic progress in India. Every variable that is used is converted to its

Table 1 Data description

Variables	Descriptions	Measurement	Web-sources
SEG	Sustainable economic growth	Annual GDP per capita growth rate	WDI (2024)
FI	Financial inclusion	Financial inclusion index	
HCD	Human capital development	Human development index	
GOVEXP	Government expenditure	Total government expenditure	

Source World Development indicators (WDI)—2024

natural logarithm. According to [35, 36], this was done to get the series to conform to normalcy. Source, measurement, and unit of measurement of the data are shown in Table 1. This research paper economic function is showed in Eq. (1):

$$\text{SEG} = f(\text{FI}_t, \text{HCD}_t, \text{GOVEXP}_t) \quad (1)$$

In Eq. (1), SEG, FI, HCD, and GOVEXP illustrate sustainable economic growth, financial inclusion, human capital development, and government expenditure-based sustainable economic growth. The following is an illustration of the study's economic model.

$$\text{SEG}_t = \alpha_0 + \alpha_1\text{FI}_t + \alpha_2\text{HCD}_t + \alpha_3\text{GOVEXP}_t + \varepsilon_t \quad (2)$$

This empirical study explains why the variables indicated above are being used. Therefore, the model given in Eq. (2) is expanded to include the interaction term (HCI * FI) in order to measure the indirect impacts of FI on sustainable economic development via the human capital development channel.

$$\text{SEG}_t = \alpha_0 + \alpha_1\text{FI}_t + \alpha_2\text{HCD}_t + \alpha_3\text{GOVEXP}_t + \alpha_4\text{HCD}_t * \text{FI}_t + \varepsilon_t \quad (3)$$

4.2 *Methodology*

4.2.1 Unit Root Test

Prior to performing additional analysis, it is imperative to review the integration series order. Consequently, we evaluate the integration characteristics of the series using unit root tests. The Phillips–Perron (PP) and traditional augmented Dickey–Fuller (ADF) unit root assessments were applied in the investigation.

4.2.2 ARDL Bounds Co-integration Test

The ARDL limits test is used to examine how government spending, financial inclusion, and human capital development relate to India's sustainable economic growth. We utilised the bounds test for ARDL developed by [37] to determine the co-integration across the sequences. It has some positive attributes: (i) the ARDL approach offers the most accurate linear and undistorted long-term relationship prediction; (ii) the effectiveness of the ARDL approach is unaffected by data size; (iii) the long-term and short-term relationships are evaluated concurrently; (iv) the ARDL approach can be applied regardless of whether the series are I(0), I(1), or both; (v) The ARDL approach resolves the issues of autocorrelation and series endogeneity; (vi) The series' lag sequences are not necessarily the same [38]; and (vii) its unique equation structure facilitates use and makes interpretations simple. Determine [39] the critical values of the limits test, which is based on the F-distribution. The limited ARDL test, which captures longer and shorter run impacts, assesses the following unrestricted error correction mechanism via ordinary least squares and is defined as follows Eq. (4).

$$\begin{aligned}\Delta \mathrm{SEG}_t &= \vartheta_0 + \beta_1 \mathrm{SEG}_{t-1} + \beta_2 \mathrm{FI}_{t-1} + \beta_3 \mathrm{HCD}_{t-1} + \beta_4 \mathrm{GOVEXP}_{t-1} \\ &\quad + \beta_5 \mathrm{HCD} * \mathrm{FI}_{t-1} + \sum_{i=1}^{t} \vartheta_1 \mathrm{SEG}_{t-i} + \sum_{i=1}^{t} \vartheta_2 \mathrm{FI}_{t-i} + \sum_{i=1}^{t} \vartheta_3 \mathrm{HCD}_{t-i} \\ &\quad + \sum_{i=1}^{t} \vartheta_4 \mathrm{GOVEXP}_{t-i} + \sum_{i=1}^{t} \vartheta_5 \mathrm{HCD} * \mathrm{FI}_{t-i} + \varepsilon_t \end{aligned} \quad (4)$$

There is neither co-integration nor evidence of co-integration, which are the alternative and null hypotheses, respectively. We cannot accept the null assumption if the F-statistic is greater than the smaller and higher bond critical thresholds. The null and alternative hypotheses are, respectively, shown by Eqs. 5 and 6:

$$H_0 = \vartheta_1 = \vartheta_2 = \vartheta_3 = \vartheta_4. \quad (5)$$

$$H_a \neq \vartheta_1 \neq \vartheta_2 \neq \vartheta_3 \neq \vartheta_4. \quad (6)$$

where H_1 represents the alternative theory and H_0 stands for the null hypothesis. The research investigation applies [40] criteria, which need the calculated the T-statistic and F-statistics to be greater than the related upper significant figures, a crucial requirement for determining on co-integration, in contrast to earlier rules that require the F-statistics to be more substantial than the greatest critical threshold for the co-integration. Furthermore, the generated p-values ought to be below the desired thresholds.

4.2.3 ARDL Approach

We applied the ARDL method once co-integration of the criteria was substantiated. Once the long-term links have been identified, short-term interconnections are examined using the error correction framework (ECM), which was developed by [41] for the evaluation of a shorter-term parameters and the Error Correction Term (ECT). To achieve this, incorporate the ECM as follows into the Model Eq. (7) ARDL framework:

$$\Delta \mathrm{SEG} = \vartheta_0 + \sum_{i=1}^{t} \vartheta_1 \Delta \mathrm{SEG}_{t-i} + \sum_{i=1}^{t} \vartheta_2 \Delta \mathrm{FI}_{t-i} + \sum_{i=1}^{t} \vartheta_3 \Delta \mathrm{HCD}_{t-i} + \sum_{i=1}^{t} \vartheta_4 \Delta \mathrm{GOVEXP}_{t-i} + \sum_{i=1}^{t} \vartheta_5 \Delta \mathrm{HCD} * \mathrm{FI}_{t-i} + \rho \mathrm{ECT}_{t-i} + \varepsilon_t \quad (7)$$

where ρ characterises the adjustment speed and ECT_{t-1} denotes the error correction term.

5 Result and Discussion

5.1 Summary of Statistics

This study explores the causal relationship between inclusion of financial, HCD, government expenditure on sustainable economic progress in India over the long-term using annual datasets from 1990 to 2023. Table 2 displays a concise statistical synopsis of the investigation. In addition, Table 2 displays the sequence's maximum, minimum, median, mean, kurtosis, and skewness. To ascertain if the variable is heavy tailed in relation to normal distributions, with the study used Kurtosis. The empirical outcomes demonstrate that each of the statistics is platykurtic since their estimates are lower than three. Moreover, moderate skewness is shown in all of the indices. Additionally, the Jarque–Bera (JB) was used in our investigation to test the parameters' normalcy properties. The results of JB p-value showed that while the financial inclusion, human capital development, government expenditure, and sustainable economic growth all follow normal distribution patterns with respect to the Jarque–Bera probability of less than 1%.

5.2 Unit Root Test

As part of this empirical investigation, the stationarity assets of the used parameters are checked first. Based on this, we utilised the standard unit root assessments, PP

Table 2 Descriptive statistics

Variables	SEG	FI	HCD	GOVEXP
Mean	4.44	39.51	103.79	3.83
Median	5.38	44.63	102.70	3.82
Maximum	8.82	54.51	119.51	5.65
Minimum	− 6.67	22.51	92.00	2.90
Std. dev.	2.85	12.37	8.65	0.59
Skewness	− 1.80	− 0.23	− 0.14	− 0.77
Kurtosis	4.55	2.79	1.51	1.62
Jarque–Bera	5.694	4.5056	2.9696	4.7074
Probability	0.2340***	0.1051***	0.2265***	0.09502***

***designates the significant level at 1%, respectively. Author calculation

Table 3 Conventional unit root analysis

Variables	At level I (0)	First difference I (1)	Conclusion
ADF-test			
SEG	− 3.5367***	− 6.2142***	I (0)
FI	− 2.1092	− 4.653***	I (1)
HCD	− 1.7499	− 5.9859***	I (1)
GOVEXP	− 0.80078	− 4.3631***	I (1)
PP-test			
SEG	− 29.91***	− 44.553***	I (0)
FI	− 5.022	− 51.581***	I (1)
HCD	− 9.1925	− 38.283***	I (1)
GOVEXP	− 13.719	− 41.315***	I (1)

The signs *** denote the statistical significance at 1% level. Author Calculation

and ADF, to determine the series' stationarity properties. The outcomes derived from the ADF and PP unit root analyses are shown in Table 3. The findings show that SEG are stagnant at level, or I (0). But all the limitations are stationary once the first-order difference, or I (1), is taken.

5.3 ARDL Method

This analysis used ARDL limits test to scrutinise the features of variables' long-term association after the reliability of each variable unit-roots was confirmed.

Table 4 exposes that the empirical outcomes of co-integration utilising the ARDL limits testing approach. According to the estimated *F*-statistic, which was higher than the upper threshold restriction, the experimental results strongly supported the existence of long-term co-integration between the parameters being studied.

After the ARDL longer and shorter-term estimates, the study determined the impact of inclusion of financial, human capital development, Government spending on SEG in India between 1990 and 2023 after establishing the longer-run co-integration amid the limitations. Table 5 displays the findings from both the longer- and shorter-term ARDL calculations. It is abundantly evident that inclusion of financial and SEG are positively connected. It can be shown from this that a 1% drop in financial inclusion corresponds to a 0.774% (long-run) and 0.759% (short-run) drop in GDP growth, holding other variables equal. This consequence indicates that FI is a factor in India's growing economy. These outcomes are reliable with those of [42, 43].

As anticipated, there is a positive nexus between economic expansion and the development of human capital. This proposes that, holding other variables constant, a 1% upsurge in human capital development is correlated with an increase in economic

Table 4 Outcomes of ARDL bound assessment

Test statistics	Value	Critical value (%)	Lower bound	Upper bound
F-statistics	4.6700	10	3.008	4.150
		5	3.710	5.018
		1	5.333	7.063

Author calculation

Table 5 Outcome derived from ARDL for long and short-run estimation

Regressors	Coefficient	Std. error	*t*-statistics	*p*-value
Longer run				
FI	0.77431	0.05051	3.1471	0.0001***
HCD	0.0496	0.0695	2.7137	0.0004***
GOVEXP	0.2781	0.9770	3.3081	0.0002***
HCD * FI	0.6030	0.0029	2.0356	0.0011***
R-squared	0.8324			
Adj-squared	0.8067			
Shorter run				
ECM (− 1)	− 0.7867	0.4829	− 3.493	0.000***
FI	0.7593	0.3327	4.282	0.000***
HCD	0.3416	0.1421	2.404	0.0086**
GOVEXP	0.38991	1.3676	3.285	0.039**
HCD * FI	0.502414	0.002666	2.906	0.0041**

Author Calculation: The symbols ** and *** denote statistical significance at the 1% and 5% levels, respectively growth of 0.049% (longer run) and 0.341% (shorter run). The consequence is that the country is better positioned to see a notable growth in the amount of economic activity as human development advances. The rationale is that as human capital increases, so do the citizens' health and educational standards. As expected, there is a strong and favourable nexus amid GOVEXP and economic expansion. Thus, in India, a 1% increase in wind energy is associated with a 0.2781% (long run) and 0.389% (short run) rise in GDP growth, leaving other variables equal. Additionally, the outcome supports comparable research conducted by [44]. According to our discoveries, the Keynesian theory of government spending in India is supported by the fact that higher government spending increases immediate production, consumption, and investment in the nation as well as short- and long-term economic growth.

Lastly, Table 5 includes the moderating effect of both FI and human capital to investigate how higher human capital may affect the effect of FI on SEG. There is a notable favourable interaction impact when FI, economic growth, and school enrolment are combined. This suggests that, if all the additional factors remain unchanged, a 1% growth in HCD * FI will lead to a 0.60% long-term and a 0.502% short-term increase in economic growth. Consequently, school enrolment and FI collaborate to support economic growth. As a result, FI uses human capital to positively influence economic growth. This validates earlier research [45] and the endogenous theory, which maintains that human capital is created domestically within nations. Consequently, improving India's human capital quality fosters financial inclusion and economic growth. Human capital reduces credit risk, which fosters FI and growth. Furthermore, it is noteworthy that the ECM (− 1) is statistically substantial and adverse, suggesting that − 0.78% of the short-term shocks will vanish within a month. The proposed regression strategy fit the dataset remarkably well, as evidenced by longer-run estimate R^2 and adjusted R^2 values of 0.8324 and 0.8067. This advises that the dependent-factors changes can be virtually entirely described by the independent-factors. Figure 3 provides a summary of the causal relationships among the variables.

5.4 Diagnostic Inspection

The empirical estimations of the different diagnostic statistics are shown in Table 6. The uniform distribution of residuals can be determined by applying the Jarque–Bera test. The problem of serial correlation was investigated using the Lagrange multiplier (LM) technique. The consequences of the LM test show that there are no problems with serial correlation in model. The forecast model's heteroscedasticity was investigated using the Breusch–Pagan–Godfrey methodology. It is clear from the results that the suggested model does not exhibit heteroscedasticity. The Ramsey reset test was employed in order to assess the correctness of model.

Additionally, the cumulative sum of recursive-residuals (CUSUM) and cumulative sum of squares of recursive-residuals (CUSUMQ) tests were used to assess the stability of the model. Figure 4 displays the CUSUMQ and CUSUM charts at a

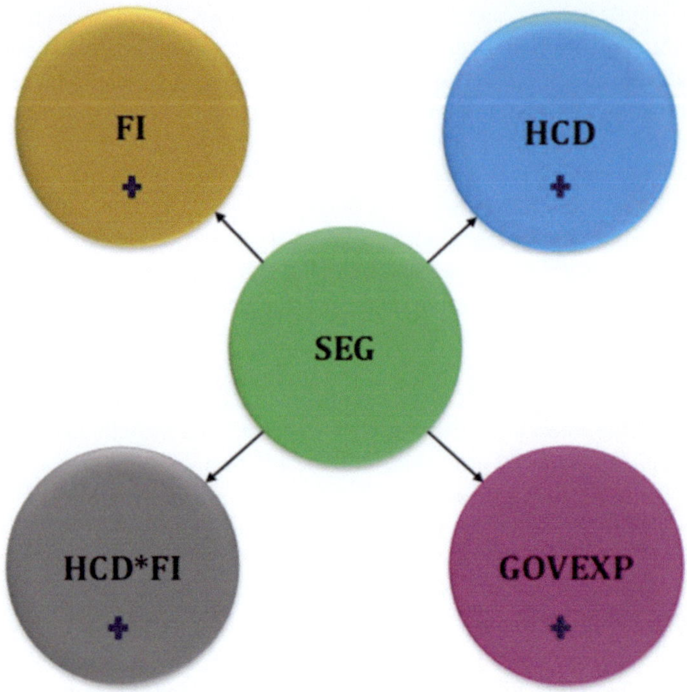

Fig. 3 An overview of the ARDL approach's outcomes (self-created)

Table 6 Result of diagnostic assessment

Diagnostic test	Test statistics	Probability value
JB test	1.41	0.70
LM ARCH test	1.13	0.99
NORMALITY test	0.64	0.73
RESET test	0.46	0.51

Author calculation

5% significant level; the red-outlines illustration the confidence-levels, and the blue-outlines represent the residual values. The estimated findings demonstrate that values of analysed residuals stay within lines of confidence at a 5% significance level, confirming the system's stability.

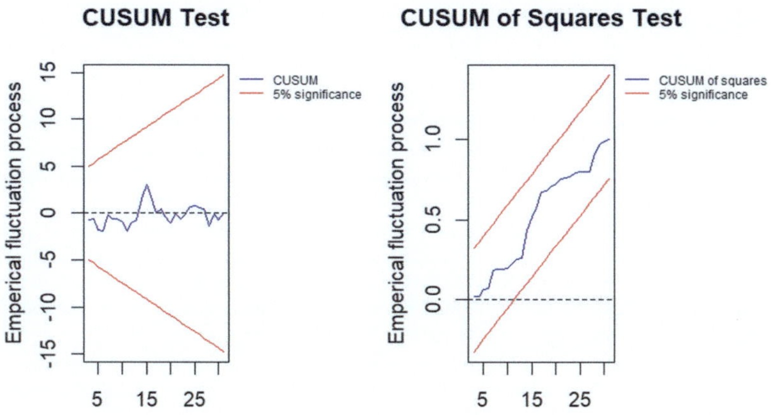

Fig. 4 Plot of stability test (self)

6 Conclusion and Policy Implications

This study examined the relationship between inclusion of financial, HCD, government expenditure on SEG and the moderating effect of HCD * FI in India from 1990 to 2023. The mechanism by which FI promotes GDP growth is human capital. While the literature has recognised the influence of FI on economic progress, empirical explore on this problem is still in its infancy and focuses on a few nations. This article aims to shed more light on the linking between FI and global economic advance. The shorter and longer-term interactions between the variables were explored using the ARDL method. We find important long- and short-term connections between government spending, HCD, inclusion of financial, and the moderating effect of HCD FI all have encouraging and substantial impacts on India's sustainable economic development. While maintaining profitability and economic growth, the Indian government should develop complementary measures to grow financial inclusion (FI) by helping financial markets reduce entry barriers and increase the penetration of financial services. It is possible to implement policies that would help financial institutions offer services in remote areas at affordable rates, hence fostering economic expansion. They must advocate for the financial system's accessibility to considerable population segments in order to positively influence the development of society and the nation overall. Managers ought to make investments in human capital in order to expand job openings, stimulate economic expansion, and raise productivity. Managers, governments, and policymakers must develop appropriate reforms and policies that address the problems on a national and regional level in order to rectify the financial system's flaws. Thus, implementing formalised policies and reforms pertaining to human capital is the main focus of the specific policy recommendations. To support FI, promote economic growth, and facilitate the rise of regional commerce, this entails enhancing education and training resources.

References

1. Collard, S. (2007). Toward financial inclusion in the UK: Progress and challenges. *Public Money and Management, 27*(1), 13–20. https://doi.org/10.1111/j.1467-9302.2007.00550.x
2. Rastogi, S., & E, R. (2018). Financial inclusion and socioeconomic development: Gaps and solution. *International Journal of Social Economics, 45*(7), 1122–1140. https://doi.org/10.1108/IJSE-08-2017-0324
3. Aggarwal, R. (2014). Financial inclusion in India: Challenges and opportunities. *International Journal of Research, 1*(4), 557–567. https://ijirt.org/publishedpaper/IJIRT150147_PAPER.pdf
4. Sharma, U., & Changkakati, B. (2022). Dimensions of global financial inclusion and their impact on the achievement of the United Nations Development Goals. *Borsa Istanbul Review, 22*(6), 1238–1250. https://doi.org/10.1016/j.bir.2022.08.010
5. Dikshit, S., & Pandey, A. C. (2021). Role of financial inclusion in realising sustainable development goals (SDGs). *International Journal of Research in Finance and Management, 4*(2), 35–39. https://doi.org/10.33545/26175754.2021.v4.i2a.103
6. Pandey, A., Kiran, R., & Sharma, R. K. (2022). Investigating the impact of financial inclusion drivers, financial literacy and financial initiatives in fostering sustainable growth in North India. *Sustainability, 14*(17), 11061. https://doi.org/10.3390/su141711061
7. Beck, T., Demirgüç-Kunt, A., & Levine, R. (2007). Finance, inequality and the poor. *Journal of economic growth, 12*, 27–49. https://doi.org/10.1007/s10887-007-9010-6
8. Sarma, M., & Pais, J. (2011). Financial inclusion and development. *Journal of international development, 23*(5), 613–628. https://doi.org/10.1002/jid.1698
9. Pearce, D. (2011). Financial inclusion in the Middle East and North Africa: Analysis and roadmap recommendations. World Bank policy research working paper, (5610). https://ssrn.com/abstract=1794915
10. Sharma, D. (2016). Nexus between financial inclusion and economic growth: Evidence from the emerging Indian economy. *Journal of financial economic policy, 8*(1), 13–36. https://doi.org/10.1108/JFEP-01-2015-0004
11. Huang, W., Gu, X., Lin, L., Alharthi, M., & Usman, M. (2023). Do financial inclusion and income inequality matter for human capital? Evidence from sub-Saharan economies. *Borsa Istanbul Review, 23*(1), 22–33. https://doi.org/10.1016/j.bir.2022.09.002
12. Ali, M., Hashmi, S. H., Nazir, M. R., Bilal, A., & Nazir, M. I. (2021). Does financial inclusion enhance economic growth? Empirical evidence from the IsDB member countries. *International Journal of Finance & Economics, 26*(4), 5235–5258. https://doi.org/10.1002/ijfe.2063
13. Rewilak, J. (2017). The role of financial development in poverty reduction. *Review of Development Finance, 7*(2), 169–176. https://hdl.handle.net/10520/EJC-fa3732772
14. Swamy, V. (2014). Financial inclusion, gender dimension, and economic impact on poor households. *World Development, 56*, 1–15. https://doi.org/10.1016/j.worlddev.2013.10.019
15. Dahiya, S., & Kumar, M. (2020). Linkage between financial inclusion and economic growth: An empirical study of the emerging Indian economy. *Vision, 24*(2), 184–193. https://doi.org/10.1177/0972262920923891
16. Mehrotra, A. N., & Yetman, J. (2015). Financial inclusion-issues for central banks. BIS Quarterly Review March. https://ssrn.com/abstract=2580310
17. Chiwira, O. (2021). The co-integrating relationship between financial inclusion and economic growth in the Southern African Development Community. *Eurasian Journal of Economics and Finance, 9*(3), 170–188. https://doi.org/10.15604/ejef.2021.09.03.003
18. Menyelim, C. M., Babajide, A. A., Omankhanlen, A. E., & Ehikioya, B. I. (2021). Financial inclusion, income inequality and sustainable economic growth in sub-Saharan African countries. *Sustainability, 13*(4), 1780. https://doi.org/10.3390/su13041780
19. Sinha, R. P. (2014). Index of financial inclusion: Some empirical results. In: Analytical issues in trade, development and finance: Essays in honour of Biswajit Chatterjee, pp. 467–478. https://doi.org/10.1007/978-81-322-1650-6_28

20. Evans, D., Green, C. J., & Murinde, V. (2000). The importance of human capital and financial development in economic growth: new evidence using the translog production function. Institute for Development Policy and Management, University of Manchester. https://citeseerx.ist.psu.edu/document?repid=rep1&type=pdf&doi=58f4d32df562be6ada468d125f458644a43debe0
21. Mohieldin, M., Hussein, K., & Rostom, A. (2019). On financial development and economic growth in Egypt. *Journal of Humanities and Applied Social Sciences, 1*(2), 70–86. https://doi.org/10.1108/JHASS-08-2019-0027
22. Abdelghaffar, R. A., Emam, H. A., & Samak, N. A. (2023). Financial inclusion and human development: Is there a nexus? *Journal of Humanities and Applied Social Sciences, 5*(3), 163–177. https://doi.org/10.1108/JHASS-11-2021-0178
23. Beck, T., Demirguc-Kunt, A., & Peria, M. S. M. (2007). Reaching out: Access to and use of banking services across countries. *Journal of financial economics, 85*(1), 234–266. https://doi.org/10.1016/j.jfineco.2006.07.002
24. Matekenya, W., Moyo, C., & Jeke, L. (2021). Financial inclusion and human development: Evidence from Sub-Saharan Africa. *Development Southern Africa, 38*(5), 683–700. https://doi.org/10.1080/0376835X.2020.1799760
25. Allen, F., Demirguc-Kunt, A., Klapper, L., & Peria, M. S. M. (2016). The foundations of financial inclusion: Understanding ownership and use of formal accounts. *Journal of financial Intermediation, 27*, 1–30. https://doi.org/10.1016/j.jfi.2015.12.003
26. Singh, B. P., Kumari, A., Sharma, T., & Malhotra, A. (2021). Financial inclusion, Pradhan Mantri Jan Dhan Yojna scheme and economic growth: Evidence from Indian states. *Economic Notes, 50*(3), e12186. https://doi.org/10.1111/ecno.12186
27. Tarique, M., Zafar, S., & Abusaad, M. (2022). An enquiry into the success of Pradhan Mantri Jan Dhan Yojana in Bihar: An empirical evidence. *Indian Journal of Economics and Development, 18*(4), 971–975. https://doi.org/10.35716/IJED/22178
28. Ray, K. K. (2024). Customer perception towards Pradhan Mantri Jan-Dhan Yojana (PMJDY): An empirical investigation from slum-dwellers perspective. *Millennial Asia, 15*(3), 487–514. https://doi.org/10.1177/09763996221118722
29. Reserve Bank of India. (2008). Report on currency and finance 2007–08. RBI. https://elischolar.library.yale.edu/ypfs-documents/12915/
30. Chavan, P. (2007). Access to bank credit: Implications for Dalit rural households. *Economic and Political Weekly*, 3219–3224. https://www.jstor.org/stable/4419871
31. Lenka, S. K., & Sharma, R. (2017). Does financial inclusion spur economic growth in India? *The Journal of Developing Areas, 51*(3), 215–228. https://doi.org/10.1353/jda.2017.0069
32. Nwafor, M. C. (2018). The nexus between financial inclusion and economic growth: Evidence from Nigeria. *International Journal of Science and Innovation in Social Science, 2*(4). http://eprints.gouni.edu.ng/id/eprint/236
33. Ahmad, M., Majeed, A., Khan, M. A., Sohaib, M., & Shehzad, K. (2021). Digital financial inclusion and economic growth: Provincial data analysis of China. *China Economic Journal, 14*(3), 291–310. https://doi.org/10.1080/17538963.2021.1882064
34. Hannig, A., & Jansen, S. (2010). Financial inclusion and financial stability: Current policy issues. https://hdl.handle.net/10419/53699
35. Adebayo, T. S. (2020). Revisiting the EKC hypothesis in an emerging market: An application of ARDL-based bounds and wavelet coherence approaches. *SN Applied Sciences, 2*(12), 1945. https://doi.org/10.1007/s42452-020-03705-y
36. Hussain, B., Naqvi, S. A. A., Anwar, S., & Usman, M. (2023). Effect of wind and solar energy production, and economic development on the environmental quality: Is this the solution to climate change? *Gondwana Research, 119*, 27–44. https://doi.org/10.1016/j.gr.2023.01.012
37. Pesaran, M. H., Shin, Y., & Smith, R. J. (2001). Bounds testing approaches to the analysis of level relationships. *Journal of applied econometrics, 16*(3), 289–326. https://doi.org/10.1002/jae.616
38. Adebayo, T. S., & Acheampong, A. O. (2022). Modelling the globalization-CO2 emission nexus in Australia: Evidence from quantile-on-quantile approach. *Environmental Science and Pollution Research, 29*(7), 9867–9882. https://doi.org/10.1007/s11356-021-16368-y

39. Pesaran, M. H., & Timmermann, A. (2005). Small sample properties of forecasts from autoregressive models under structural breaks. *Journal of Econometrics, 129*(1–2), 183–217. https://doi.org/10.1016/j.jeconom.2004.09.007
40. Kripfganz, S., & Schneider, D. C. (2023). ARDL: Estimating autoregressive distributed lag and equilibrium correction models. *The Stata Journal, 23*(4), 983–1019. https://doi.org/10.1177/1536867X231212434
41. Engle, R.F. and Granger, C.W. (1987). Co-integration and error correction: representation, estimation, and testing. *Econometrica: Journal of the Econometric Society*, 251–276. https://doi.org/10.2307/1913236
42. Huang, R., Kale, S., Paramati, S. R., & Taghizadeh-Hesary, F. (2021). The nexus between financial inclusion and economic development: Comparison of old and new EU member countries. *Economic Analysis and Policy, 69*, 1–15. https://doi.org/10.1016/j.eap.2020.10.007
43. Van, L. T. H., Vo, A. T., Nguyen, N. T., & Vo, D. H. (2021). Financial inclusion and economic growth: An international evidence. *Emerging Markets Finance and Trade, 57*(1), 239–263. https://doi.org/10.1080/1540496X.2019.1697672
44. Nyarko-Asomani, A., Bhasin, V. K., & Aglobitse, P. B. (2019). Government capital expenditure, recurrent expenditure and economic growth in Ghana. *Ghanaian Journal of Economics, 7*(1), 44–70. https://hdl.handle.net/10520/EJC-1b2f6ea97f
45. Beck, T., Demirgüç-Kunt, A., & Pería, M. S. M. (2011). Bank financing for SMEs: Evidence across countries and bank ownership types. *Journal of Financial Services Research, 39*, 35–54. https://doi.org/10.1007/s10693-010-0085-4

Green Bonds in India: Evolving Practices, Perspectives, and Challenges

Neetha Mary Avanesh and N. Usha Devi

Abstract Green bonds have brought about a phenomenal growth in sustainable finance worldwide and India is no exception. Increasing focus on Sustainable Development Goals, rising climate awareness, renewable energy targets, technological advancements, decline in the cost of renewable energy technologies, government interventions and policy support, regulatory framework by the Securities and Exchange Board of India, and the need for individual investors going for investment portfolio diversification have contributed to the green bonds growth trajectory. Nevertheless, green bonds are faced with challenges relating to high verification costs, risk of greenwashing and a standardized global framework in order to ensure regulatory clarity. The present chapter focuses on the current state of green bonds adoption in India, the various types of green bonds and their role in sustainable debt financing, the key players involved, the major challenges being faced, remedial mechanisms, and the way forward. A few case studies on companies offering green bonds have also been included to give a wholistic perspective to the topic in-hand. The results of the study are expected to deliver valuable insights to myriad stakeholders, thereby expanding the reach of sustainable debt financing.

Keywords Green bonds · Sustainable finance · Sustainable debt finance · Renewable energy · Individual investors

1 Introduction

Green bonds have been developed to fund environment-friendly projects. These bonds are essential for addressing climate change, supporting renewable energy, and achieving sustainability goals. The European Investment Bank heralded the issuance

N. M. Avanesh (✉)
School of Business and Management, Christ University, Bangalore, India
e-mail: neetha.mary@christuniversity.in

N. Usha Devi
Vivekananda Degree College, Bangalore, India
e-mail: ushakrish51@gmail.com

K. Khanna et al. (eds.), *From Poverty to Prosperity*, Information Systems Engineering and Management 64, https://doi.org/10.1007/978-3-031-99166-0_12

of green bonds in 2007, for initiatives revolving around renewable energy sources. In 2008, the World Bank followed suit, with its green bond emphasizing on transparency and accountability. With the formulation of standard guidelines like the Climate Bonds Standard, the investors became more confident to put their money in green bond projects. Later with the corporate participation through companies like Apple and Toyota and global expansion, the popularity of green bonds transcended boundaries with China emerging as the largest issuer of green bonds by 2023. According to S&P Global [1], the Global sustainable bonds issuance is expected to exceed $900 Billion.

In India, green bonds have been strategically integrated into the country's decarbonization plans, helping fund large-scale projects in renewable energy, clean transportation, and carbon-intensive industry transitions [2]. The issuance of green bonds also brings 'greenium' benefits, offering slightly lower yields compared to conventional bonds due to their environmental appeal. However, challenges such as liquidity, green washing concerns, and the need for robust governance remain. Addressing these issues requires clear regulations, risk mitigation strategies, and enhanced investor awareness. India's experience with green bonds demonstrates their potential to mobilize significant capital for sustainability while providing lessons for other emerging markets [3]. By focusing on domestic and international standards, these bonds can contribute meaningfully to the global net-zero transition.

2 Evolution and Growth

The evolution of green bonds in India dates back to 2015, when Yes Bank issued India's first green bond worth ₹1000 crore to fund renewable energy projects. This was followed by IDBI Bank, Axis Bank, and Exim Bank issuing green bonds to support renewable energy and sustainable infrastructural projects. The role played by the apex bank of India namely, the Reserve Bank of India (RBI) in exploring and promoting sustainability initiatives, and the regulations on green bond disclosures issued by The Securities and Exchange Board of India (SEBI) in 2017 had been influential in boosting investor confidence in green projects. In January 2023, India issued its first-ever sovereign green bonds worth ₹16,000 crore ($2 billion), which focused on funding clean energy, biodiversity conservation, and climate adaptation projects. These bonds adhere to a framework designed to ensure transparency and credibility, aligning with international standards. The funds sourced through these bonds were allocated to government projects within 24 months of issuance, targeting clean energy infrastructure and emissions reduction efforts. This approach has not only attracted investment but also fostered the growth of a sustainable financial market domestically. According to the policy brief issued by the Energy and Resources Institute, New Delhi [4], India had issued approximately $21 billion in green bonds, as of February 2023.

3 Types of Green Bonds

Green bonds had evolved over time to cater to the myriad financing needs for environmentally sustainable proposals. The prominent types of green bonds issued in India are listed below (Fig. 1):

i. Corporate Green Bonds: These bonds are issued by private companies to raise funds for specific projects aimed at sustainability and green practices. It mainly included projects relating to renewable energy sources [5]. Nevertheless, projects relating to energy efficiency upgrades within industries were also considered.
ii. Sovereign Green Bonds: These bonds are issued by the Government of India to fund environmentally sustainable projects. The earnings generated from these bonds are allocated to green initiatives which includes renewable energy, biodiversity conservation, efficient waste management, ensuring clean transport systems, and so on.
iii. Municipal Green Bonds: The debt instruments issued by local governments like the municipal corporations to fund sustainability initiatives within the city fall under this category [6].
iv. Masala green bonds are rupee-denominated bonds issued in international markets to finance environmentally sustainable projects in India. These bonds can be issued by a variety of entities including corporates, banks and financial institutions, Public Sector Undertakings, Municipal Corporations and Special Purpose Vehicles (SPVs), provided they comply with regulations set by the

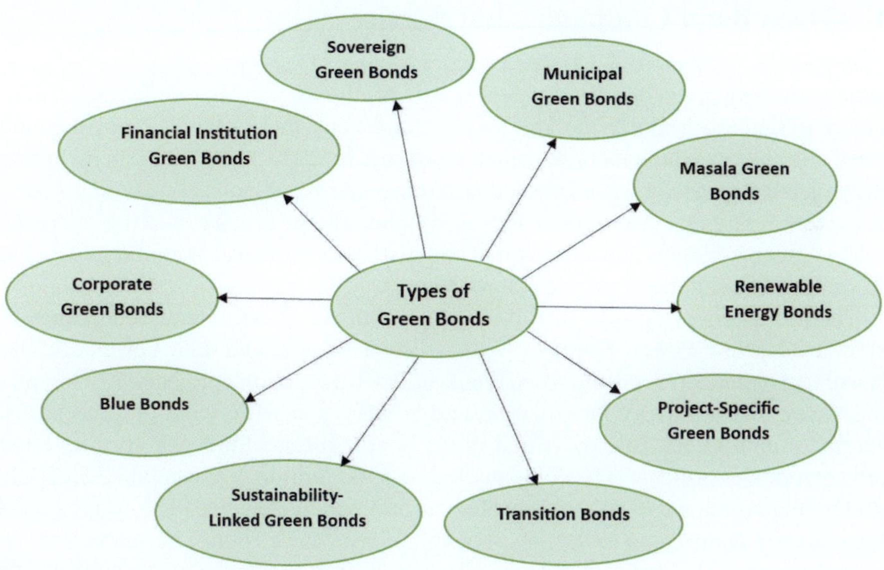

Fig. 1 Types of green bonds. *Source* Developed by Authors

Reserve Bank of India (RBI) and the Securities and Exchange Board of India (SEBI). SPVs refer to those entities which have been established to implement specific projects, such as solar parks, wind farms, and so on. The major advantage of masala green bonds is that it allows Indian issuers to raise funds globally while transferring the currency risk to foreign investors.

v. Financial Institution Green Bonds: Financial Institutions like banks and non-banking financial companies (NBFCs) also issue bonds to raise finance for green projects. These funds are specifically beneficial to small developers and businesses engaged in green projects.

vi. Renewable Energy Bonds: These are dedicated to funding renewable energy initiatives such as solar, wind, hydroelectric, and biomass energy projects.

vii. Project-Specific Green Bonds: These are issued to fund an individual project or a series of related projects. These bonds provide clarity to the investors as to fund allocation and the expected environmental impact.

viii. Sustainability-Linked Green Bonds: These bonds are linked to specific sustainability targets such as reduce water consumption per unit of production or carbon emission by a specified percentage over a defined period of time and so on. These bonds are meant to generate variable returns based on performance.

ix. Transition Bonds: The issuance of such bonds is done by companies in carbon-intensive industries to finance the transition to greener operations.

x. Blue Bonds: These bonds are focused on financing ocean and water-related projects such as coastal management, sustainable fishing, marine biodiversity conservation, and so on.

4 Green Bond Companies and Services

India's growing green bond ecosystem boasts of diverse players promoting sustainability initiatives. Majorly the key sectors funded by these players revolve round renewable energy which includes, solar, wind, and biomass; transportation including electrification of railways, and clean urban transport; water management comprising of wastewater treatment and water supply systems; energy efficiency initiatives in the form of green buildings and smart grids; and sustainable urban development focusing on smart cities and green infrastructure (Fig. 2).

The prominent companies involved in the issuance of corporate green bonds in India are ReNew Power, Greenko, Adani Green Energy, and Tata Power. ReNew Power had issued $375 million green bonds in 2019 to fund solar projects [7]. Greenko and Adani Green Energy have also issued bonds for specific wind or solar parks. Further, Adani Green Energy issued green bonds to raise funds for specific wind and solar power projects. The company raised $500 million in green bonds in 2020, which helped finance its expansion of solar and wind capacities. The funds raised have directly contributed to the development of renewable energy infrastructure.

In January 2023, India government issued its first sovereign green bond worth ₹16,000 crore ($2 billion) to finance green projects. This issuance is part of India's

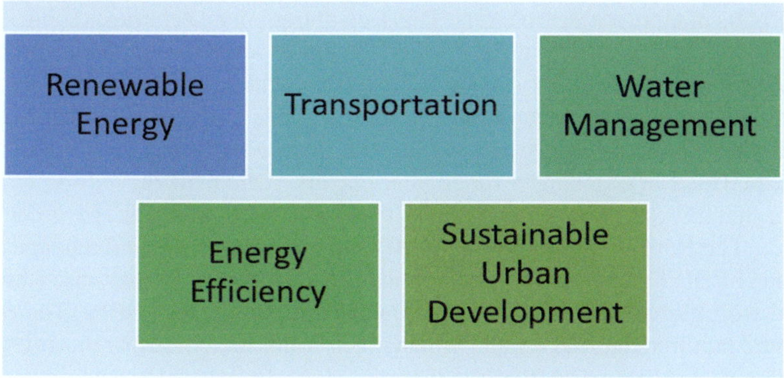

Fig. 2 Key sectors funded by green bonds. *Source* Developed by Authors

broader strategy to meet its climate goals and enhance green infrastructure. The funds from the sovereign green bond were allocated to clean energy, biodiversity conservation, and climate adaptation projects. Notably, the proceeds were expected to be used within 24 months of issuance, which indicates a commitment to quickly deploy the capital for green project. For the Indian government, the green bond issuance not only raised crucial funds but also demonstrated the country's commitment to sustainability and its ability to attract international investors. The issuance of sovereign green bonds enhances India's credibility in the global green finance ecosystem. The issuance of sovereign green bonds has helped to attract both domestic and international investments, contributing to the growth of India's green bond market and enabling large-scale sustainable development projects.

Sovereign green bonds have been issued by National Thermal Power Corporation (NTPC) to source clean energy projects. The Indian Railway Finance Corporation (IRFC) had also issued green bonds for sustainable transportation projects. The Pune Municipal Corporation had issued municipal green bonds in 2017 to fund a 24 × 7 water supply project. NTPC had issued masala green bonds in 2016 to finance renewable energy projects. Yes Bank had also issued masala bonds to finance green energy projects. Financial Institutions such as Yes Bank, Axis Bank, ICICI Bank, and HDFC have issued green bonds to fund solar and wind energy projects. The State Bank of India (SBI) had issued $650 million green bonds in 2018 for renewable energy projects.

India has set ambitious renewable energy goals, such as achieving 500 GW of non-fossil fuel capacity by 2030. Green bonds are expected to play a critical role in meeting these targets. As of February 2023, India had raised over $21 billion through green bonds, a substantial portion of which has been allocated to renewable energy infrastructure.

Nevertheless, India faces unique challenges in its green bond market compared to mature markets like those in Europe and the United States. For instance, green bonds in India suffer from lower trading volumes, which makes them less liquid than their counterparts in more mature markets. This lack of liquidity discourages some institutional investors from participation. Further, the funds raised through green bonds are not used for environmentally sustainable projects, which is a concern in emerging markets, including India. Again, issuing green bonds involves significant upfront costs for certification, third-party verification, and compliance with international standards. This can be a barrier for smaller players in the market.

India has introduced various regulatory frameworks to encourage green bond issuance. The Reserve Bank of India (RBI) and the Securities and Exchange Board of India (SEBI) have developed guidelines for green bond disclosures, which are aligned with global standards like the Green Bond Principles (GBP). The SEBI's disclosure requirements have made Indian green bonds more transparent and reliable for investors. Further, to address the liquidity issue, India has allowed green bonds to count toward statutory liquidity ratios (SLRs) for banks and has facilitated repo transactions with green bonds. This incentivizes banks to hold green bonds in their portfolios, thus improving market liquidity.

India has taken steps to raise investor awareness about green bonds through education campaigns and collaboration with international bodies like the Climate Bonds Initiative. These efforts aim to enhance understanding of green bonds' potential for long-term sustainable returns. Again, India has aligned its green bond standards with international frameworks like the Green Bond Principles (GBP). This has improved investor confidence and helped India become an attractive market for international green bond investors.

5 Tool for Sustainable Debt Financing

Green bonds are debt instruments created to secure funding for projects that promote environmental sustainability [8, 9]. These bonds channel investments into climate-friendly and sustainable initiatives [10]. The positive role of green bonds in sustainable debt financing is as follows:

i. Purpose-Driven Financing: Green bonds focus on projects that constitute to eco conscious projects especially those pertaining to clean energy, sustainable water management, green infrastructure, and climate change adaptation. For instance, the World Bank funds have been directed toward solar energy projects in India [11] and wind energy farms in Argentina. These projects contribute directly to renewable energy generation and promote sustainable development.
ii. Encouraging Private Sector Participation: Green bonds provide a secure way for private investors to support sustainability, enabling capital to flow from the private sector into green initiatives and reducing reliance on public funding.
iii. Lower Borrowing Costs (Greenium): The term 'greenium' was derived from 'green premium', which refers to the reduced borrowing costs that issuers of green bonds may enjoy in relation to conventional bonds due to high demand from environmentally conscious investors. An example of this phenomenon

can be seen when the Government of India had released sovereign green bonds in 2023.

iv. Transparency and Accountability: Green bonds require rigorous reporting and verification to ensure that the funds are used exclusively for their intended green projects, fostering greater accountability and boosting investor confidence. In India, Yes Bank was the first Indian bank to issue green bonds in 2015, raising ₹1000 crore to fund wind and solar power projects. The bonds adhered to the globally recognized Green Bond Principles (GBP), ensuring clear guidelines for the use of proceeds. Yes Bank not only provided detailed disclosures on how the bond proceeds were allocated, but also focused on systematic reporting and independent project verification by reputed agencies to ensure alignment with the green criteria specified during the bond issuance. This commitment to transparency ensures that the green bond proceeds were being used for their intended purposes, enhancing trust among investors.

v. Mitigating Climate Risks: Green bonds are meant to support proposals that reduce carbon emissions and enhance climate resilience, helping countries and companies address the risks posed by climate change. India has issued green bonds to finance solar and wind energy projects, which in turn would reduce fossil fuel usages.

vi. Stimulating Economic Growth: Green bonds promote job creation in green sectors, contributing to economic development while reducing the carbon foot print. ReNew Power, which had raised $475 million for the development and expansion of solar and wind energy projects through green bond issuance in 2017, had added hundreds of megawatts to India's green energy grid, supporting the country's 500 GW renewable energy target by 2030. Further it also led to infrastructural development and job creation in areas where the project was implemented.

vii. Supporting Sustainable Development Goals: These debt instruments help in aligning the country's financing mechanisms with global sustainability frameworks, such as the Green Bond Principles (GBP) and Sustainable Development Goals (SDGs).

viii. Promoting Innovation in Financial Markets: The Sovereign Green Bonds, issued by the Government of India in 2023, had been a catalyst in the Indian green bond segment. It enabled further innovations in green products like sustainability-linked bonds and transition bonds [12].

ix. Attracting International and Domestic Investments: The issuance of masala green bonds had been instrumental in attracting foreign institutional investors to the Indian debt market.

6 Obstacles in Green Bond Adoption and Strategies to Overcome

The following case studies and examples provide insights into the challenges faced by the green bond market [13] and the strategies being employed to address them:

i. Limited Liquidity in the Market

Challenge: Green bonds often suffer from lower trading volumes, making them harder to buy or sell. This liquidity issue can deter institutional investors.

Strategy-1: Denmark introduced twin bonds, which are essentially two interchangeable bonds—one green and one conventional bond, thus improving flexibility and market liquidity. This allows investors to easily switch between the two types of bonds.

Strategy-2: In India, green bonds have been allowed to count toward statutory liquidity ratios (SLRs) for banks, and repo transactions (repurchase agreements) are also permitted. This move encourages banks and financial institutions to hold green bonds as part of their portfolios.

ii. Risk of Green Washing

Challenge: Green washing occurs when the capital sourced from green bonds are not used for truly sustainable or green projects, undermining investor confidence.

Strategy-1: To avoid green washing, having strict guidelines and standard operating procedures is paramount. The Green Bond Principles (GBP) by the International Capital Market Association (ICMA) is one of those principles which ensure transparency and accountability.

Strategy-2: Regular Audits and Public Disclosures has to be mandatory. The issuers of green bonds are increasingly required to undergo independent verification and audits.

iii. Higher Initial Costs and Complex Issuance Process

Challenge: Green bond issuance often involves significant upfront costs for project verification, third-party certifications, and compliance with international standards.

Strategy-1: Here again the standardized frameworks like GBP can simplify the issuance process by providing clear guidelines on eligible projects, environmental impact, and reporting.

Example: The European Investment Bank (EIB) uses the GBP to streamline its green bond issuance process, making it easier and less costly to issue green bonds.

Strategy-2: Governments or multilateral organizations can subsidize the initial costs of issuance or provide guarantees to reduce the financial burden.

Example: The World Bank and the United Nations Environment Programme (UNEP) have provided advisory support and funding to help developing countries to reduce the upfront costs of issuing green bonds.

iv. Lack of Investor Awareness and Confidence

Challenge: Many investors are unfamiliar with green bonds and the specific benefits they provide, which limits the pool of potential investors.

Strategy-1: Some governments and financial institutions have launched investor education campaigns to raise awareness about the funding avenues and advantages associated with green bonds.

Example: The Climate Bonds Initiative conducts regular webinars and produces reports to educate investors about the benefits of green bonds and their role in achieving the sustainability objectives.

Strategy-2: Some jurisdictions have introduced tax incentives (e.g., tax exemptions, credits) to attract investors to green bonds.

Example: California offers tax exemptions for green bond investors to encourage participation in green bond markets, fostering greater investor interest. Similar efforts can be made by the Indian government.

v. Limited Pool of Eligible Projects

Challenge: Developing economies may struggle to identify and implement projects that meet green bond criteria.

Strategy-1: Some countries have broadened the scope of eligible green projects to include projects that may not strictly fall under the traditional green bond criteria but still have positive environmental impacts.

Example: Mexico expanded its eligible project categories to include not just renewable energy but also sustainable agriculture and water conservation initiatives to address the needs of developing regions.

Strategy-2: Capacity-building programs help developing countries plan and execute projects that qualify for green bond funding.

Example: The Asian Infrastructure Investment Bank (AIIB) offers technical assistance to countries like India and the Philippines to develop pipelines of green projects that can be financed through green financial instruments.

vi. Currency and Macroeconomic Risks

Challenge: Issuing eco bonds in foreign currencies can expose countries to exchange rate volatility, which increases financial risk.

Strategy-1: To mitigate currency risks, some countries issue green bonds in their local currencies.

Example: Nigeria issued its first green bond in 2017 in Nigerian Naira, allowing the country to avoid foreign exchange risks associated with global bond markets.

Strategy-2: Fostering Domestic Markets Strengthening local investor markets helps reduce dependency on foreign currency funding.

Example: The Brazilian Development Bank (BNDES) has worked to foster a local green bond market, which has led to increased participation from domestic investors and a reduction in currency risks.

vii. Lack of Bench marking and Market Standards

Challenge: The absence of universally accepted standards and benchmarks can make green bond pricing complex and inconsistent, leading to confusion among investors.

Strategy-1: Harmonizing Benchmarks Efforts are underway to harmonize green bond standards and create consistent benchmarks to simplify the issuance process.

Example: The Climate Bonds Initiative (CBI) works with international standard-setting bodies to develop benchmarks for green bonds, which enhances transparency and investor confidence.

Strategy-2: Issuing sovereign green bonds can help establish a benchmark pricing model for the broader market.

Example: France issued the first sovereign green bond in 2017, setting a pricing benchmark for other green bond issuers and establishing the French government's commitment to climate finance.

To summarize, in mature markets like Europe and the U.S., green bond issuance is more standardized and streamlined. However, India is gradually catching up by establishing clearer regulations and creating frameworks for transparency and accountability. In countries like the U.S., institutional investors are highly active in the green bond market. While this is slowly gaining momentum in India, more efforts are required to build confidence among institutional investors. Mature markets have larger, more liquid green bond markets. India is still building the infrastructure for a large-scale, liquid green bond market, but the success of recent sovereign green bond issuances shows significant progress.

India is making significant strides in addressing the challenges faced by its green bond market. With support from regulatory bodies, greater transparency, investor education, and improvements in liquidity, India is on track to develop a more robust green bond market. Compared to mature markets, India's green bond market is still in a growth phase, but the country is rapidly closing the gap. Addressing these challenges requires multi-faceted strategies, including regulatory support, market infrastructure development, and enhanced transparency [14]. By implementing these strategies, the green bond market can continue to grow and attract a wider pool of investors, while ensuring that funds are used effectively to address global environmental challenges. The examples provided highlight how various stakeholders are working to overcome the obstacles and advance the green bond market.

7 Future Prospects

The future potential of sustainable bonds are extremely promising [7], with their role as a key financial instrument in combating climate change becoming even more significant. The future prospects relating to green bonds are incorporated under the pointers listed below:

i. Growth in Emerging Markets: Green bonds are gaining traction in emerging markets as countries seek financing for renewable energy, clean transportation, and sustainable infrastructure projects. These markets are becoming more attractive to investors due to increasing climate commitments and international support.
ii. India's Climate Goal Achievement: India's commitment toward net-zero emissions by 2070 will require significant capital allocation in renewable energy, green infrastructure, and sustainable urbanization, which in turn could be sourced with the help of eco-friendly bonds.
iii. Increased Participation from the Private Sector: As private corporations align their business strategies with sustainability, they are issuing green debt instruments to fund environmentally friendly projects [15].
iv. Growth of Municipal Green Bonds: Successful issuances of municipal green bonds by Pune and Ghaziabad municipal corporations have paved the way for other Indian cities to take up the green initiative.
v. Policy and Regulatory Enhancements: Guidelines on green bonds, coupled with government subsidies for green bond issuers, improve market transparency and encourage issuance
vi. Integration with Global Standards: Adoption of GBP and Climate Bonds Standard has enhanced India's credibility in the global green finance ecosystem. Nevertheless, partnerships with international development banks and other such bodies could help in scaling up the bond issuances. As global standards become more integrated, investors will have greater confidence in comparing green projects across borders.
vii. Technological Advancements: Advancements in green technologies, including smart grids and renewable energy storage, will present new opportunities for green bond financing. Leveraging blockchain for green bond issuance and tracking can improve transparency and investor confidence.
viii. Expanding Project Categories: The projects categories in future are found to expand beyond renewable energy sources to include climate-resilient agriculture, sustainable water management, circular economy, and waste management initiatives, to name a few.
ix. Rising Investor Demand: The growing sustainability consciousness among institutional and retail investors would further boost the appetite for sustainable investment bonds.

By addressing the existing challenges and leveraging opportunities, the Indian green debt securities market could be instrumental in building a sustainable nation, thereby contributing to a resilient global economy.

8 Conclusion

Green bonds in India have evolved as a transformative instrument for securing funds toward green developmental initiatives. The green bonds have supported the nation's renewable energy targets and have enabled numerous green infrastructure projects across different sectors, thereby acclimatizing the masses on creating a green eco space. Nevertheless, there has been challenges relating to awareness, accessibility, greenwashing, and so on. Examples cited above have demonstrated how green bonds have successfully merged financial returns with environmental impact, addressing challenges like greenwashing and market liquidity, and playing a crucial role in scaling up global climate action. Further, many evolving practices, including regulatory frameworks, sovereign green bonds, innovative financial instruments, investors increasingly prioritize ESG compliance, and issuers recognize the long-term benefits of sustainable financing, have opened up a plethora of opportunities for growth and expansion. Through efforts like enhanced transparency, investor education, regulatory frameworks, and the collaborative efforts of the government, corporates and global stakeholders, green bonds in India are becoming a more reliable and attractive investment option. As their adoption continues to grow, green bonds will remain a cornerstone of the global financial system, driving the transition to a more eco-friendly and resilient future.

References

1. Cochelin, P., et al. (2023). Global sustainable bonds 2023 issuance to exceed $900 Billion. S&P Global.
2. Abhilash, et al. (2023) Green bond as an innovative financial instrument in the indian financial market: Insights from systematic literature review approach. *SAGE Open 13*(2), 21582440231178783.
3. Gupta, V., Kumar, A. (2024). Green bonds: Accelerating green finance towards sustainable economic development. In *Green Innovations for Industrial Development and Business Sustainability* (pp. 127–145). CRC Press.
4. Singh, U. V., & Shrivastava M. K. (2024). Accelerating the growth of green bonds in India. The Energy and Resources Institute.
5. Sreenu, N. (2024). The impact of fintech and green bonds on the Indian renewable energy production. *Renewable Energy*, 121807.
6. Tiwari, D., & Singh, A. (2023). Exploring the role of municipalities in promoting sustainable development with special reference to green bonds in India. *International Journal of Environmental Sciences, 9*(2).
7. Sarma, P., & Roy, A. (2021). Green financial instruments in India: A study on its current status and future prospects. *International Journal of Business Innovation and Research, 26*(2), 194–218.
8. Maltais, A., & Nykvist, B. (2020). Understanding the role of green bonds in advancing sustainability. *Journal of sustainable finance & investment*, 1–20.
9. Park, S. (2019). Green bonds and beyond: Debt financing as a sustainability driver. In *Cambridge handbook of corporate law, corporate governance and sustainability*. Cambridge University Press.

10. Tuhkanen, H., & Vulturius, G. (2022). Are green bonds funding the transition? Investigating the link between companies' climate targets and green debt financing. *Journal of Sustainable Finance & Investment, 12*(4), 1194–1216.
11. Manaktala, S. (2020). Green bonds in sustainable finance: exploring the case of India. Available at SSRN 3644116.
12. Prakash, N., & Sethi, M. (2022). A review of innovative bond instruments for sustainable development in Asia. *International Journal of Innovation Science, 14*(3/4), 630–647.
13. Jain, K., Gangopadhyay, M., & Mukhopadhyay, K. (2024). Prospects and challenges of green bonds in renewable energy sector: Case of selected Asian economies. *Journal of Sustainable Finance & Investment, 14*(3), 708–731.
14. Bhutta, U. S., et al. (2022). Green bonds for sustainable development: Review of literature on development and impact of green bonds. *Technological Forecasting and Social Change, 175*, 121378.
15. Kumar, P., Taneja, S., Ozen, E. (2024). Exploring the influence of green bonds on sustainable development through low-carbon financing mobilization. *International Journal of Law and Management*.

Unlocking Potential: Financial Inclusion Strategies for Women and Marginalized Communities

Priyanka Chadha, Gitanjali Upadhaya, and Pallavi Dua

Abstract The main focus of this chapter is to discuss different mechanisms and initiatives that can be used as avenues to enhance financial inclusion in the capital markets space, specifically for women and underserved communities. This chapter shows how various key strategies can empower women and underrepresented groups. Additionally, it also delivers practical insights that encourage gender equality by focusing on reducing current barriers. Several key solutions have been highlighted to promote diversity and financial inclusion in the workplace. The chapter has demonstrated that financial literacy must be promoted among marginalized groups. So that people are aware of investing, saving, and managing wealth. Moreover, diversity-based funds, favorable policy reforms, access to credit, and public–private partnerships to address systemic disparities are also needed. The chapter concludes with a plea for further research to measure the long-term effects of these approaches. These measures can be implemented specifically in digital finance and alternative credit models to refine and continually expand inclusive capital market practices worldwide.

Keywords Financial inclusion · Women participation · Minorities · SDGs

P. Chadha (✉) · P. Dua
Manav Rachna International Institute of Research and Studies, Faridabad, India
e-mail: priyankachadha94@gmail.com

P. Dua
e-mail: duapalllavi@gmail.com

G. Upadhaya
Central University of Himachal Pradesh, Dharamshala, Himachal Pradesh, India
e-mail: gitanjali.cuhp@hpcu.ac.in

K. Khanna et al. (eds.), *From Poverty to Prosperity*, Information Systems Engineering and Management 64, https://doi.org/10.1007/978-3-031-99166-0_13

1 Introduction

Economic empowerment is a key driver of social equity and sustainable development. For women and marginalized communities, access to financial resources represents a means of economic survival and an opportunity for social mobility, autonomy, and improved quality of life. Capital markets offer transformative potential for these groups as critical mechanisms for resource allocation, investment, and wealth generation. However, systemic barriers rooted in historical, socioeconomic, and institutional biases continue to limit their full participation. Addressing these challenges requires a multifaceted approach that combines policy reform, innovation, and education tailored to the unique needs of these communities (Seven Clean Seas—October 2023. https://www.sevencleanseas.com/post/why-are-women-most-affected-by-the-environmental-crisis).

Historically, financial systems have been structured in ways that exclude or marginalize certain groups, particularly women and economically disadvantaged populations. Cultural norms and institutional policies have often perpetuated gender biases, leading to unequal access to financial services and capital markets. Studies indicate that women, especially in developing economies, are less likely to own bank accounts, access credit, or participate in investment activities than men [1]. These inequalities are further aggravated by socioeconomic factors such as poverty, illiteracy, and lack of digital access.

Marginalized communities such as ethnic minorities, rural populations are the ones which have faced the similar exclusion. Poor income distribution and disadvantages with respect to their geography limit their contact with the primary financial institutions. Therefore, these communities tend to practice the informal financial institutions that seems to be more susceptible, inadequate growth and wealth-building mechanisms. Superseding against such obstacles requires a more nuanced consideration of the socio-cultural and economic situations that inform these systemic barriers and inequities or how this affect inequity.

Financial inclusion is an important determinant of economic empowerment. It provides the access and equivalence opportunity to access financial services. Without the accessibility of accounts, insurance, savings, and other opportunities, one cannot manage risk, invest in any business, or build revenue throughout their lifetime. Hence, financial inclusion shows the path to overcoming the obstacles for women and marginalized communities, thereby succeeding in economic and social equity.

Over the preceding few periods, worldwide projects like the United Nations' Sustainable Development Goals (SDGs) [2] have highlighted the role of financial inclusion systems that foster inclusive development. The part of inclusive finance in accomplishing SDGs has received more consideration in recent years, as SDG 5 (Gender equality) and SDG 10 (Reduced inequalities) aim to reduce gender and socioeconomic differences in financial accessibility. However, considerable gaps exist notwithstanding those measures, especially in low- and middle-income countries. Institutional strategies for enhancing women's economic empowerment: One of the key strategies for enhancing the financial inclusion of women and marginalized groups is developing customized financial instruments. Women entrepreneurs, to some extent, have benefited from innovations like microfinance, which, unlike traditional banking systems, provides small loans without collateral requirements. Microfinance has proved to be a well-known engine of entrepreneurship credits to organizations like Grameen Bank. Organizations such as Grameen Bank have demonstrated the effectiveness of microfinance in fostering entrepreneurship and economic independence among women [3].

Other innovative financial instruments like gender bonds and community investment funds have also demonstrated potential, in addition to microfinance. Unionization at work and gender bonds that direct capital specifically to projects that benefit women. These are the ways toward the accessibility of financial markets that can further help in achieving social equity objectives. For instance, community investment funds can be used to pool resources among marginalized groups for investing in projects that benefit their own community members. The participation of the marginalized group in this process can be considered as one aspect of financial empowerment. It can also be seen as an instrumental enabler to split the fundamental hurdles and progress an inclusive financial ecosystem. To uplift marginalized communities, gender-sensitive and inclusive policies must be developed. They must address the sensitivity and underlying inequalities prevailing in the financial systems. Frameworks that inculcate equal opportunity of access to financial services further raise women's representation within financial institutions. It also helps fight discrimination and promotes a better financial environment. Additionally, legal contexts that confirm property rights and inheritance laws for women are mandatory for supporting their financial autonomy. Research has shown that women who own possessions are more likely to approach credit and contribute to investment activities, leading to their economic empowerment [4]. Policymakers must also focus on the digital divide, as digital financial services have become progressively significant for financial inclusion in the modern era.

Financial literacy and capacity must be built to leverage the capital markets among women and marginalized communities. Most of these groups have less financial knowledge in order to access the resources effectively. To ensure the maximum participation from these groups, education programs that reflect the understanding of financial planning, investment strategies, and effective use of digital tools should be delivered. Hence, individuals will be furnished with the knowledge and skills to create confidence to participate in financial markets. Specific programs like the Global Financial Literacy Excellence Center (GFLEC) [5] have indicated that targeted financial education programs for women and underrepresented groups can meaningly close knowledge gaps, contribute to better decision-making, and enhance financial health. Such programs benefit individuals and society at large by enabling informed participation in global financing systems.

Digital technology has changed the realm of financial inclusion. All the hurdles regarding accessibility and affordability are getting ironed out by Digital Continue. Mobile banking, digital wallets, and fintech platforms have enabled access to financial services in remote and underserved communities. Mobile money is one example of this disruption, where services like M-Pesa in Kenya have enabled financial access for millions of people, notably women, who may not have had the means to access traditional banking services [6]. Digital innovation also presents opportunities to build inclusive capital markets. As an example, blockchain technology can democratize access to investment opportunities while lowering the cost of transactions and increasing transparency. Nonetheless, this is not without challenges, such as digital literacy, cybersecurity, and access to good internet infrastructure, which necessitate addressing digital equity.

Women and marginalized communities are dealing with multiple problems; thus, an intersectional perspective is necessary. Various challenges faced by individuals in financial inclusion are race, ethnicity, disability, geographic concerns, etc. The listed barriers often intersect with gender to create unique obstacles to financial inclusion. Therefore, strategies should be adopted, keeping in consideration the unique needs and experiences of various subgroups within these communities. Execution of intersectional approaches requires good collaboration among the government, financial institutions, private sector, and NGOs. For this reason, public–private partnerships should be leveraged in order to pool expertise and resources from different participants and create and extend inclusive financial solutions.

Marginalized groups and financial inclusion are seen as a holistic process to unlock the potential of underrepresented groups. A multi-pronged approach must combine customized financial services, policy adjustment, education, and digital breakthroughs. A more inclusive and sustainable economic future can be built for marginalized groups by tackling general obstacles and advancing equitable access to financial markets. When it is made accessible and inclusive, capital markets can act as a powerful engine for this empowerment. With innovative strategies, collaboration, and commitment, this can be achieved through financial inclusion and equity.

1.1 Flow Chart of Chapter

The flow chart of the book chapter is as follows. The first section discusses the introduction, followed by flow chart of the chapter. The second section shows overview of current barriers in capital markets. The third section discusses the case studies of successful initiatives in overcoming barriers to capital markets. The fourth section covers the diverse key strategies for empowerment. The fifth section discusses the role of government and regulatory bodies. The following section, i.e., the sixth, covers the challenges and considerations. Furthermore, conclusion is discussed in the seventh section. Lastly, the eighth section covers the future research direction part. Figure 1 shows the flow chart of the chapter.

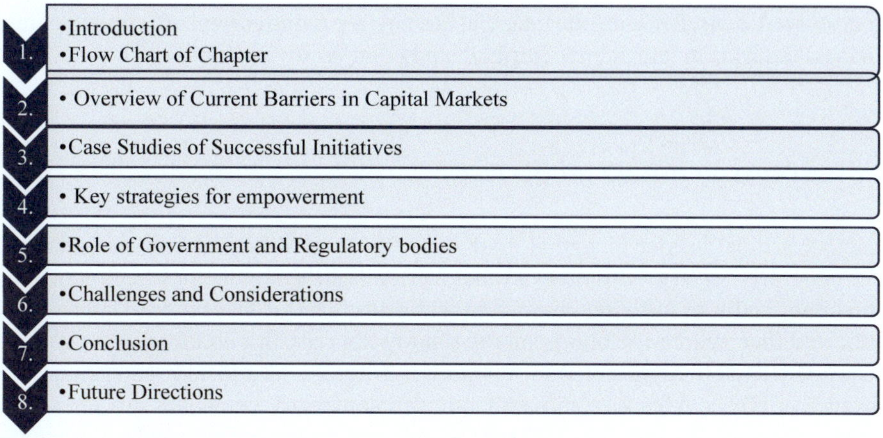

Fig. 1 Flow chart of book chapter. *Source* Authors Compilation

2 Overview of Current Barriers in Capital Markets

To drive the economic development and distribution of wealth, capital markets play an important role. Nevertheless, operational inequalities within these markets frequently exclude women and marginalized communities, prolonging cycles of poverty and restricting their economic enablement. These barriers include financial literacy gaps, restricted access to financial services, socioeconomic constraints, discriminatory lending practices, cultural and social norms, and geographic limitations. Addressing these problems is decisive for creating reasonable access to financial opportunities.

2.1 Financial Literacy Gaps

Lack of financial literacy among women and underprivileged communities is depicted as one of the most critical barriers to capital market participation. Financial literacy refers to the information and skills required to comprehend and effectually use financial resources. According to Lusardi and Mitchell [7], "financial illiteracy is more pronounced among women, particularly in developing countries." Without satisfactory financial education, one struggle is to understand investment options, risk management, and the potential profits of appealing with capital markets. This outcomes in inferior participation rates, as potential stockholders perceive capital markets as multifaceted and unapproachable.

For example, [8] revealed that financial education programs significantly increased women's likelihood of opening savings accounts or investing in assets. Though, such initiatives remain infrequent and underfunded, especially in rural and underserved areas. Bridging the financial literacy gap requires total efforts to integrate financial education into school curricula and community programs.

2.2 Access to Financial Services

Another critical barrier in capital markets is the accessibility of financial services. The underprivileged groups many a times face complications opening bank accounts, obtaining credit, or utilizing investment platforms. The Global Findex Database [1] indicated that nearly 1.4 billion adults worldwide remain unbanked, with women disproportionately represented in this group. In rural and underserved regions, the absence of physical banking infrastructure supplementary intensifies this problem.

Microfinance institutions (MFIs) have emerged as a possible solution, predominantly for women. Institutions like Grameen Bank have established the positive impact of microloans in empowering women economically. Nevertheless, MFIs often function on a restricted scale and cannot fully address the structural barriers to financial inclusion. Digital financial services, such as mobile banking, show capacity to overcome these contests, but their adoption is delayed by restricted digital literacy and internet access in side-lined communities [9].

2.3 Socioeconomic Constraints

Socioeconomic factors extensively inhibit women's involvement in capital markets. To engage in investment activities, women and marginalized groups often have an unavailability of independent income or assets. According to the United Nations [2] women earn only 77 cents for every dollar men earn globally. This income difference leaves women with fewer investment resources, even if they possess the essential financial knowledge.

Furthermore, legal and established frameworks in many countries limit women's property rights and access to inheritance. For instance, in some areas, women require male relatives' permission to open bank accounts or manage assets, restricting their financial autonomy. Highlighting these socioeconomic barriers requires policy improvements to safeguard gender equivalence in income and asset ownership. Schemes such as conditional cash transfers and government-backed savings schemes have demonstrated possibilities in developing women's monetary freedom [10].

2.4 Discriminatory Lending Practices

Discrimination within financial institutions represents a significant hurdle as women entrepreneurs face more hurdles than their counterparts. A study by Beck et al.

[11] also highlighted that women entrepreneurs face more challenges concerning interest rates and collateral requirements. Underrepresented groups sometimes get outright rejection of credit because of partialities about gender and ethnic concerns. Often, it results in higher borrowing costs. These kinds of methods lead to labels that specify and perceive women and marginalized individuals as high-risk borrowers. This obstacle should be highlighted to control oversight and comprehensive guidelines that further obligate non-discriminatory practices in lending. Certain initiatives, like Ethiopia's Women's Entrepreneurship Development Project (WEDP), can be an example for this. It provides loans to female entrepreneurs without traditional collateral requirements. This highlights the prospect of besieged interferences to overcome these barriers [1].

2.5 Cultural and Social Norms

Cultural and social norms influence the financial behavior of women. Due to these norms, it has a significant impact on underprivileged communities. They play an important role in shaping their behavior. These societal expectations have largely limited their financial contribution. It shows them predominantly as custodians rather than economic funders. Such frameworks do not permit women's competence to be independent regarding their financial responsibilities, and thus, they are being excluded from capital markets.

For example, in South Asia and parts of Sub-Saharan Africa, women's empowerment is often hampered due to cultural systems. These customs have generated obstacles for marginalized individuals, making it difficult for them to approach financial institutions. Moreover, they cannot even appear on financial literacy platforms. There is a strong need to modify these deeply rooted norms. It would require long-term efforts involving community engagement, education, and advocacy. Certain empowerment programs must be run to inspire cultural change and challenge the perceived norms. They can further showcase successful women investors and entrepreneurs to challenge the stereotypes.

2.6 Geographic Constraints

Geographic barriers intensify exclusion from capital markets, in particular for remote or rural individuals. The absence of financial institutions in these regions compels people to travel long distances to access banking services, creating high transaction costs and time constraints. Geographic barriers have a positive outcome with digital skills. Modern financial applications such as mobile money platform M-Pesa in Kenya facilitate financial inclusion by enabling users to make transactions without the need for physical bank branches. However, the viability of these solutions is also contingent on accessibility to a digital foundation and low-cost internet access.

Governments and private sector shareholders need to be involved in helping improve connectivity.

2.7 Addressing the Barriers

It requires a method that includes policy interventions, technological interferences, and community-driven proposals to overcome the barriers. Key recommendations to overcome these obstacles include:

1. **Enhancing Financial Literacy**: To enhance financial literacy among women and marginalized communities, the government should run several financial education programs. Governments and non-governmental organizations (NGOs) should cooperate together to implement the programs to make the task smoother. Such initiatives will further help these groups gain more financial knowledge. In addition, these programs will influence digital tools to reach broader spectators and include culturally relevant content.
2. **Expanding Access to Financial Services**: Financial institutions should implement comprehensive procedures to expand access to financial services to the underrepresented groups. The reach must be extended to the rural and underserved areas. It must be confirmed that digital financial services are affordable and secure for these regions. Regulatory bodies should create a relevant structure to manage and implement such frameworks.
3. **Promoting Gender Equality**: To empower women financially, social protection programs must be designed. Therefore, certain focus should be given to policy reforms that specifically deal with income disparities. Hence, women's equal property and inheritance rights can be safeguarded.
4. **Combating Discrimination**: To prevent discrimination among underprivileged communities, anti-discrimination laws should be available to them. It should be made mandatory by the regulatory bodies that anti-discrimination laws should be implemented within financial institutions. To moderate the unconscious biases, capacity-building programs should be run for bankers and financial service providers
5. **Challenging Cultural Norms**: Community engagement and advocacy are pivotal to challenging limited cultural norms. Media promotions and role models can substantially reshape societal attitudes toward women's economic involvement.
6. **Investing in Digital Infrastructure**: Governments and private sector stakeholders must highlight investments in digital connectivity to ensure that rural and remote regions can access financial services. Public–private partnerships can go faster the deployment of affordable Internet and mobile banking policies.

3 Case Studies of Successful Initiatives in Overcoming Barriers to Capital Markets

It is very important to know the role models to overcome the barriers. To gain constructive comprehension, it is necessary to understand the strategies that have efficiently addressed the barriers. Repeating these models in other areas is equally essential to control the barriers. Hence, this section demonstrates key initiatives that have permitted women and marginalized societies to contribute to capital markets. It can include women-focused investment funds, social impact bonds, microfinance institutions, crowdfunding platforms, and digital financial services.

3.1 Women-Focused Investment Funds

Investment funds focusing on women have emerged as a strong tool to tackle the gender gaps in capital markets. Such fund totally offers resources exclusively for women-led startups. It advances the ability of women entrepreneurs to break the pressures of things like limited access to credit and prejudicial lending practices. A great example is the SEAF Women's Opportunity Fund, which provides financial and technical assistance to women entrepreneurs in developing markets. Traditional financing can be the hardest for female entrepreneurs to access. Studies like [11] show that capital access and performance are affected by funds, which creates more barriers for women entrepreneurs.

Research by Brush et al. [11] also points to the convenience of such funds, as female entrepreneurs tend to face lower access to traditional resources. These funds add to the economic incentive for women-led businesses and contribute to job creation and community progress. The success of women-focused investment funds highlights the significance of targeted financial devices in addressing structural inequalities.

3.2 Social Impact Bonds (SIBs) and Blended Finance

Social Impact Bonds (SIBs) and blended finance structures are new gatekeepers for using private capital in public projects for social good. SIBs, or pay-for-success contracts, are private investors who fund projects with significant social outcomes. This "pay-for-success" modeling means investors are only paid back by governments or other entities if results are achieved, which de-risks investments.

Blended finance is a strategy for targeting capital to projects that benefit marginalized communities that combines philanthropic or public funding with private investment. By altering risks for private investors, blended finance underwrites projects that might otherwise be thought too hairbrained to pursue. An example is the Uganda Renewable Energy Fund, which blended public grants with private capital to deliver solar energy solutions to rural households to improve energy access and economic opportunities [12].

SIBs and blended finance have demonstrated that it is helpful in establishing paths to market contribution for underprivileged groups. Their success emphasizes the potential of public–private partnerships in addressing systemic obstacles to financial inclusion.

3.3 Microfinance Institutions

Microfinance institutions (MFIs) have transformed credit approaches for women and low-income entrepreneurs. By proposing small loans with minimal collateral conditions, MFIs authorize entities to create or develop establishments, fostering financial independence and economic encouragement. Grameen Bank in Bangladesh is a groundbreaking instance of this method.

Research by Armendáriz and Morduch [13] demonstrates the transformative influence of microfinance on women's incomes. Women who collect microloans are more likely to invest in their children's education and healthcare, creating a ripple effect of positive conclusions. While barriers such as high interest rates and repayment pressures endure, the overall success of microfinance highlights its potential as a tool for financial inclusion.

3.4 Crowdfunding Platforms

Crowdfunding platforms have appeared as an alternative means of financing for women and underprivileged populations. Programs like Kiva and Milaap [14] permit persons globally to participate in lesser amounts of income to fund business ventures. This method goes around traditional banking limitations and connects funders and beneficiaries.

For instance, Kiva [14] has enabled over $1.5 billion in loans to underserved industrialists worldwide, with a considerable proportion directed toward women-led businesses [14]. Crowdfunding provides financial resources and builds a sense of community and trust among participants. However, ensuring transparency and accountability on these platforms remains a challenge.

3.5 Digital Financial Services

Digital financial services have been instrumental in bridging access gaps for individuals without physical access to banking services. Mobile banking platforms, particularly in African countries, have transformed how people save, borrow, and transfer money. M-Pesa, a mobile money service in Kenya, is a prime example of this success.

According to Suri and Jack [15] M-Pesa has significantly increased financial inclusion in Kenya, lifting approximately 2% of households out of poverty. The platform's success lies in its simplicity, affordability, and widespread adoption, even in remote areas. Digital financial services not only enhance access but also offer opportunities for innovation in financial products tailored to the needs of marginalized communities.

4 Key Strategies for Empowerment

Numerous strategies can be used to encourage more comprehensive capital markets. These tactics address the fundamental obstacles and inspire the underprivileged and women groups to actively encourage women and marginalized groups to contribute to financial ecosystems.

4.1 Financial Literacy and Education Programs

Financial literacy programs must be made to address women and marginalized communities, especially if we want economically marginalized groups to grow. Tailored financial literacy programs enhance women's financial behavior regarding ownership of banks, savings, and expenditures [16]. According to Jarecke et al. [17] women tend to have a preference for learning from experts and in group settings. Thus, the study calls out on the need to localize financial literacy education to local culture and involve local organizations for effective delivery [18]. On the other hand, inclusive pedagogies in financial literacy workshops also bring forth the challenge of reconstituting traditional understandings of finance alongside an essential critique of those marginalized [19]. What these findings emphasize is the necessity for targeted approaches that will address specific challenges encountered by women and marginalized communities in various contexts to boost their financial literacy levels and economic participation.

Programs should be designed to address the needs of a practical knowledge base and skills, such as mobile investment applications or online resources. NGO partnerships and community-based organizations will ensure that these programs reach rural and underserved areas. Financial literacy enhances self-confidence and equips individuals with the skills to manage complex financial systems.

4.2 Gender and Diversity-Focused Funds

Governments and financial institutions can facilitate diversity with funds dedicated to women and other marginalized communities. These funds, particularly those that target diversity, will help promote inclusivity in financial markets by allocating investment toward women-operated enterprises or companies within underrepresented demographics. Examples include venture capital funds that target women entrepreneurs and grants for minority-owned businesses. This type of program covers the continual funding gap, enabling businesses to scale and creating a positive economic enablement impact. Bestowing social pressure by measuring and announcing the results of the capital disbursed can also propel investments in inclusive financial products and services.

4.3 Policy Reforms for Equal Access

Inclusive financial systems lead to economic growth, poverty reduction, and access to chance [20]. Gender-sensitive credit scoring, flexible loan periods, and subsidies for poorly served groups are policies that governments can also adopt to bring about more financial inclusion [21]. While increased financial inclusion nurtures inclusive banking sectors and capital markets in emerging economies, implying a unidirectional relationship with stock markets [22], and capital market-centered banks counterpart. To ensure the reach of financial services, removing barriers such as transaction costs, perceived risks, and information asymmetries are effective strategies [23]. Better access to financial services drives trialability, innovation, and business entry and growth among small firms and poor households [23]. Ongoing investment in data gathering and impact evaluation to ensure policy effectiveness to promote broader economic development.

4.4 Improved Access to Credit and Financial Services

Many of these marginalized individuals have no traditional collateral means, which makes obtaining the credit impossible. Capital markets can be opened up further through microcredit programs, less stringent collateral requirements, and new credit scoring models that incorporate alternative forms of data.

For example, mobile phone usage or social network information can be used in credit scoring models to more accurately measure creditworthiness. Access can be expanded by reaching financial service networks to rural areas and by simplifying application procedures.

4.5 Technological Integration and Digital Platforms

Technology has the potential to support regions that are under or underserved, such as remote locations. Financial services can be further provided through mobile banking, digital investment platforms, and blockchain technology, which give more power and access. For example, women living in rural areas can use mobile applications for accessing loans, opening accounts, and joining investment platforms.

Financial transactions through digital platforms are easier and cheaper for both service providers and users, promoting transparency and efficiency. The private sector and governments should allocate resources toward building digital infrastructure as well as teaching people how to use it to ensure that technology integration is beneficial.

4.6 Public–Private Partnerships

Governments and private companies and NGOs need to work together, for example, to create infrastructure, education, and financial products that target underserved populations. The former use resources to apply the latter solutions toward long-term, sustainable empowerment initiatives.

Joint ventures are the perfect example, whether they want to establish rural banking facilities or partnerships which offer direct delivery of financial literacy campaigns. These partnerships harness the expertise of each sector to tackle complex problems more effectively.

5 The Role of Government and Regulatory Bodies

The role of government in establishing policies is inevitable. They help in encouraging inclusive capital markets. Some key actions driven by government and regulatory bodies are listed below.

5.1 Creating Enabling Environments

The government can be a key player in inclusive capital markets. They have helped create an atmosphere where women can invest further. There are other parameters like incentives, subsidies, and grants provided by the government to encourage inclusiveness. Such policies advance the cause of financial institutions engaging more members of disenfranchised communities. For instance, in the industry, tax incentives for banks with diversity-lending targets can catalyze greater inclusion. Public funding for both capacity-building programs and financial awareness campaigns is equally important. Such initiatives empower women and underrepresented communities to invest more in closing the knowledge gaps and gaining confidence in market participation.

5.2 Strengthening Financial Regulations

The desire to normalize this concept of bias in the capital markets runs contrary to the goal of being transparent and inclusive by defining general standards before engaging. Some, such as fair lending policies and practices, should be mandatory. By providing equal opportunities for credit, regardless of gender or culture, discrimination could be reduced even more. To make the corporate boards and investment funds diverse, it is important to have a representation of these underrepresented groups in the financial decision-making rooms. However, among those, in order to make the policies truly effective, enforcement mechanisms like penalties in case of non-compliance as well as incentives in favor of inclusive practices must be preserved. In terms of increasing accountability, it is going to be really important to have regular audits and public reports.

5.3 Data Collection and Monitoring

Robust data collection and monitoring systems enable governments and regulatory bodies to track the participation rates of women and marginalized groups in capital markets. In case of any disparity, this data can be further analyzed. With the help of profound data, areas that require intervention can be evaluated, and further policies can be implemented.

For example, gender-disaggregated data on loan approvals, interest rates, and investment participation can highlight systemic biases. Making the data available to the general public promotes transparency. Furthermore, this will help in evidence-based policymaking. Collaborative efforts with academic institutions and international organizations can enhance data quality and utilization.

5.4 Promoting Public Awareness and Advocacy

Government agencies can run campaigns to promote public awareness. These awareness programs can further inform people about the benefits of inclusive capital markets. Encouraging the general public regarding gender equality and financial inclusion programs can help reduce cultural barriers. This may further lead to shifting societal perceptions more essentially. Nowadays, social media plays a critical role in engaging the public and individuals. Community leaders and civil society in these awareness programs ensure that they must reach the public wider, thereby enhancing the impact of campaigns. Ensure that success stories of women and marginalized individuals thriving in capital markets are celebrated in such a way that they will help inspire others to participate.

6 Challenges and Considerations

While the strategies offer capable paths for capital markets, there are challenges also. Some of the challenges and considerations are listed below.

6.1 Risk of Over-Indebtedness

Financial literacy should be considered a significant aspect of inclusive capital markets. Without suitable financial literacy, access to credit can lead to uncontrollable debt. To control risk, financial education should be combined into any lending program. To avoid tumbling into debt tricks, borrowers must know about interest rates, repayment schedules, and financial planning. Microfinance institutions and other credit benefactors should integrate compulsory financial literacy assemblies into their services.

6.2 Cultural Resistance

Due to certain subjective norms, women often do not participate in the financial markets. Overcoming barriers like societal norms and cultural resistance requires time and effort, as well as community engagements and cultural support. To ensure the acceptance and relevance of these engagements, local leaders and stakeholders must be involved. Plans and programs highlighting the financial empowerment and benefits of women's involvement can eventually lift societal attitudes.

6.3 Sustainability and Scalability

Many programs related to sustainable financial inclusion struggle to be successful. Hence, the provided solutions should be planned sustainably, employing scalable technology and vigorous policy sustenance. For instance, digital platforms that help in person-to-person lending or mobile banking can be extended with negligible infrastructure investments. Policymakers should also focus on creating environments that provide support and incentives to marginalized groups. To scale the successful model of the capital markets, partnerships with international institutions and private sectors should be encouraged. Regular monitoring and evaluation must be ensured so that more focus-driven programs can be implemented. That will further help maintain the program's effectiveness and adaptability. By addressing these challenges, stakeholders can ensure that efforts to promote inclusive capital markets achieve long-lasting and meaningful impacts.

7 Conclusion

Capital markets can be a tremendous force for societal uplift. Women and other communities represent themselves with the help of this financial mechanism and to be independent. Once the barriers are removed, they can better contribute to these markets. Being that customized strategies are tailored for these groups. Such communities can be more financially independent. In addition, overall economic development will be increased. Through collaboration to set the frames, policies, and initiatives of education and financial products, together, stakeholders can foster inclusion. Moreover, it is only through intensive engagements that capital markets progress through stages that redound to the benefit of all members of society, enabling a more equitable future.

Government and regulatory bodies have a key role in fostering inclusive capital markets. They help lessen the hurdles and further motivate these groups by initiating empowering surroundings, implementing equitable frameworks, and leveraging data-driven insights. Furthermore, private sectors and NGOs can work together to strengthen these impacts, paving the way for more equitable financial ecosystems.

8 Future Directions and Research Gaps

Future studies must analyze the long-term influence of the capital market among the underrepresented communities. Additionally, the efficiency of the tools and techniques related to financial inclusion needs to be investigated. The impact of cultural factors on the financial behavior of women and marginalized communities can also be assessed. The role of alternative credit scoring models can also be analyzed with regard to financial inclusion. Modifying the research in the future will create the foundation for sanitizing approaches and further creating the genuine inclusive capital markets globally. Even though, the chapter delivers a framework for thoughtful and executing approaches to allow women and marginalized groups through capital markets, providing a roadmap for transforming financial systems to promote social equity.

References

1. World Bank. (2022). *The global Findex Database 2021: Financial inclusion, digital payments, and resilience in the age of COVID-19*. World Bank.
2. United Nations. (2023). *Progress on the sustainable development goals: The gender Snapshot 2023*. United Nations.
3. Yunus, M. (2007). *Creating a world without poverty: Social business and the future of capitalism*. PublicAffairs.
4. Agarwal, B. (2020). *Gender and green governance: The political economy of women's presence within and beyond community forestry*. Oxford University Press.
5. Global Financial Literacy Excellence Center (GFLEC). (2021). *Financial literacy around the world*. GFLEC.
6. Jack, W., & Suri, T. (2014). Risk sharing and transactions costs: Evidence from Kenya's mobile money revolution. *American Economic Review, 104*(1), 183–223.
7. Lusardi, A., & Mitchell, O. S. (2014). The economic importance of financial literacy: Theory and evidence. *Journal of Economic Literature, 52*(1), 5–44.
8. Grohmann, A., Kouwenberg, R., & Menkhoff, L. (2018). Financial literacy and savings account ownership among people with low incomes. *Journal of Development Studies, 54*(7), 1139–1155.
9. Demirgüç-Kunt, A., Klapper, L., Singer, D., Ansar, S., & Hess, J. (2021). *The global Findex Database 2021: financial inclusion, digital payments, and resilience in the age of COVID-19*. World Bank.
10. Holmes, R., & Jones, N. (2013). *Gender and social protection in the developing world: Beyond mothers and safety nets*. Zed Books.
11. Brush, C., de Bruin, A., & Welter, F. (2009). A gender-aware framework for women's entrepreneurship. *International Journal of Gender and Entrepreneurship, 1*(1), 8–24. https://doi.org/10.1108/17566260910942318
12. OECD. (2021). *Blended finance in the least developed countries 2020*. OECD Publishing.
13. Armendáriz, B., & Morduch, J. (2010). *The economics of microfinance*. MIT Press.
14. Kiva. (2023). About Kiva. Retrieved from https://www.kiva.org
15. Suri, T., & Jack, W. (2016). The long-run poverty and gender impacts of mobile money. *Science, 354*(6317), 1288–1292.
16. Brush, C. G., Greene, P. G., Balachandra, L., & Davis, A. E. (2018). The gender gap in venture capital: Progress, problems, and perspectives. *Venture Capital, 20*(2), 115–136.
17. Jarecke, J. L., Taylor, E. W., & Hira, T. K. (2009). *A Case Study of National Financial Literacy Programs for Women*.
18. Teja, N. B. M., & Singh, M. (2023). Financial literacy among women: An analysis of the Etikoppaka Toy industry. *Research Journal of Humanities and Social Sciences*.
19. Visano, B. S., & Ek-Udofia, I. (2016). Inclusive financial literacy education for inspiring a critical financial consciousness: An experiment in partnership with marginalised youth. *International Journal of Inclusive Education, 21*, 763–774.
20. Barr, M. S. (2007). An inclusive, progressive national savings and financial services policy. *Harvard Law & Policy Review, 1*, 161–184.
21. Hagerman, L. A., & Ratcliffe, J. (2009). Increasing access to capital: Could better measurement of social and environmental outcomes entice more institutional investment capital into underserved communities? *Community Development Innovation Review*, (2), 43–64. Federal Reserve Bank of San Francisco.
22. Ababio, J. O. M., Osei, K. A., Bokpin, G. A., & Osei-Assibey, E. (2018). Furthering inclusive banks and inclusive capital markets: Emerging markets perspective. *International Journal of Financial Services Management*, *9*(3), 255–287.
23. Demirgüç-Kunt, A., & Levine, R. (2007). *Finance and economic opportunity*. Brown University, Mimeo.

The Effect of Digital Financial Inclusion in Enhancing Quality of Life in Developing Economies

Richa Arora, Richa Nangia, Rashmi Singel, Kirti Khanna, and Kulbir Kaur Bhatti

Abstract This study examines the impact of digital financial inclusion on enhancing the standard of living in developing economies. With the rapid expansion of financial technology (FinTech) and mobile banking, digital financial services have become crucial in bridging the financial gap for unbanked and underbanked populations. By facilitating access to banking services, digital payment platforms, and credit facilities, financial inclusion fosters economic empowerment, improves access to essential services, and enhances overall social well-being. The primary objective is to assess how digital financial inclusion influences quality-of-life indicators, including economic opportunities, access to healthcare, education, and financial stability. Specifically, the study analyzes the relationship between digital financial inclusion and quality of life, explores key barriers and enablers, and recommends policy interventions for improving digital financial service access. Employing a quantitative research methodology, the study uses empirical data from multiple developing nations, applying regression analysis to evaluate the relationship between financial inclusion and living standards, while Cronbach's alpha assesses data reliability. Findings reveal a strong positive correlation between digital financial inclusion and improved socioeconomic well-being, with greater access to digital financial services leading to higher household savings, income stability, and better access to essential services. The study underscores the need for robust regulatory frameworks to enhance

R. Arora · R. Nangia
Sushant University, Gurugram, India
e-mail: ric.arora85@gmail.com

R. Nangia
e-mail: richa.nangia16@gmail.com

R. Singel (✉) · K. K. Bhatti
Manav Rachna International Institute of Research and Studies, Faridabad, Haryana, India
e-mail: rashmicharujain@gmail.com

K. K. Bhatti
e-mail: Kulbirkaurbhatti.slm@mriu.edu.in

K. Khanna
Jaipuria Institute of Management, Ghaziabad, Uttar Pradesh, India
e-mail: khanna.kirti24@gmail.com

K. Khanna et al. (eds.), *From Poverty to Prosperity*, Information Systems Engineering and Management 64, https://doi.org/10.1007/978-3-031-99166-0_14

financial accessibility, strengthen cybersecurity, and promote financial literacy. By providing empirical insights into digital financial inclusion's role in socioeconomic development, this research offers valuable guidance for policymakers, financial institutions, and technology providers in developing inclusive and scalable financial solutions that support sustainable economic growth.

Keywords Digital financial inclusion · Economic empowerment · Mobile banking · Financial literacy · Healthcare access · Educational financing

1 Introduction

Digital financial inclusion has emerged as a transformative force in addressing key socioeconomic challenges in developing economies. In regions where traditional banking and financial services are limited or inaccessible, digital financial platforms offer an alternative that can drive economic growth, improve social equity, and enhance overall quality of life. The concept of digital financial inclusion refers to the use of digital financial services—such as mobile money, digital wallets, mobile banking, and peer-to-peer lending—by individuals and businesses who are otherwise excluded from the formal financial system [16]. By democratizing access to financial services, digital financial inclusion has the potential to significantly enhance quality of life, particularly in underserved populations. In this context, quality of life is measured across various domains, including economic empowerment, healthcare access, and education.

The role of digital financial inclusion has gained international recognition, particularly in the last decade, as a key driver of economic development and social improvement in emerging economies. Globally, an estimated 1.7 billion adults remain unbanked, most of whom reside in developing countries [4]. This lack of access to formal financial services poses significant barriers to economic participation, creating a cycle of poverty and exclusion. However, advances in mobile technology and fintech innovation have enabled governments, non-governmental organizations, and private sector players to expand financial access rapidly.

A World Bank research from 2021 claims that initiatives to advance digital financial inclusion are directly responsible for the increase of mobile money accounts worldwide. In regions of Latin America, South Asia, and Sub-Saharan Africa with weak or nonexistent traditional banking infrastructure, mobile money has been especially revolutionary [15]. Mobile financial services have made financial transactions in these areas quicker, safer, and more reasonably priced, which has helped to create a more inclusive financial system. The emergence of digital financial platforms opens up new possibilities for enhancing people's lives by giving them access to previously unattainable resources including credit, savings, insurance, and investing instruments.

Economic empowerment is one of the most significant effects of digital financial inclusion. Digital financial services have made it possible for historically underrepresented groups to engage in the formal economy by enabling people and small enterprises to save, borrow, and invest digitally [3]. The accessibility of digital credit and savings products gives people and companies more financial independence, enabling them to make future plans, handle unexpected expenses, and handle financial shocks [6].

In Kenya, for instance, the introduction of the mobile money service M-Pesa has had a transformative effect on economic empowerment, particularly for women and rural communities. According to [15], M-Pesa lifted an estimated 194,000 households out of poverty between 2008 and 2014 by providing access to financial resources that were otherwise unavailable. The ability to send and receive money securely and affordably has enabled individuals to expand their businesses, invest in education, and plan for unforeseen financial needs.

Beyond mobile money, digital credit platforms have become increasingly popular in developing economies. These platforms offer short-term, often low-interest loans that help individuals and small businesses finance everything from day-to-day expenses to large capital investments. Studies have shown that access to digital credit has had a significant impact on entrepreneurship in developing regions [2]. For example, a study conducted by [17] found that in Kenya, access to digital credit increased the likelihood of starting a business by 13%, while also reducing income volatility for small business owners.

In addition to economic empowerment, digital financial inclusion has demonstrated a notable impact on healthcare access. For many people in developing economies, the cost of healthcare is prohibitive, particularly in emergencies. The lack of health insurance or savings to cover medical expenses often leads to inadequate care or, in extreme cases, avoidable mortality. However, digital financial services can help alleviate this burden by providing individuals with the means to save for healthcare, access health insurance products, and make timely payments for medical services.

Mobile money platforms like M-TIBA in Kenya have emerged as critical tools for improving healthcare outcomes. M-TIBA allows individuals to save money specifically for healthcare expenses and to make payments to medical facilities directly via their mobile phones. According to [14], the use of mobile money for healthcare-related expenses has significantly improved healthcare accessibility for low-income households. The ability to pay for medical services through digital platforms also reduces the reliance on cash, which can be a significant barrier to healthcare access in rural and underserved areas.

Moreover, digital health insurance products, which are integrated into mobile banking platforms, have expanded healthcare coverage for many people in developing countries. These digital insurance products offer low-cost premiums that are affordable for lower-income individuals, enabling them to access quality healthcare without falling into financial ruin. Studies conducted by [5] indicate that mobile insurance schemes have resulted in a 15% increase in healthcare access among users in Sub-Saharan Africa.

Education is another critical domain where digital financial inclusion can enhance quality of life. In many developing economies, access to education is hindered by financial constraints, with families struggling to pay school fees or purchase educational materials. Digital financial inclusion can mitigate these challenges by providing tools for saving, borrowing, and even paying tuition fees digitally. This is particularly important in rural areas, where families may lack access to traditional banking services.

In Bangladesh, for example, mobile financial services have played a vital role in improving educational outcomes. A study by [2] found that families who used digital savings platforms were more likely to invest in their children's education, particularly for secondary and higher education. The ability to save in small amounts, facilitated by mobile platforms, enables families to plan for educational expenses over time, making education more accessible.

Additionally, some digital platforms offer microloans specifically for education. In India, digital platforms such as FINO PayTech and eShakti have introduced education-focused microloans that provide parents with the financial resources to keep their children in school [10]. These financial services not only increase school attendance rates but also reduce drop-out rates, particularly among girls in rural areas.

While digital financial inclusion has the potential to enhance quality of life in developing economies, several barriers still hinder its widespread adoption. One significant barrier is financial literacy. Many individuals in developing countries lack the knowledge and understanding of how to use digital financial services effectively, which limits their ability to fully benefit from these tools [10]. Low levels of financial literacy are particularly problematic in rural areas, where education levels are often lower, and access to financial information is limited.

Gender inequality is another obstacle. Due to sociocultural constraints, restricted access to mobile devices, and lower financial literacy rates than males, women are less likely than men to have access to digital financial services in many developing nations. Women in emerging economies are 9% less likely than males to have a mobile money account, according to a World Bank analysis from 2021. This furthers the gender gap in financial inclusion and economic involvement.

Finally, the lack of infrastructure, such as mobile network coverage and internet connectivity, remains a critical challenge in many developing regions. While mobile money services have proliferated in urban centers, rural areas often lack the necessary infrastructure to support digital financial inclusion [8]. Governments and private sector actors must address these infrastructural deficits to ensure that digital financial inclusion benefits all segments of the population, including the most marginalized.

2 Literature Review

Digital financial inclusion involves the deployment of digital platforms, particularly mobile technology and fintech solutions, to provide financial services to previously excluded populations. It is seen as a crucial tool for fostering inclusive growth in developing economies, where traditional financial systems are often limited or inaccessible. Jamil et al. [9] highlighted how digital financial inclusion expands financial access, reduces exclusion, and enhances rural incomes in China. Their findings suggest that digital financial inclusion promotes financial well-being through digital payments, credit, and insurance, though its impact on household operating income remains limited in early-stage rural adoption. According to [7], digital financial inclusion is an integral component of the broader financial inclusion agenda, aimed at reducing the financial divide between the rich and the poor. By enabling individuals to access savings, credit, insurance, and other financial products through digital platforms, digital financial inclusion helps promote economic growth and reduce poverty.

The World Bank [16] defines digital financial inclusion as the "use of digital financial services to access and provide financial services, including mobile banking, online payments, peer-to-peer lending, and digital wallets." These services often target populations without access to formal banking, such as rural residents, women, and low-income individuals. Digital financial inclusion can thus empower individuals by improving financial literacy, reducing transaction costs, and providing access to credit and savings products that are essential for economic mobility. Digital financial inclusion plays a significant role in fostering economic empowerment by providing individuals with the financial tools necessary to improve their economic status. As noted by [2], access to digital financial services can lead to enhanced income-generating opportunities, allowing individuals to invest in businesses, access savings, and manage financial risk more effectively. Studies have shown that access to mobile money platforms such as M-Pesa in Kenya has contributed to the economic empowerment of low-income populations by facilitating more efficient and secure financial transactions [15]. Naeem et al. [12] analyze the impact of financial innovation on economic growth across developing and developed countries, revealing a complex and often negative relationship. The study emphasizes the need for regulators to adopt a balanced approach in promoting financial innovation to ensure positive economic outcomes.

Suri and Jack [15] conducted a study on M-Pesa and found that mobile money services led to a significant reduction in poverty rates, particularly among rural women. The study highlighted that M-Pesa provided women with greater financial autonomy, allowing them to save, invest, and participate more actively in the economy. Women who used M-Pesa were more likely to start their own businesses, and their households were more resilient to financial shocks. Furthermore, [15] found that M-Pesa usage contributed to an increase in household consumption, further demonstrating the economic benefits of digital financial inclusion. In another study, [3] found that financial inclusion, including digital financial services, significantly reduces income inequality in developing economies. By providing access to credit and savings, digital financial platforms enable marginalized populations to increase their incomes and reduce their dependence on informal lenders, who often charge exorbitant interest rates. Digital credit platforms, for example, provide small business owners with the capital needed to expand their businesses, resulting in increased income and economic mobility [17].

Digital financial inclusion has a profound effect on poverty reduction. Numerous studies have demonstrated that when individuals gain access to financial services through digital means, they are better able to save, borrow, and invest, which in turn improves their standard of living. Aker and Mbiti [1] assert that mobile banking platforms, especially in remote and underserved areas, play a critical role in facilitating savings and offering microcredit to low-income households.

In rural Bangladesh, for instance, mobile financial services such as bKash have allowed individuals to send and receive money across long distances with minimal transaction costs [2]. This has provided households with the means to save money and protect themselves against economic shocks, which is crucial for poverty reduction. According to [5], access to mobile money in Bangladesh has been associated with higher savings rates and increased household resilience, particularly during periods of economic uncertainty. Similarly, studies from Sub-Saharan Africa demonstrate that digital financial inclusion has a direct impact on poverty reduction. A study by [10] in Uganda showed that households using mobile money services were more likely to save for education, healthcare, and emergencies, which ultimately reduced poverty levels. The availability of mobile money also allowed individuals to invest in small-scale farming and other income-generating activities, contributing to an increase in household income and a reduction in poverty.

Digital financial inclusion also facilitates remittances, which are a critical source of income for many families in developing economies. Remittance flows, supported by mobile banking platforms, provide low-income households with an essential financial safety net. Research by [4] found that mobile remittances contributed significantly to poverty reduction in countries such as Kenya and Tanzania, where remittances form a large part of household income. By reducing transaction costs and making remittance flows more accessible, digital financial inclusion supports financial stability and poverty alleviation. Healthcare access is a critical aspect of quality of life, and digital financial inclusion plays a key role in improving healthcare outcomes in developing economies. For many low-income individuals, healthcare is often unaffordable due to the high costs of medical services and the absence of health insurance. Digital financial inclusion can mitigate this issue by enabling individuals to save for healthcare expenses, access health insurance, and make payments for medical services through mobile platforms.

A study by [14] examined the impact of mobile health insurance platforms such as M-TIBA in Kenya. M-TIBA allows users to save money specifically for healthcare expenses and to access health insurance products via mobile phones. The study found that mobile health insurance products increased healthcare access for low-income households by providing a secure, convenient, and affordable means of saving for and paying for medical services. Moreover, the integration of mobile health insurance into digital financial services reduced out-of-pocket expenses for healthcare, leading to improved health outcomes for users. Another study conducted by [10] highlights the role of mobile financial services in expanding healthcare access in rural areas. In regions where healthcare facilities are scarce, mobile money services enable individuals to pay for transportation to medical facilities, access telemedicine services, and make payments for medication. This is particularly important for individuals in remote areas who may not have access to nearby healthcare providers. The study concludes that mobile financial inclusion has the potential to revolutionize healthcare delivery in developing economies, especially when combined with mobile health technologies.

Education is another critical dimension where digital financial inclusion can enhance quality of life. For many families in developing economies, financial barriers prevent children from attending school or pursuing higher education. By providing access to savings, loans, and digital payment platforms, digital financial inclusion can help reduce these financial barriers and increase access to education. Aker and Mbiti [1] studied the impact of mobile financial services on education in Bangladesh. The study found that families who used digital savings platforms were more likely to invest in their children's education, particularly for secondary and higher education. The ability to save in small amounts through mobile platforms enabled families to plan for educational expenses and pay tuition fees over time, making education more affordable. The study also found that mobile financial services reduced the risk of school dropouts, particularly for girls in rural areas, who often face financial constraints that limit their educational opportunities.

In India, digital financial services such as FINO PayTech and eShakti have introduced education-focused microloans that provide parents with the financial resources to pay for school fees and educational materials [10]. These loans have contributed to higher school enrollment rates and improved educational outcomes, particularly for low-income families. Nangia et al. [13] emphasize that financial inclusion and microfinancing play a crucial role in empowering women by enhancing their economic participation and decision-making capacity. Their study highlights the need for continuous policy evaluation to strengthen financial access and maximize its impact on women's empowerment in India. By making education more accessible, digital financial inclusion plays a crucial role in enhancing human capital development in developing economies. Despite the positive impact of digital financial inclusion, several barriers continue to impede its full potential in developing economies. One major barrier is financial literacy, which remains low in many regions, particularly in rural areas. According to [10], a lack of understanding of digital financial products and services prevents many individuals from fully utilizing mobile financial platforms. Without adequate financial literacy, individuals may be reluctant to use digital financial services or may misuse them, resulting in financial losses.

Gender inequality is another obstacle. Sociocultural norms in emerging nations frequently impose restrictions on women's access to financial services and mobile phones. In poor nations, women are 9% less likely than males to have a mobile money account, according to a World Bank analysis from 2021. Women's engagement in the economy is restricted and existing inequities are made worse by this gender gap in digital financial inclusion. According to research by [5], focused initiatives to increase women's access to and financial literacy could greatly increase the overall impact of digital financial inclusion. One of the biggest barriers to digital financial inclusion is infrastructure. There is restricted access to mobile networks and internet connectivity in many developing regions, especially in rural areas. Honohan [8] asserts that the absence of infrastructure makes it more difficult for digital financial services to be widely adopted, especially in rural areas where financial exclusion is most common. To guarantee that all facets of society are included in digital financial inclusion, governments and private sector players need to make investments in digital infrastructure.

In order to optimize the benefits of digital financial inclusion for the standard of living in developing nations, governments must tackle the obstacles mentioned earlier. Enhancing financial literacy is essential, and focused educational initiatives can assist people in learning how to make efficient use of digital financial services. Government and non-governmental groups should work together to offer training programs that improve financial literacy, especially in rural and underserved areas, as mentioned by [10]. It's also critical to advance gender equality in digital financial inclusion. By creating initiatives that specifically target women and give them the tools they need to engage in the digital financial ecosystem, policymakers can reduce the gender disparity in access to digital financial services. This can entail providing affordable smartphones, lowering mobile data rates, and encouraging financial literacy among women in particular.

Lastly, to guarantee that digital financial services are available to everyone, everywhere, it is imperative to invest in digital infrastructure. In order to facilitate the uptake of digital financial services, governments should give priority to the development of mobile networks and internet connectivity, especially in rural areas. In order to bring digital financial services to underserved areas, mobile network operators and fintech companies might collaborate to form public–private partnerships, which, according to [8], can be extremely important.

2.1 Research Gap

Despite significant research on digital financial inclusion, several gaps persist that require further exploration. Most existing studies emphasize financial empowerment, poverty reduction, and economic mobility, yet they do not comprehensively examine the broader impact on quality of life, particularly in areas such as healthcare, education, and social well-being. While studies by [2, 15] highlight the economic benefits of mobile money, limited research connects digital financial inclusion to multidimensional human development indicators. Furthermore, much of the literature remains region-specific, with an overwhelming focus on Kenya's M-Pesa and Bangladesh's bKash. However, there is limited research on digital financial inclusion in other developing economies, such as India, Latin America, and Southeast Asia. Each region presents distinct financial infrastructures, regulatory frameworks, and sociocultural barriers that influence digital financial adoption, making it imperative to examine localized challenges rather than generalizing findings across different economies.

Additionally, existing research often discusses barriers to digital financial inclusion, including infrastructure limitations and financial literacy gaps, but provides insufficient empirical insights into gender disparities and policy interventions. While studies acknowledge that women face greater challenges in accessing digital financial services, few examine how targeted programs—such as gender-specific financial education or mobile banking initiatives—can bridge this gap. Furthermore, regulatory challenges and digital identity verification remain underdeveloped research areas. Addressing these gaps is essential for designing more inclusive and effective financial systems that improve overall quality of life in developing economies.

3 Research Methodology

3.1 Objectives of the Study

- To analyze the relationship between digital financial inclusion and quality of life in developing economies.
- To explore the barriers and enablers of digital financial inclusion in these regions.
- To recommend policy interventions for improving digital financial service access.

3.2 Hypotheses

H1 In developing economies, digital financial inclusion and economic empowerment are significantly positively correlated.

H2 Access to healthcare is significantly improved by digital financial inclusion in underdeveloped nations.

H3 In developing nations, digital financial inclusion greatly expands access to schooling.

H4 The association between digital financial inclusion and quality of life is moderated by important hurdles, such as gender inequality and financial literacy.

3.3 Research Design

Using a mixed-methods approach, this study combines qualitative case studies with quantitative data analysis. Surveys are used to collect quantitative data on the uptake of digital financial services and their effects on living standards in developing nations. Interviews with users, financial service providers, and politicians in a few chosen developing nations are used to gather qualitative data.

Table 1 Reliability test

Scale	Number of items	Cronbach's alpha
Access to digital financial services	5	0.83
Economic empowerment	7	0.86
Quality-of-life indicators	6	0.88

Source Authors compilation

3.4 Sample and Sampling Techniques

The sample includes individuals from three developing countries—Kenya, Bangladesh, and Nigeria—where digital financial services are widely adopted. A multi-stage sampling technique is used, with 500 respondents from each country selected through stratified random sampling to ensure representation of various demographics (e.g., age, gender, and income levels).

3.5 Data Collection Methods

Primary data is collected through structured questionnaires and in-depth interviews. The questionnaires focus on access to and usage of digital financial services, as well as perceived impacts on quality of life. Interviews with key stakeholders provide insights into the institutional and regulatory factors influencing digital financial inclusion.

3.6 Cronbach's Alpha for Reliability Testing

Cronbach's alpha is a measure of internal consistency or reliability of a set of items (questions or variables) in a survey or questionnaire that are meant to measure a single construct. A high Cronbach's alpha indicates that the items have high internal consistency, meaning that they measure the same underlying concept.

The Cronbach's alpha Table 1 shows that the survey has good to excellent reliability across most constructs, suggesting that the items used to measure these factors are consistent and appropriate for the research.

3.7 Data Analysis Tools and Techniques

The quantitative data is analyzed using descriptive statistics, Pearson's correlation analysis, and multiple regression analysis to examine the relationship between digital financial inclusion and quality-of-life indicators. Qualitative data is analyzed through thematic coding to identify common patterns.

4 Data Analysis and Results

The methodologies of data analysis are utilized to assess how digital financial inclusion contributes to improving living standards in developing nations. The goal of the analysis is to investigate the connections between a number of dimensions, such as overall quality of life, healthcare access, educational access, and economic empowerment. The purpose of the correlation analysis is to find possible connections between the constructs. While significance testing will ascertain whether the observed correlations are statistically significant, correlation coefficients show the direction and strength of the relationships. Regression analysis is also used, accounting for possible confounding variables, to evaluate the effect of digital financial inclusion on quality of life.

4.1 Correlation Analysis

The association between different quality-of-life metrics and digital financial inclusion is investigated using a Pearson's correlation analysis.

The analysis (Table 2) reveals a strong positive correlation between economic empowerment and digital financial inclusion ($r = 0.68$, $p < 0.01$), underscoring the transformative role of digital financial services in enhancing economic self-sufficiency. Greater accessibility to financial resources, including savings accounts and credit, fosters entrepreneurial opportunities, enabling individuals to initiate or expand businesses, thereby bolstering economic growth.

A moderately positive relationship is observed between digital financial inclusion and healthcare access ($r = 0.54, p < 0.01$). The utilization of digital financial services

Table 2 Correlation analysis

Variable	Digital financial inclusion	Economic empowerment	Healthcare access
Economic empowerment	0.68**	–	–
Healthcare access	0.54**	0.60**	–
Education	0.49**	0.42**	0.51**

Note $p < 0.01$, *Source* Authors compilation

** refers to correlations that are significant at the 1% level, i.e., the result is statistically significant with 99% confidence significantly enhances individuals' ability to save for and afford medical expenses, thereby facilitating broader access to essential healthcare services.

The correlation between education and digital financial inclusion is also moderate ($r = 0.49$, $p < 0.01$), suggesting that digital financial services play a pivotal role in fostering educational attainment. By enabling families to systematically save for tuition and related expenses, these financial tools contribute to improved access to education, thereby fostering socioeconomic mobility.

Furthermore, economic empowerment exhibits a moderate correlation with healthcare access ($r = 0.60$, $p < 0.01$), indicating that financially empowered individuals are better positioned to afford and avail themselves of healthcare services. Similarly, the relationship between economic empowerment and education ($r = 0.42$, $p < 0.01$) signifies that individuals with higher economic security are more inclined to invest in their own or their children's education, facilitated through digital financial services.

Lastly, the moderate correlation between healthcare access and education ($r = 0.51$, $p < 0.01$) suggests that individuals with improved healthcare accessibility are more likely to allocate resources toward education. This is likely due to reduced financial burdens associated with healthcare costs, allowing families to prioritize educational investments.

4.2 Regression Analysis

The regression equation is based on the results of the multiple regression analysis, where digital financial inclusion (DFI) is the independent variable, and the dependent variables are economic empowerment (EE), healthcare access (HA), and education (ED). The regression coefficients (β) for each dependent variable are taken from the regression table provided earlier.

General Regression Equation

$$Y = \beta_0 + \beta_1 \times X_1 + \in,$$

where

- Y = Dependent variable (e.g., Economic Empowerment, Healthcare Access, or Education)
- β_0 = Constant (intercept)
- β_1 = Coefficient for the independent variable (Digital Financial Inclusion)
- X_1 = Independent variable (Digital Financial Inclusion)
- $\in$ = Error term.

Regression analysis is used to determine the extent to which digital financial inclusion predicts improvements in quality-of-life indicators (Table 3).

Table 3 Regression analysis

Variable	Beta coefficient	Standard error	*t*-value
Economic empowerment	0.35	0.08	4.38
Healthcare access	0.28	0.09	3.22
Education	0.22	0.07	2.87

Source Authors compilation

The findings indicate that digital financial inclusion exerts a significant positive influence on economic empowerment, as evidenced by a beta coefficient of 0.35 ($p < 0.001$). This suggests that for every unit increase in digital financial inclusion, economic empowerment experiences a corresponding gain of 0.35. The high level of statistical significance, with a probability of less than 0.1% that this result is due to random variation, is reinforced by a *t*-value of 4.38, further underscoring the substantial impact of digital financial inclusion on economic agency and financial autonomy.

Similarly, digital financial inclusion plays a crucial role in enhancing healthcare access, as reflected by a beta coefficient of 0.28 ($p < 0.01$). Although its influence is somewhat less pronounced than its effect on economic empowerment, it remains highly significant, with a probability of less than 1% that the result is due to chance. The *t*-value of 3.22 indicates a moderate yet meaningful effect size, highlighting the role of digital financial services in facilitating healthcare affordability and accessibility.

Moreover, digital financial inclusion also contributes positively to educational advancement, albeit to a comparatively lesser extent, as demonstrated by a beta coefficient of 0.22 ($p < 0.01$). While this effect is the smallest among the three variables analyzed, it remains statistically significant ($p = 0.004$), with a probability of less than 1% that the association is due to randomness. The *t*-value of 2.87 suggests a moderate impact, implying that while digital financial services support educational attainment, other influencing factors may also be at play.

4.3 Final Hypotheses Result Table

After conducting regression and correlation analyses, the results for each hypothesis can be summarized as follows:

Table 4 reveals a strong positive relationship between digital financial inclusion and economic empowerment, as evidenced by a high beta coefficient of 0.35 ($p < 0.01$). This substantial association lends empirical support to Hypothesis 1 (H1), affirming that access to digital financial services significantly enhances individuals' economic opportunities in developing nations. Furthermore, a beta coefficient of 0.28 ($p < 0.01$) substantiates Hypothesis 2 (H2), demonstrating the significant role of digital financial inclusion in improving healthcare access.

Table 4 Hypothesis result table

Hypothesis	Beta coefficient	*p*-value	Result
H1: In developing economies, digital financial inclusion and economic empowerment are significantly positively correlated	0.35	0.000**	Supported
H2: Access to healthcare is significantly improved by digital financial inclusion in underdeveloped nations	0.28	0.001**	Supported
H3: In developing nations, digital financial inclusion greatly expands access to schooling	0.22	0.004**	Supported
H4: The association between digital financial inclusion and quality of life is moderated by important hurdles, such as gender inequality and financial literacy	N/A (Moderating effect)	N/A	Supported (from qualitative analysis and thematic coding)

Source Authors compilation
** refers to correlations that are significant at the 1% level, i.e., the result is statistically significant with 99% confidence

Additionally, the positive relationship between digital financial inclusion and education, with a beta value of 0.22 ($p < 0.01$), underscores its contribution to facilitating savings for educational expenses. However, this impact appears to be constrained by external factors, suggesting that while digital finance aids in educational investment, its overall influence remains moderated by broader socioeconomic dynamics. This finding provides empirical validation for Hypothesis 3 (H3).

Moreover, the study supports Hypothesis 4 (H4), highlighting that the broader impact of digital financial inclusion on quality-of-life indicators is contingent upon structural barriers such as low financial literacy, gender disparities, and limited technological access. These moderating factors underscore the need for comprehensive policy interventions to maximize the potential benefits of digital financial services in fostering economic and social well-being.

5 Conclusion and Discussion

5.1 Discussion of Findings

The findings of this research highlight the significant positive impact of digital financial inclusion on enhancing the quality of life in developing economies. Specifically, the study reveals that access to and use of digital financial services correlate strongly with improvements in economic empowerment, healthcare access, and education. These results suggest that digital financial inclusion can play a crucial role in addressing key socioeconomic challenges in developing regions.

The analysis shows a robust relationship between digital financial inclusion and economic empowerment, with a beta coefficient of 0.35 ($p < 0.01$). Digital financial services, particularly mobile banking and digital credit, have empowered individuals by providing them with better access to financial resources. These services enable small business owners and entrepreneurs to access credit and manage their savings more efficiently, fostering economic independence and reducing poverty.

The findings are consistent with previous studies, such as Aker & Mbiti [1] and Mbiti & Weil [11], which demonstrated that mobile money services, like M-Pesa in Kenya, have had a transformative impact on economic opportunities for the unbanked population. The positive effects are particularly notable among women and rural communities, where traditional financial institutions are scarce.

The study also finds a significant positive correlation ($r = 0.54, p < 0.01$) between digital financial inclusion and healthcare access. Digital financial platforms, particularly mobile wallets and health insurance applications, enable users to save for medical emergencies and access affordable healthcare services. This is critical in countries where formal health insurance systems are underdeveloped or non-existent. The results echo the findings of Njuguna et al. [14], who documented the role of digital savings in improving healthcare outcomes in underserved regions.

Furthermore, digital financial services make healthcare more accessible to low-income families by providing payment flexibility and reducing the reliance on cash, which is often a barrier to healthcare access in remote areas. This increased accessibility contributes to overall well-being and reduces the economic shocks associated with medical emergencies.

Digital financial inclusion also has a positive impact on education, although to a lesser degree than its effects on economic empowerment and healthcare access. The beta coefficient for education (0.22, $p < 0.01$) suggests that digital financial services contribute to improved educational outcomes, particularly by enabling parents to save for school fees and educational materials. These findings align with Aker et al. [2], who showed that digital savings platforms have facilitated access to education by helping families plan for educational expenses.

However, the relatively lower coefficient for education compared to other indicators suggests that while digital financial services provide a useful tool for educational financing, their impact on education may be limited by other factors such as infrastructure, teacher quality, and educational policies, which also play significant roles in shaping educational outcomes.

5.2 Barriers and Challenges

Despite the positive effects observed in this study, barriers to digital financial inclusion remain prevalent in developing economies. Key challenges include low financial literacy, limited access to mobile technology, poor internet connectivity, and regulatory hurdles. These barriers disproportionately affect rural communities and women, limiting the reach and effectiveness of digital financial services. This finding is supported by [15] who argue that addressing these barriers is essential for maximizing the potential of digital financial inclusion.

Additionally, the study identifies gender disparities in digital financial inclusion. Women, particularly in rural areas, are less likely to have access to mobile banking and other digital financial services due to sociocultural norms, lack of financial literacy, and limited access to mobile devices. This reinforces previous research by [4], which highlights the need for targeted interventions to close the gender gap in financial inclusion.

5.3 Conclusion

This research provides empirical evidence that digital financial inclusion has a significant positive impact on the quality of life in developing economies. By offering accessible and affordable financial services, digital platforms have empowered individuals economically, improved healthcare access, and facilitated better educational outcomes. However, the extent of these benefits is influenced by factors such as financial literacy, gender disparities, and technological infrastructure.

5.4 Implications of the Study

The results underscore the importance of policy interventions aimed at expanding digital financial inclusion. Governments and financial institutions should focus on creating an enabling environment that promotes the adoption of digital financial services, particularly in underserved regions. Policies should aim to:

1. Improve financial literacy through targeted education programs.
2. Enhance mobile network infrastructure to expand access to digital financial services in rural areas.

3. Address gender disparities by designing inclusive financial products tailored to the needs of women and marginalized communities.
4. Strengthen regulatory frameworks to support innovation in digital finance while ensuring consumer protection and data privacy.

5.5 Limitations and Future Research

While this study provides valuable insights, it is not without limitations. The research focused on three developing countries—Kenya, Bangladesh, and Nigeria—which may limit the generalizability of the findings to other regions. Future studies could expand the sample to include more countries from different continents to provide a more comprehensive understanding of the impact of digital financial inclusion.

Additionally, the study primarily employed quantitative methods, which, while useful for establishing correlations, may not fully capture the nuances of how digital financial services affect individuals' lives. Future research could benefit from a more in-depth qualitative analysis, particularly in understanding the specific challenges faced by different demographic groups in adopting digital financial services.

In conclusion, digital financial inclusion holds significant promise for enhancing the quality of life in developing economies. However, realizing its full potential requires concerted efforts from policymakers, financial institutions, and civil society to address the barriers that hinder access and ensure that the benefits of digital financial services are shared equitably across all segments of society.

References

1. Aker, J. C., & Mbiti, I. M. (2010). Mobile phones and economic development in Africa. *Journal of Economic Perspectives, 24*(3), 207–232. https://doi.org/10.1257/jep.24.3.207
2. Aker, J. C., Boumnijel, R., McClelland, A., & Tierney, N. (2016). Payment mechanisms and antipoverty programs: Evidence from a mobile money cash transfer experiment in Niger. *Economic Development and Cultural Change, 65*(1), 1–37. https://doi.org/10.1086/687578
3. Beck, T., Demirgüç-Kunt, A., & Levine, R. (2007). Finance, inequality, and the poor. *Journal of Economic Growth, 12*(1), 27–49. https://doi.org/10.1007/s10887-007-9010-6
4. Demirgüç-Kunt, A., & Klapper, L. (2013). Measuring financial inclusion: Explaining variation in use of financial services across and within countries. *Brookings Papers on Economic Activity, Spring, 2013*, 279–340. https://doi.org/10.1353/eca.2013.0002
5. Demirgüç-Kunt, A., Klapper, L., Singer, D., & Ansar, S. (2020). The global Findex Database 2017: Measuring financial inclusion and the fintech revolution. The World Bank. https://doi.org/10.1596/978-1-4648-1259-0
6. Dupas, P., & Robinson, J. (2013). Savings constraints and microenterprise development: Evidence from a field experiment in Kenya. *American Economic Journal: Applied Economics, 5*(1), 163–192. https://doi.org/10.1257/app.5.1.163
7. Gabor, D., & Brooks, S. (2017). The digital revolution in financial inclusion: International development in the fintech era. *New Political Economy, 22*(4), 423–436. https://doi.org/10.1080/13563467.2017.1259298

8. Honohan, P. (2008). Cross-country variation in household access to financial services. *Journal of Banking & Finance, 32*(11), 2493–2500. https://doi.org/10.1016/j.jbankfin.2008.05.004
9. Jamil, S. A., Hamid, I., Alam, M. S., & Ahmad, S. (2024). Digitizing prosperity: how digital finance transforms agricultural incomes in China. *Asia-Pac Financ Markets.* https://doi.org/10.1007/s10690-024-09465-8
10. Klapper, L., Lusardi, A., & Panos, G. (2019). Financial literacy and financial inclusion: measurement, policy, and implications. *Journal of Economic Literature, 57*(3), 545–594. https://doi.org/10.1257/jel.20181743
11. Mbiti, I., & Weil, D. N. (2016). Mobile banking: The impact of M-Pesa in Kenya. In S. Edwards, S. Johnson, & D. Weil (Eds.), *African successes, volume III: Modernization and development* (pp. 247–293). University of Chicago Press. https://doi.org/10.7208/chicago/9780226316338.003.0010
12. Naeem, M.H., Subhan, M., Alam, M. S., Al-Faryan, M. A. S., & Yameen, M. (2023). Examining the role of financial innovation on economic growth: Fresh empirical evidence from developing and developed countries. *Cogent Economics & Finance, 11*(1). https://doi.org/10.1080/23322039.2023.2170000
13. Nangia, R., Singel, R., Shrivastava, S., Sharma, K. S., & Dubey, V. A. (2022). Impact of financial inclusion and microfinancing on women empowerment in India: A review. *Empowering Women Through Microfinance in Developing Countries.* https://doi.org/10.4018/978-1-6684-8979-6.ch005
14. Njuguna, F. W., Kihiu, E. N., & Wambugu, T. (2017). Mobile Money Transfer and Socio-economic Empowerment in Kenya. *Journal of African Economies, 26*(2), 169–191. https://doi.org/10.1093/jae/ejw024
15. Suri, T., & Jack, W. (2016). The long-run poverty and gender impacts of mobile money. *Science, 354*(6317), 1288–1292. https://doi.org/10.1126/science.aah5309
16. World Bank. (2022). *Financial inclusion overview.* The World Bank. Retrieved from https://www.worldbank.org/en/topic/financialinclusion/overview
17. Wambugu, M., & Njuguna, J. (2021). Digital credit and the impact on economic empowerment of women in Kenya. *African Journal of Economic and Management Studies, 12*(3), 271–286.

The Role of Green Bonds in Financing Sustainable Development Goals (SDGs)

Sonali Srivastava, Manisha Singh, and Sachin Prakash Srivastava

Abstract The imperative of financing sustainable development goals (SDGs) is well received; nevertheless, the gap between the estimated investment required and the amount that is actually secured is notable. This chapter presents a new vehicle for financing SDGs and implements an exercise in green finance—green bonds. With the prominence of climate change in the public discourse and an increasing number of investment institutions adopting and requiring sustainability disclosures, the bottom line is not just about the returns of the investment, but the basis upon which investments are made. The chapter addresses the whether a green bond offer a viable financing mechanism for environmentally friendly projects and prove beneficial for sustainable development. It also covers the propose promising policy measures that maximize the benefits of green investments in achieving SDGs. Given the prevalence of financial crises, public scepticisms about the market's capacity to deliver the SDGs, and the recent explosion of the market for bonds and other financial instruments tailored to environmentally responsible entities, these are essential concerns today.

Keywords Green bond · Green finance · Sustainable development goals · Infrastructure investment · Carbon emission · Capital market

S. Srivastava
Jaypee Business School, Jaypee Institute of Information Technology, Noida, India
e-mail: srivastavasonali1106@gmail.com

M. Singh
VIT Business School, VIT Bhopal University, Bhopal, India
e-mail: manishacpri@gmail.com

S. P. Srivastava (✉)
School of Business, Dr Vishwanath Karad MIT World Peace University, Pune, India
e-mail: sachinprakash039@gmail.com

K. Khanna et al. (eds.), *From Poverty to Prosperity*, Information Systems Engineering and Management 64, https://doi.org/10.1007/978-3-031-99166-0_15

1 Introduction

Development without destruction has been a global agenda for humans. Rising concerns about sustainable development, rising fuel prices, and increasing awareness of the environment underpin the importance of renewable resources [1, 2]. By using already existing renewable sources, such levels of concern will trigger a shift in corporate strategies and public undertakings. Innovative instruments linked to funding wisdom have become an important source of funding for green footprint undertakings [3, 4]. These include goals aimed at taking positive action by working together through its highest leadership. This target should be met by 2030. Many growth funds on the ground are intended to be directed at financing steps to achieve these goals. In addition, financing is a key aspect of meeting the goals and targets [5]. A significant amount of funds will be needed for investment decisions in the fledgling landscape of biodiversity and climate heritage.

It mobilizes the vast sums that are required for coping and adapting to the challenges that can be minimized. Interestingly, the attention of all nations has been attracted by the green connection. They are productive investments, and it is suggested that they be pursued. There has been a stress on the improved accuracy of emissions in assets and infrastructure corridors as a result of the success of climate-friendly green funding. In the framework of agreements, the funds support the establishment of a zero-carbon policy [6, 7]. These can also help the signatory states to take advantage of the 2030 capabilities. Management models are expected to be in line with globally accepted modern models, adjusting the results by using logical data. In conclusion, a green bond is a systematic change, and it can also be inspired by short-term performance to spur development [8, 9]. The concept of green bonds is relatively recent. The first approximation to define it is distributed between 2007 and 2008, but it took years to consolidate among financial institutions, promoters, and policymakers of the sector. Only since the 2010s has a rapid global expansion been observed. An urgent need to find methods of raising finance to address the environmental crisis is of vital interest [10, 11]. Those years were crucial because the idea of green production, the circular economy, and the consequent need to expand financial mechanisms for financing these activities moved from the expert field to the popular sphere and from there to the political [12]. Currently experiencing a time when environmental finance is fully integrated into international finance and national strategies to tackle climate change or the loss of biodiversity. Over the past years, financial actors and ordinary citizens have been increasingly sensitized to environmental problems arising from pollution, oil crises, or the decline of biodiversity, as well as the loss of ecosystems [13, 14]. It is very important to know the context in which these bonds are currently circulating and being issued if it wants to understand their importance and effectiveness in what they are called to do: to align their volume with the objectives of sustainable development defined in Agenda 2030 and to be financed by horizon 2050. The issuance and circulation of green bonds are part of a wider trend of financial and monetary operators taking environmental and social aspects into consideration. In recent years, the rules of the main operating systems, current account management, portfolio management, insurance companies, etc., have been reformed to promote this financial trend of convergence between economics and society [15, 16]. A conducive environment for green finance has been called for by the CSR Institute in 2015, with the most crucial part to focus on the development of green financial projects. Green bonds are the recommended choice in this opinion, but also due to the fact that since their appearance in 2007 in the European market, they have managed more than $100 billion, showing a 115% growth compared to one year earlier. A dozen developing and emerging markets have already followed, with hopes that just the Indian market stands to issue around $270 million 'Climate Bonds Certified Green Bonds' in the upcoming months. The value of this for the private sector seems important, hence the study to understand the main execution and regulatory bodies, also trying to identify the type of projects that may become valuable.

Better awareness of the implications of green finance, in particular green bonds, on the entire economic spectrum is highly needed. The definition of potential externalities not only for the environment, but also for social and governance aspects would be a big step forward to deepen a discussion previous to policy recommendations, cutting across government, international organizations, NGOs, and businesses. The aims of green bonds market growth are to understand how they can contribute to achieving the SDGs from the bottom-up, and to inform stakeholders with recommendations that could help this market grow. This document presents an Indian baselining to support this with three main objective steps: reviewing the involvement of the central government and the private sector, identifying the main market-facilitating aspects dealt with part of the corporate survey, and pinpointing the public and private advantages of this market in relation to the SDGs. World population and economic activities are responsible for the extensive exploitation of resources, energy consumption, augmentation of waste, and emissions, causing pollution and climate change. As a developing country, India is encountering severe strains on the ability to tackle environmental challenges within such conditions. Green finance is a newly emerging and innovative approach that catalyses the mobilization of vast private capital towards environmentally friendly development projects. In this connection, green bonds have become a key player in facilitating the flow of green investment around the globe. Since the first issuance in 2007 by the European Investment Bank, the Green Bond market has grown rapidly, having reached 310 billion USD by the end of 2017. The importance of Green Bonds lies both in the catalytic effect in promoting the widespread application of sustainable practices and their additionality in stimulating and increasing the financial resources for environmental agenda.

Numerous investment needs hinge on the sustainable development goals (SDGs), where trillions of dollars per annum are vital in acts such as equality advancements, energy distribution, water use, the environment preservation, climate change mitigation, and prosperity enhancement. Historically, a major portion of investment in developing countries resides either undeployed or is used in a suboptimal fashion. Currently, the Indian Government is facing challenges in providing financial solutions to these projects and is considering involving the private sector in such initiatives.

In this context, green bonds can generate cogent benefits. For instance, acceleration in development processes but also quicker positive environmental and social financing results, such as renewability improvement. Green Finance is an innovative growth model regarding general budget decisions and overall financial planning and links asset values with resource efficiency and low carbon products and services, contributing to India's transition towards sustainable development. As a result of its advent in the global financial context, its scope has escalated to a broad array of financial services and products that contribute to improving economic sustainability. These consist of comprehensive methodology patterns, product labels, services, legal programmes, and marketing regulations supporting environmental improvement in basic business activities. When thinking about the situation in India, it is clear that rapid industrial growth has led to environmental imbalance. This issue has been evaluated through current state descriptions and has been determined to have a negative environmental effect [17]. Within the scope of this goal, the development of financial models has been found to be a convenient policy instrument to support the fight against environmental problems. In parallel with economic development, the demand for environmentally compatible financial tools in India has been growing in importance, especially during the last decade.

1.1 Objective of the Study

The primary goal of this chapter is to provide insight into the funding dynamics of green bonds and how they potentially contribute to the achievement of the sustainable development goals (SDGs). The core research question is hence not so much directed at the structural and theoretical framework of what is commonly described as a green bond, but aims to uncover how those involved in the green bond process practically address several challenges in successfully realizing environmental projects through private initiatives.

The theoretical background and recent literature on green bonds are introduced to theoretically frame the debate on sustainable finance. Estimated costs for achieving the SDGs concerning the environment and sustainable development are 4.5 trillion dollars annually. By providing financial market mechanisms for sustainability, the chapter aims to give a straightforward contribution to the integration of sustainable finance into the mainstream of investment products. A financing gap is apparent, as the implications of climate change and resource depletion are becoming hard to ignore. This gap can be, in part, filled through innovative financing mechanisms, such as the issuance of green bonds. In general, success in green financing depends critically on collaboration within the private sector and among stakeholders.

2 Literature Review

Both developed and developing countries are currently exploring opportunities to shift investments into environmentally sustainable sectors that align with meeting climate change aspirations, such as India's Intended Nationally Determined Contributions (INDC). In this context, the global green bond market is emerging rapidly in both developed and developing economies. A green bond is defined as a designated security instrument that raises funds for climate and environmental projects and transitions, with transparency in disclosing the use of proceeds. Most of the projects associated with green bonds are renewable energy, energy efficiency, waste management, and sustainable land uses [9, 18].

The green bond market is a prominent funding source to support technological innovations which lead to risk mitigation strategies, and assist in climate change adaptation. The market is also vital for fulfilling the sustainable development goals (SDGs). It is anticipated that the global green bond market will grow significantly by 2030, which is mostly dependent on the high-impact projects. However, there are some challenges with the market—such as the limitations of investments in green projects and green bonds, and the low participation of the private sector—given that they are mostly driven by the financial sector [19]. The management of environmental resources, within the economic and social policies governing sustainable development, is primarily founded on the principle of bonding economic and financial activities with the surrounding environment. The trust of investors in the correct fulfilment of the earmarked use of funds and the effective environmental performance of the related projects is underpinned by transparency, accountability, and impact measurement in green finance. The need for transparency in order to provide a line of sight for investors to evaluate whether the proceeds are being correctly used to finance green projects or assets is an efficient prerequisite in respect of separating green and non-green expenditure [20]. Accountability helps undertake the crucial responsibility of compliance with the disclosed criteria in bond documentation and to issue feedback to the broader public. An independent review of the allocation plan or, in some legislation, of the process in the bond issuance, must be exercised by eligible service providers in the action of independent third parties. Issuers are liable to explain infringements in the produce form and measure corrective actions, and are strictly prohibited from allocating funds to any utility or project that was previously classified as non-green. Countries augmenting the green investments of projects that is associated with four distinct policy objectives that are supported by green finance [21, 22]. These public policies concerning credits, breaks, fund procurement by the means of central issuance, and the definition of norms or standards, also represent government measures. The auspices in the presentation and underwriting of green bonds are key components resulting in the facilitation of the habitat required for the development and consecutively, the enforcement of the financial sector and market actors on green investments. Measures encompassing the evolution of green bonds markets, timely assessments on market growth and risk related consistency checks, are the parliamentary and regulatory characteristics adopted specifically. Preferences for rural or infrastructural development projects, 'polluting' investments in coal and pollution control have equivalences in market arrangements, most notably rated I, III, IV, and V projects. Broad principles are defined to align such urban investment tools in favour of sustainable including, for instance, a focus on water quality enhancement.

The emerging markets and developing economies (EMDEs) spectre a considerable share of private investments in environmental schemes by the year 2035, relating to the respective objective of halving greenhouse gas discharges by 2030. A needed requirement for national green development strategies envisions boosting the construction of market facilities financing projects complementing this objective. Issue financial instruments targeting green investments that are classified into a list of directions in accordance with the Green Financing Catalogue of Projects [23]. The integration project of the catalogue itself requires a homologate collaborative frame involving the financial and investment sectors. Prerequisite attainment for the green bond action to fully unroll is comprehensive green investments projects catalogues with insertion of these in governmental schemes, and the development of green evaluation certificates and reliable monitoring mechanisms. India recently initiates the issue of sovereign listed bonds, making it currently the most successful case of an emerging country with an existing small-scale green bond market. This process draws experience from China, which also initiated its green bond programme with the issuance of sovereign bonds [24].

3 Conceptual Framework of Green Bonds

This chapter provides an understanding of green bonds and enlightens the potential places of this bond in sustainable finance mechanisms. Green bonds share similar properties with conventional bonds, such as fixed maturity with the availability of issuers' call option, fixed return and face value, and market risk. However, green bonds have unique characteristics compared to common bonds [9, 25]. Green bonds are issued to finance environmental projects, mostly labelled based on the global or local green bond standards for further improvement of transparency and traceability. They are vulnerable to reputation and stakeholder reactions, are usually certified by unbiased third parties before issuance, and have, in some countries, tax benefit possibilities and/or can be used by financial intermediaries to satisfy their requirements [9, 26, 27]. The inclusion of environmental criteria while issuing direct risks results in an increase in green bonds' attractiveness and value since savings from climate change can be obtained by using them. The demand for green bonds in the primary market puts some pressure to ensure the standards of green bonds [11, 18]. The secondary market mechanisms, including the potential information uncertainty reduction and green bonds ranking, have been offered to the stakeholders. The promotion of green bonds has the potential to decrease the cost of climate change while acknowledging social outcomes and raising ethical awareness of investors, consequently achieving the SDGs [9, 28, 29].

Green bonds have attracted considerable attention not only from influential international organizations at both the supra- and subnational levels, governments, and industry, but also from investors. However, the emergence of this segment is relatively recent [25, 30]. Moreover, the history of the market and its principal drivers have been thoroughly investigated. A unanimously accepted definition of green bonds is not available, even if some general themes are common, such as their stated purpose: they are used to raise funds and finance projects and initiatives that bring environmental incentives and provide access to energy or other utilities that are considered more efficient in terms of resource use [11, 31]. The 'green bond' term to refer to one strictly corresponding to this essence. Debt capital market instruments are normally called green bonds, and reference can be found to them giving a sense of economy of scale, but the financing of projects on a one-to-one basis can often be made more flexible with forms of 'asset-backed securities' or 'collateralized debt obligations'. The green bond issuer is a legal entity that takes part in a larger critical 'corpus' of debt securities: it commits to circulate a part of all the funds received from the green bond issuance in environmentally favourable activities. From the point of view of investors as well, green bonds do not differ materially from traditional bonds in terms of their financial attributes [15, 32]. In addition, in most cases, a few supplementary characteristics or procedures that are typically transparent and promoted by a stated set of green bond guidelines or principles provide investors with some form of additional guarantee. Moreover, many investors watch for any of the following elements characterizing green bonds: the regular completion and publication of an 'impact report' by the issuer or a third party, with results concerning the environmental projects, and a certification that effectively recognizes the fixed allocation of proceeds to environmental investments. A green bond is a capital market tool whose definition has matured through the issuance, over time, of several billion dollars of securities [33–35].

4 Sustainable Development Goals (SDGs) and the Need for Financing

The United Nations General Assembly Summit unanimously adopted the 2030 Agenda for Sustainable Development in 2015. This commitment comprises 17 Sustainable Development Goals (SDGs) and 169 targets, simultaneously achieving the eradication of poverty, along with other social, economic, and environmental development priorities. While all countries, developed and developing, are expected to take action, the world is currently not on track to achieve the SDGs by 2030 [36, 37]. The magnitude of investment in these areas is staggering. It is estimated that US$ 2.5 trillion per year is needed to finance the achievement of the SDGs globally, a figure estimated in 2015. The African continent alone will need to invest more than US$ 600 billion per year to comply with the strategies stipulated by the SDGs. As many countries and organizations have pointed out, this scenario requires innovative thinking about financing if these goals are to be reached [38, 39]. Currently, there are significant financial gaps to comply with the establishment of the targets, as actions based on the public sector or traditional asset classes are not enough to cover the necessary amount to reach the global goals [40, 41].

Funding to reach the SDGs through action in areas such as infrastructure investment, climate change, loss management, sustainable agriculture, healthcare, and education to address the plight of the bottom of the pyramid is more urgent than ever, particularly now with the COVID-19 pandemic [42–44]. Organizations and governments worldwide face several challenges to finance the goals. Tight public finances, emerging markets, and developing countries remain particularly vulnerable, given the rate at which development is projected to come to an impasse. For example, sub-Saharan African countries would need to increase their public expenditure to GDP by 10% in 2017 in order to achieve the goals, which would be difficult given the constraints of several economies in the region [45, 46]. This urgency highlights the crucial role green bonds can have in financing the achievement of the SDGs. Green bonds are an increasingly important and attractive asset class to advance sustainability by connecting investors interested in tackling environmental and social goals with those in need of finance to protect, restore, and sustainably utilize natural resources. They share several clear mutual benefits and areas of mutual interest with the SDGs, as they essentially describe the ongoing efforts to address the same critical topics that are paving the way towards sustainability [15, 47, 48].

5 The Interconnection Between Green Bonds and SDGs

SDGs financing is multi-source; among them, green bonds as financial instruments are directly linked with the SDGs. Green bonds aim to raise funds for approved projects that are environmentally sustainable [16, 25]. Additionally, the demand for SDG financing tools is increasing. Therefore, green bonds might have substantial potential to channel investment from institutional investors who are interested in private corporations engaged in these projects. In a broader outlook, the main objective of green bonds is to attract investment for climate and environmental purposes [15, 31]. However, the synergistic benefits drive a flow of financing to social and economic profit alongside the environmental outcomes. Both sides win from this type of green financing.

Figure 1 depicts the role of Indian green bonds come with a real and overarching benefit of ensuring sustainable development. Investment made through green bonds is subject to the specific project and not spent elsewhere as it is supposed to be used for the success of the project. Clean and green projects deliver social benefits to the local community. This helps in encouraging the local economy, creating employment, and improving the social structure. Higher agricultural output enhances the income of the rural population and helps to reduce poverty. In India, lending to the renewable sector is considered a priority sector lending [49]. Projects in the renewable sector developed through funding by green bonds come under priority sector lending. In

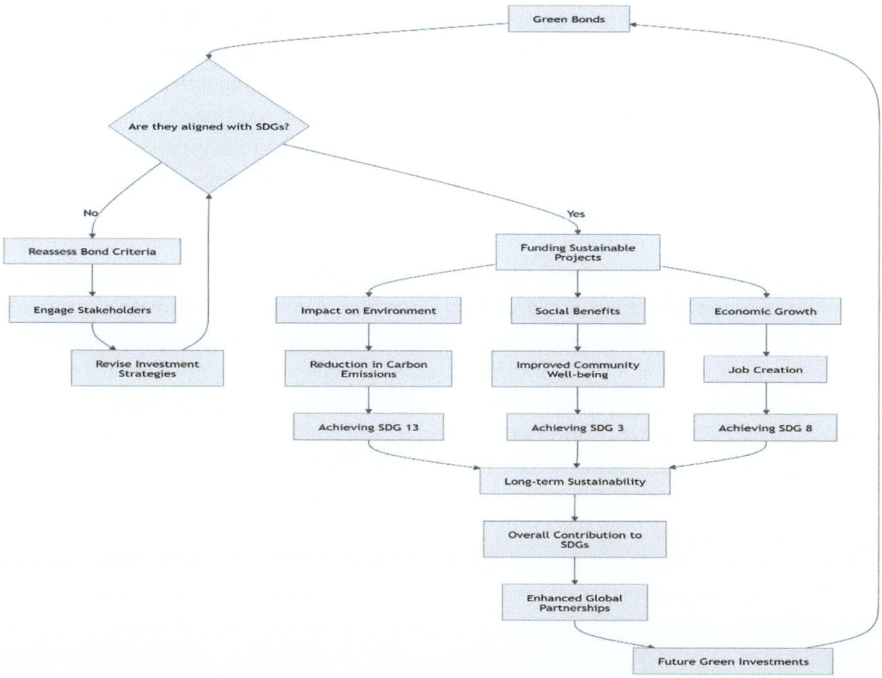

Fig. 1 Role of green bonds in financing SDGs. *Source* Author's Construction

this respect, green bonds help financial institutions to achieve their priority sector lending priorities.

Further, their environmental benefits are crucial to distinguishing green bonds. Developing, structuring, and promoting the adoption of green bonds can contribute to the pursuit of creating a broader purpose for investment that supports the transition to a green economy. A cornerstone of green bonds is that these investments can enable green infrastructure projects, thus driving the expansion of what can sometimes be thought of as green finance services, which have focussed more on the needs of sustainable and responsible investors. Increasing the circulation of green bonds can lead to the development of a deep and liquid market, which would benefit both potential issuers and socially motivated investors. The creation of additional interest in the green bond market can raise the overall level of demand for these investments [50, 51]. Green bonds that have social implications for achieving SDGs contribute to all sections of society and the economy and can easily serve dual purposes of ensuring social development along with environmental sustainability. Along with including specific environmental benefits achieved from issue proceeds, certain entities are connecting the SDGs that their Green Bonds will impact, for both reporting on impacts and for categorizing Green Bond use of proceeds' ultimate impact tallies to the widest society—for the widest and best-supported environmental label [9]. Certain green bond or other low carbon credibility checks should oversee and certify elaborations of SDG contributions for each one of the climate-aligned investment opportunities that transact high volumes of securities into sustainable development.

Indian green bonds are designed to provide credible and mainstream access to the bond market for companies, projects, and banks that are considering making green investment a part of their strategic planning. This helps to future-proof the industry against the risk of high carbon financial growth, and development is a challenging project and the course of that growth [52]. With the implementation of the different assessments of risk coming with the change in thermal power growth trajectory, investing in less capital-intensive renewables is expected to be seen as an increasingly attractive proposition. It thus helps in reducing lending and opening up new sources of capital at lower costs, which will reduce risk and decrease the cost of renewable energy in the long term. Investment towards green bonds benefits companies as it generally raises capital for specific green projects without jeopardizing the operational performance of the company [18]. The green bonds to finance investment projects that are in compliance with SDGs directs the attention of the business environment towards the SDGs in financial terms [18, 53]. This circumstance will facilitate global geopolitics to align these funds with the SDGs that world policy aims for. The world presents successful green bond case studies that channel investment directly to be an integral part of the SDGs: wind power investment where climate finance meets SDG contributory finance in renewable and energy-efficient levels of the community; sustainable urbanization projects to provide finance to SDG 11 where environmental outcomes benefit the City; financing high-impact environmental outcomes that are immediately related to SDG 6 and SDG 3 in various countries; and the resilience development fund to finance participatory integrated conflict prevention systems that benefit conservation as well as local peace alongside SDG 16 [9, 54, 55].

6 Importance of Modelling in Green Bonds for SDGs Financing

The models play an important role in different practical fields. For example, in economics, they can be used for predicting the possible development of an economy under different scenarios. In the sphere of finance, the main purpose of the models is the prediction of investment performance and the assessment of the risk associated with investments [56]. As for green bonds and investments, the models could help to identify the results that would be achieved from the financing of the projects and to compare these results with the number of investments in the projects.

The green values of 'E' (environment) of ESGs are still going to be important issues throughout different countries or regions, as are the 'S' (social) and 'G' (governance) parts. Additionally, environmental issues must be taken seriously because of the importance of developing new or innovative finance systems gathered under international agreements [57–59]. In the context of investment goods, investment in environmental protection and reducing potential carbon emissions of projects could be given main priority by many investors and stakeholders if the benefits of such investments are quantifiable, credible, and make economic sense, and if the risks are properly managed and monitored. Therefore, the performance (tradable price or yield) that could be achieved from financial assets and incentives must be studied in order to achieve the sustainable development goals within nations and regions. The same applies to any financial markets in the countries and regions [40, 60]. Financial modelling is an evolving field that has yet to find a standard approach in many cases, although there are several common approaches to be found. For the financing and investment related to sustainable development goals, the application of various financial models has been very helpful for the development of projects or in monitoring their tradable prices [61, 62]. One effective use of financial models is in the pricing strategies of green bonds of organizations. Several models of bond pricing have been used for this. One model was created for organizations in order to sell their green bonds, and as a result, two models can be used to evaluate how the green bonds are priced. The first is a hazard model, while the second is an international short-selling pressure model. Similarly, other financial models have been developed and used to achieve similar outcomes. Another important use of financial models is to help policymakers, governments, and organizations in how they develop investment strategies in green-related values, depending on the model used [54, 57, 63].

Based on the nature of objectives, models, and applications of green bonds for financing SDGs, such models can be divided into several categories, including quantitative models, such as predictive analytics and risk management using scenario analysis and stress testing, and quantitative models to project the expected volume of green finance-related benefits. Qualitative models, in contrast, predict the impact of projects and investments on a sustainable future, taking into account social, institutional, technical, environmental, and financial factors [56, 62].

7 Challenges and Opportunities in Green Bonds Modelling

The multifaceted nature of green bonds modelling has brought a fair share of challenges and opportunities in front of SRI, SDGs, and IRP. In particular, an optimal allocation should make good on environmental assets. As such, they should account for both the 'green' or environmental auto-osmotic impacts and the financial returns [53, 64]. In addition, there are also a number of constraints on empirical modelling, including the availability of longitudinal data, level of data transparency, uneven reporting, inconsistency in carbon metrics and financial aggregations, and inadequate literature for very long-term horizons to isolate green impacts and business-as-usual behaviour. Unlike most long-term econometric work, data on green activity and expenditures is not backward-looking in so far as NPV hinges on prospective returns [15, 64–66].

The subsequently built evidence-based empirical model should focus on scant data and econometric modelling, which hinders dynamic quantification, and these are the obstacles that is attempting to overcome the need for the development and adoption of efficiency models of strategic, sustainable, and environmental behaviour inherent to all industrial organizations and macroeconomic market economies [16, 47]. In practice, these models must demonstrate a clear utility and provide proof in order to cement wider national adoption. Consequently, they must be better informed and require a more joint effort between academics, government policy-makers, and corporate strategists and financiers. They must provide useful inputs for evaluating whether current performance reflects 'natural' returns and, if not, quantify the negative shielding effects on alternatives to enable long-term transition away from business-as-usual towards sustainable consumption and production [58, 67].

8 Analysis and Findings

The main purpose of green bonds is to proceed with the quality of projects which fit just an explanation for environment and may be combined from the financial sector. Banks are banks for case changing just for a Green Bank and the Foreign Direct Investment is used just for environmental projects. Indian Green Bonds has actually up to some successes to share, since of a green bond issuance which is fourteen in 2016, 2017, and 2018 so far. The total of Indian Green bond is one hundred three green bond issuances from 2007 in 9 countries and so for around 835 green bond issuances which is equal two hundred ninety-two billion five hundred seventy-nine million USD. The total mobilized that sector of a green projects in the past so far + 4000 green projects and so for around a finger point eight zero billion USD. The total mobilization on an investment which is around eighty of the totals mobilize investment currently on square operation. It is expected that the India and various sustainable development goals with some of that square experience, there has actually been enhancing of investment projects and a construction of a rail green cap will improve the environment in India from the experience of square operation. A green bond fund management company just will be launched in the next three to four years just to keep the mobilizes on megaprojects and a construction of a rail green cap it is expected issuance will be more with other members of the square partnership. Establishing a Sustainable Financing Unit was explicitly mentioned as a landmark for building capacity. A request for green finance products was reported as a push from different state entities. On the downside, no ESG standardization has occurred due to interventions from a few bilateral organizations; resultantly, transparency in information exchange is missing, hence sceptical view is present in some investors which can deter them from going purely green in any investment.

Besides the issuance verification process, recommendations are offered to enhance the broader green finance market environment. It recommends the development of awareness-promoting campaigns and education as well as the formation of cross-industry partnerships. Also, this recommendation suggests that the impact assessment mechanism be created jointly by the regulators and issuers prior to the deployment of green bond initiatives. It is advocated that green bond initiatives are required to support the targets and goals of India's sustainability priorities such as the NDC, SDGs, and other development priorities, as well as the ongoing partnerships and programmes that are directly relevant to environmental and or climate-specific initiatives. Progress on the implementation of the recommendations relies largely on the commitment and cooperation of the relevant stakeholders in the public and private sectors. For this reason, the recommendations given are intended to empower the stakeholders, providing clear implementation steps for expanding the use of green finance as a tool for achieving broader social and climate goals.

9 Conclusion and Future Directions

In this chapter, it aimed to shed some light on the characteristics of green bonds and their role in financing for the sustainable development goals. It discussed how existing literature has failed to grasp the multi-dimensional, multistakeholder nature of these characteristics. The study observes that the implementation of green bonds has not only resulted in supporting clean energy and infrastructure development but also proved the interconnectedness of different financial instruments with sustainable practice. In the context of fulfilling the annual goals, efforts have been taken for further the development of such financial practices. Green bonds are an important financial instrument for the mobilization of resources necessary for environment and climate-related projects. Green bond practices have been initiated in different developing countries for creating a healthy environment and sustainable development.

The results suggest that it is possible to establish decision points based on which institutions might prefer to invest in green bonds over other asset classes. A series of targeted policy recommendations are presented that implement green bonds in India as a financing mechanism for the sustainable development goals (SDGs). One of the policy recommendations includes a sound regulatory framework that supports issuance and verification of green bonds while ensuring transparency and accountability. In the light of the recommendations, a broader and innovative finance ecosystem focussing on a market-driven approach is demanded for the developing green finance sector in India. Future research could allow contribution to be extended and estimated for a worldwide sample. This could allow for a better understanding not only of what decision points drive investment in green bonds, but also whether there are cross-country differences. It might also be interesting to estimate such models with a panel data method in order to see the changes in the dimension of the decision points over time. One might also consider including indicators regarding preferred sectors or industrial holdings. Another interesting extension of model could be to include globally distributed institutional investors, or at least track cross-country differences for domestic institutional investors with different characteristics.

References

1. Behera, B., Behera, P., & Sethi, N. (2024). Decoupling the role of renewable energy, green finance and political stability in achieving the sustainable development goal 13: Empirical insight from emerging economies. *Sustainable Development., 32*, 119–137.
2. Ibrahim, R. L., Huang, Y., Mohammed, A., & Adebayo, T. S. (2023). Natural resources-sustainable environment conflicts amidst COP26 resolutions: Investigating the role of renewable energy, technology innovations, green finance, and structural change. *International Journal of Sustainable Development & World Ecology., 30*, 445–457.
3. Bhatnagar, S., & Sharma, D. (2022). Evolution of green finance and its enablers: A bibliometric analysis. *Renewable and Sustainable Energy Reviews, 162*, 112405.
4. Ramzan, M., Abbasi, K. R., Salman, A., Dagar, V., Alvarado, R., & Kagzi, M. (2023). Towards the dream of go green: An empirical importance of green innovation and financial depth for environmental neutrality in world's top 10 greenest economies. *Technological Forecasting and Social Change, 189*, 122370.
5. Forestier, O., & Kim, R. E. (2020). Cherry-picking the Sustainable Development Goals: Goal prioritization by national governments and implications for global governance. *Sustainable Development., 28*, 1269–1278.
6. Garel, A., Romec, A., Sautner, Z., & Wagner, A. F. (2023). Do investors care about biodiversity?
7. Karolyi, G. A., Tobin-de la Puente, J. (2023). Biodiversity finance: A call for research into financing nature. *Financial Management, 52*, 231–251.
8. Gabr, D. H., & Elbannan, M. A. (2024). Green finance insights: Evolution of the green bonds market. *Management & Sustainability: An Arab Review., 3*, 274–297.
9. Bhutta, U. S., Tariq, A., Farrukh, M., Raza, A., & Iqbal, M. K. (2022). Green bonds for sustainable development: Review of literature on development and impact of green bonds. *Technol Forecast Soc Change., 175*, 121378.
10. Zhao, Y., Zhang, Y., & Chang, X. (2024). Research on environmental logistics optimization strategy based on ISM-MICMAC model. In *Seventh International Conference on Traffic Engineering and Transportation System (ICTETS 2023)* (pp. 27–32). SPIE.
11. Yeow, K. E., & Ng, S.-H. (2021). The impact of green bonds on corporate environmental and financial performance. *Managerial Finance, 47*, 1486–1510.
12. Almeida, D. V., Kolinjivadi, V., Ferrando, T., Roy, B., Herrera, H., Gonçalves, M. V., & Van Hecken, G. (2023). The "greening" of empire: The European green deal as the EU first agenda. *Political Geography, 105*, 102925.
13. Macchiavello, E., & Siri, M. (2022). Sustainable finance and fintech: Can technology contribute to achieving environmental goals? A preliminary assessment of 'green fintech'and 'sustainable digital finance.' *European Company and Financial Law Review., 19*, 128–174.
14. Liang, H., Renneboog, L. (2020). Corporate social responsibility and sustainable finance: A review of the literature. *European Corporate Governance Institute–Finance Working Paper*.
15. Alamgir, M., & Cheng, M.-C. (2023). Do green bonds play a role in achieving sustainability? *Sustainability, 15*, 10177.
16. Jonsdottir, G. E., Sigurjonsson, T. O., Alavi, A. R., & Mitchell, J. (2021). Applying responsible ownership to advance SDGs and the ESG framework, resulting in the issuance of green bonds. *Sustainability, 13*, 7331.
17. Rasoulinezhad, E., & Taghizadeh-Hesary, F. (2022). Role of green finance in improving energy efficiency and renewable energy development. *Energy Efficiency, 15*, 14.
18. Ning, Y., Cherian, J., Sial, M. S., Álvarez-Otero, S., Comite, U., & Zia-Ud-Din, M. (2023). Green bond as a new determinant of sustainable green financing, energy efficiency investment, and economic growth: A global perspective. *Environmental Science and Pollution Research., 30*, 61324–61339.
19. Kumar, S. (2022). A quest for sustainium (sustainability Premium): Review of sustainable bonds. *Academy of Accounting and Financial Studies Journal., 26*, 1–18.

20. Agu, J. C., Nkwo, F. N., & Eneiga, R. U. (2024). Governance and anti-corruption measures in Nigeria: Strategies for enhancing transparency, accountability and public trust. *International Journal of Economics and Public Policy., 8*, 1–15.
21. Michalski, D., & Hawranek, P. (2023). Financial instruments of green transition. *internetowy Kwartalnik Antymonopolowy i Regulacyjny, 12*, 8–28.
22. Ahmed, R., & Charafeddin, M. (2021). The reallocation of capital towards green investments: A study on the EU taxonomy regulations.
23. Chen, J. M., Umair, M., & Hu, J. (2024). Green finance and renewable energy growth in developing nations: A GMM analysis. *Heliyon, 10.*
24. Damodaran, A., & van den Heuvel, O. (2023). India's low carbon value chain, green debt, and global climate finance architecture. *IIMB Management Review., 35*, 97–107.
25. Maltais, A., & Nykvist, B. (2020). Understanding the role of green bonds in advancing sustainability. *Journal of sustainable finance & investment*, 1–20.
26. Wang, T., Liu, X., & Wang, H. (2022). Green bonds, financing constraints, and green innovation. *Journal of Cleaner Production, 381*, 135134.
27. Gibon, T., Popescu, I. -Ş, Hitaj, C., Petucco, C., & Benetto, E. (2020). Shades of green: Life cycle assessment of renewable energy projects financed through green bonds. *Environmental Research Letters, 15*, 104045.
28. Zhao, L., Chau, K. Y., Tran, T. K., Sadiq, M., Xuyen, N. T. M., & Phan, T. T. H. (2022). Enhancing green economic recovery through green bonds financing and energy efficiency investments. *Economic Analysis and Policy, 76*, 488–501.
29. Sartzetakis, E. S. (2021). Green bonds as an instrument to finance low carbon transition. *Economic Change and Restructuring, 54*, 755–779.
30. Sangiorgi, I., & Schopohl, L. (2021). Why do institutional investors buy green bonds: Evidence from a survey of European asset managers. *International Review of Financial Analysis, 75*, 101738.
31. Fatica, S., & Panzica, R. (2021). Green bonds as a tool against climate change? *Business Strategy and the Environment, 30*, 2688–2701.
32. Gyura, G. (2020). Green bonds and green bond funds: The quest for the real impact. *The Journal of Alternative Investments, 23*, 71–79.
33. Piñeiro-Chousa, J., López-Cabarcos, M. Á., Caby, J., & Šević, A. (2021). The influence of investor sentiment on the green bond market. *Technological Forecasting and Social Change, 162*, 120351.
34. Russo, A., Mariani, M., & Caragnano, A. (2021). Exploring the determinants of green bond issuance: Going beyond the long-lasting debate on performance consequences. *Business Strategy and the Environment, 30*, 38–59.
35. Doronzo, R., Siracusa, V., & Antonelli, S. (2021). Green bonds: The sovereign issuers' perspective. *Bank of Italy Markets, Infrastructures, Payment Systems Working Paper*.
36. Tsalis, T. A., Malamateniou, K. E., Koulouriotis, D., & Nikolaou, I. E. (2020). New challenges for corporate sustainability reporting: United Nations' 2030 Agenda for sustainable development and the sustainable development goals. *Corporate Social Responsibility and Environmental Management, 27*, 1617–1629.
37. Yuan, M. (2021). Geographical information science for the United Nations' 2030 agenda for sustainable development.
38. Park, J. (2022). How can we pay for it all? Understanding the global challenge of financing climate change and sustainable development solutions. *Journal of Environmental Studies and Sciences, 12*, 91–99.
39. Kulkarni, S., Hof, A., Ambrósio, G., Edelenbosch, O., Köberle, A. C., van Rijn, J., & van Vuuren, D. (2022). Investment needs to achieve SDGs: An overview. *PLOS Sustainability and Transformation, 1*, e0000020.
40. Van Niekerk, A. J. (2024). Economic inclusion: Green finance and the SDGs. *Sustainability, 16*, 1128.
41. Zhan, J. X., & Santos-Paulino, A. U. (2021). Investing in the Sustainable Development Goals: Mobilization, channeling, and impact. *Journal of International Business Policy, 4*, 166.

42. Mukarram, M. (2020). Impact of COVID-19 on the UN sustainable development goals (SDGs). *Strategic Analysis, 44*, 253–258.
43. Fulzele, R., Fulzele, V., & Dharwal, M. (2022). Mapping the impact of COVID-19 crisis on the progress of sustainable development Goals (SDGs)-a focus on global environment and energy efficiencies. *Materials Today: Proceedings, 60*, 873–879.
44. Arora, R. U., & Sarker, T. (2022). Financing for sustainable development goals (SDGs) in the era of COVID-19 and beyond. *European Journal of Development Research, 35*, 1.
45. Bernhard, O.I., & Owen Affor, M. (2022). Health care expenditure-GDP nexus in Sub-Saharan Africa. *Journal of Positive School Psychology*, 1649–1664.
46. Odhiambo, N. M. (2021). Health expenditure and economic growth in sub-Saharan Africa: An empirical investigation. *Development Studies Research, 8*, 73–81.
47. Ahmed, R., Yusuf, F., & Ishaque, M. (2024). Green bonds as a bridge to the UN sustainable development goals on environment: A climate change empirical investigation. *International Journal of Finance & Economics, 29*, 2428–2451.
48. Lukšić, I., Bošković, B., Novikova, A., & Vrbensky, R. (2022). Innovative financing of the sustainable development goals in the countries of the Western Balkans. *Energy, Sustainability and Society, 12*, 15.
49. Tuhkanen, H., & Vulturius, G. (2022). Are green bonds funding the transition? Investigating the link between companies' climate targets and green debt financing. *Journal of Sustainable Finance & Investment, 12*, 1194–1216.
50. Khamis, M., & Aassouli, D. (2023). The eligibility of green bonds as safe haven assets: A systematic review. *Sustainability, 15*, 6841.
51. Abhilash, Shenoy, S. S., Shetty, D. K., Lobo, L. S., N, S. K. (2023). Green bond as an innovative financial instrument in the Indian financial market: Insights from systematic literature review approach. *Sage Open 13*, 21582440231178784.
52. Nayak, P., Taher, Z., & Maria, A. V. (2024). G20 and sustainable finance: A case for GSS+ bonds. In *NDIEAS-2024 International Symposium on New Dimensions and Ideas in Environmental Anthropology-2024 (NDIEAS 2024)* (pp. 182–196). Atlantis Press.
53. Lee, J. W. (2020). Green finance and sustainable development goals: The case of China. *Journal of Asian Finance Economics and Business, 7*, 577–586.
54. Jain, G. (2021). Green bonds: A path to greener pastures for emerging countries. *Emerging Market Multinationals Report*, 98–108.
55. Zhang, X., Song, Y., & Zhang, M. (2023). Exploring the relationship of green investment and green innovation: Evidence from Chinese corporate performance. *Journal of Cleaner Production, 412*, 137444.
56. Monaco, S. (2024). SDG 17. Strengthen the means of implementation and revitalize the global partnership for sustainable development. In *Identity, territories, and sustainability: challenges and opportunities for achieving the UN sustainable development goals* (pp. 167–175). Emerald Publishing Limited.
57. Muhammad, A., Abasido, A. U., Idris, M. B., Aliu, J. N., Adesugba, A. K., & Adedokun, T. (2022). Green bonds and the financial revolution: Facilitating the shift towards a sustainable future. *Journal of Environmental and Energy Economics, 1*, 1–10.
58. Restrepo-Ochoa, D. C., Restrepo-Castro, L. F., Lozada, J. M., Aguilera, C. A., Franco, J. F., Pinela, S., Costa, L. (2020). The potential of the green bond markets in Latin America and the Caribbean. EU-LAC Foundation.
59. Koundouri, P., Anquetil-Deck, C., Becchetti, L., Berthet, E., Borghesi, S., Cavalli, L., Chioatto, E., Cruickshank, E., Devves, S., Dibattista, I. (2023). Transforming our world: interdisciplinary insights on the sustainable development goals. In *Transforming our world: interdisciplinary insights on the sustainable development goals* (pp. 1–125). UN Sustainable Development Solutions Network.
60. Sreenu, N. (2024). Examining the dynamics of risk associated with green investment in India: A study on fintech and green bonds for clean energy production. *Journal of Economic Studies*.
61. Ghaemi Asl, M., Smutka, L., Nasr Isfahani, M., Raza, S. A., Vasa, L. (2024). Can multifarious types of green bonds be accused of greenwashing with a durative analysis? Insights from

a permanent causality vs. temporary causality phenomenon. *Environment, Development and Sustainability*, 1–25.
62. Mertzanis, C. (2023). Energy policy diversity and green bond issuance around the world. *Energy Economics, 128*, 107116.
63. Prakash, N., & Sethi, M. (2022). A review of innovative bond instruments for sustainable development in Asia. *International Journal of Innovation Science, 14*, 630–647.
64. Reichelt, H., Allen, D. P., & Cantor, S. M. (2023). Mainstreaming climate action in public and private investments: Mobilizing finance towards sustainable investments through the bond markets. In *Handbook of business and climate change* (pp. 430–457). Edward Elgar Publishing.
65. Rizzello, A., & Kabli, A. (2020). Sustainable financial partnerships for the SDGs: The case of social impact bonds. *Sustainability, 12*, 5362.
66. Chygryn, O., Pimonenko, T., & Lyulyov, O. (2022). Green investment as an economic instrument to achieve SDGs. In *Reducing inequalities towards sustainable development goals* (pp. 69–90). River Publishers.
67. Nasir, N., & Ahmed, W. (2024). Green finance initiatives and their potential to drive sustainable development. In *Climate change and finance: navigating the challenges and opportunities in capital markets* (pp. 3–29). Springer.

The Rise of Social, Green, and Sustainable E-Financial Instruments

Neha Kamboj and Vinita Choudhary

Abstract This study investigates the evolution, characteristics, and impact of social, green, and sustainable e-financial instruments in addressing environmental and social challenges. The primary objectives are to assess the effectiveness of these instruments in promoting sustainability and corporate social responsibility (CSR) and to identify the barriers to their widespread adoption. Utilizing a mixed-methods approach over a six-month period, the research employs both qualitative and quantitative analyses, including case studies and regulatory reviews. Key findings reveal that digital platforms significantly improve the accessibility and transparency of these financial instruments, facilitating broader stakeholder engagement. However, the study also identifies persistent challenges, including market volatility, greenwashing practices, and regulatory inconsistencies that undermine trust. Based on these findings, the research advocates for the development of stronger governance mechanisms, clear regulatory frameworks, and enhanced digital infrastructure to bolster the effectiveness of e-financial instruments in achieving sustainable development goals. Ultimately, this study contributes to the literature by providing insights into the dynamic interplay between technology, finance, and sustainability, and offers practical recommendations for policymakers and practitioners.

Keywords Social · Green · Sustainable · E-financial · Instruments · Artificial intelligence · Block chain

---

N. Kamboj
School of Management, IILM University, Gurugram, Haryana, India
e-mail: nehakamboj2000@gmail.com

V. Choudhary (✉)
School of Leadership and Management, Manav Rachna International Institute of Research and Studies, Faridabad, Haryana, India
e-mail: postvinita@gmail.com

K. Khanna et al. (eds.), *From Poverty to Prosperity*, Information Systems Engineering and Management 64, https://doi.org/10.1007/978-3-031-99166-0_16

1 Introduction

1.1 Overview of E-Financial Instruments

E-financial instruments refer to financial products and services delivered and managed through digital platforms. They represent a significant evolution within the finance sector, leveraging technology to enhance the way individuals and organizations manage, invest, and transact money. These instruments include digital payment systems, online trading platforms, cryptocurrencies, and blockchain-enabled financial tools.

The digital nature of e-financial instruments stems from their reliance on technology, enabling instant transactions, real-time updates, and global accessibility [1]. This innovation has been transformative, especially in an era marked by the proliferation of smartphones, high-speed internet, and advanced financial technologies [2]. Historically, traditional financial systems relied heavily on paper-based processes, brick-and-mortar institutions, and intermediaries. However, the rise of e-financial instruments has streamlined these operations, making financial transactions more transparent, efficient, and accessible to a broader audience [3].

The transition from traditional to digital financial platforms has brought significant benefits. Accessibility has increased as individuals in remote and underserved areas now have the tools to participate in the financial ecosystem. Transparency has improved, with digital platforms offering detailed, real-time tracking of transactions and data analytics. Efficiency has been maximized, reducing the time and costs associated with financial operations. This shift has set the stage for integrating broader financial innovations such as green, social, and sustainable e-financial instruments into the mainstream [4].

Objectives of the Study

1. To explore the evolution of E-financial instruments.
2. To analyze the types of sustainable E-financial instruments.
3. To evaluate the impact and benefits of sustainable E-financial instruments.
4. To identify challenges in adopting E-financial instruments.
5. To envision the future of sustainable finance.

2 Understanding Social Finance, Green Finance, and Sustainable Finance

Social finance, green finance, and sustainable finance are novel ways to incorporating ESG (environmental, social, and governance) variables into the financial system. These methods seek to achieve two goals: creating financial rewards and tackling key global issues such as poverty, climate change, and inequality.

1. **Green Finance** focuses on environmental objectives, such as financing renewable energy projects, reducing carbon emissions, and promoting energy efficiency. Green bonds, a prominent example, channel investments into projects with measurable environmental benefits, directly aligning with climate action goals outlined in the sustainable development goals (SDGs) [5].
2. **Social Finance** targets projects that yield positive social outcomes. This includes funding affordable housing, education initiatives, and healthcare improvements, especially in underprivileged regions. Social bonds, for instance, enable organizations to raise capital for projects that address societal challenges and foster inclusive growth [6].
3. **Sustainable Finance** combines green and social finance under a broader umbrella, emphasizing continuing value creation through the integration of ESG principles into investment decisions and financial strategies. These instruments reflect corporate social responsibility (CSR) efforts, where businesses align their objectives with broader societal and environmental goals [7]. The rise of e-financial instruments has amplified the reach and impact of these financial mechanisms. Digital platforms facilitate the issuance, monitoring, and trading of green and social bonds, making them available to a diverse variety of investors, from individuals to institutions. Moreover, the data-driven nature of digital finance enhances accountability by enabling transparent reporting on the impact of these instruments.

3 Importance of Sustainable Finance in the Modern World

In today's interconnected and rapidly changing world, sustainable finance has become a cornerstone of addressing global challenges. Climate change, one of the most pressing issues of our time, requires urgent and substantial investment in renewable energy, carbon reduction projects, and climate-resilient infrastructure. Sustainable finance plays a critical role in mobilizing these funds, bridging the gap between environmental needs and financial resources [8].

Social inequality is another global concern that sustainable finance seeks to address. Millions of people worldwide lack access to basic needs such as clean water, quality education, and healthcare. Social finance mechanisms enable targeted investments in projects that uplift communities, create jobs, and reduce disparities [9].

Global financial inclusion is yet another area where sustainable finance is making a difference. By leveraging digital platforms, e-financial instruments extend financial services to unbanked and underbanked populations. This ensures that marginalized groups have access to credit, savings, and insurance, empowering them economically and socially [10].

Several key drivers underscore the growing importance of sustainable finance:

1. **Investor Demand**: Modern investors increasingly prioritize ESG-focused investments, recognizing the potential for long-term returns and positive societal

impact. This shift has led to a surge in demand for green bonds, sustainable mutual funds, and similar products [11].
2. **Regulatory Support**: Regulatory support: The government and regulatory bodies throughout the world are enacting measures to promote sustainable finance. Strategies such as tax breaks for green investments and obligatory ESG reporting criteria are accelerating the widespread implementation of sustainable financial practices [12].
3. **Public Awareness**: The rising awareness among consumers about environmental and social issues is influencing financial markets. People are demanding transparency and accountability from corporations, pressuring them to adopt sustainable practices [12].

By aligning economic growth with environmental and social sustainability, sustainable finance offers a pathway to a more resilient and equitable future. The integration of e-financial instruments into this framework further accelerates progress, ensuring that the benefits of sustainable finance are far-reaching and inclusive.

4 Types of Sustainable E-Financial Instruments

4.1 Green Bonds

Green bonds are debt instruments issued to raise funds specifically for projects with environmental benefits. These bonds have gained global traction as a means to finance initiatives such as renewable energy generation, clean transportation, energy efficiency, and pollution control [13].

4.1.1 Purpose and Impact

The core purpose of green bonds is to channel investments into projects that mitigate climate change and promote environmental sustainability. Governments, corporations, and multilateral institutions issue green bonds to fund projects like solar and wind farms, green buildings, and water conservation systems. For example, a green bond might finance the installation of solar panels in underserved regions, reducing dependence on fossil fuels while providing energy access.

4.1.2 Role of Digital Platforms

The issuance and trading of green bonds have been revolutionized by digital platforms. E-platforms enable issuers to streamline the bond creation process, provide real-time updates, and enhance transparency for investors. Digital tools also facilitate the efficient distribution of green bonds to a global investor base, making it easier for individuals and institutions to contribute to environmentally beneficial projects.

Moreover, blockchain technology plays a critical role in verifying the use of proceeds from green bonds. By providing an immutable ledger of transactions, blockchain ensures that funds raised through green bonds are allocated exclusively to approved environmental projects, thereby fostering trust among investors.

4.2 Social Bonds

Social bonds are financial instruments issued to fund projects with positive social outcomes, addressing challenges such as affordable housing, healthcare, education, and job creation. These bonds are aligned with broader societal goals, aiming to foster inclusivity and reduce socio-economic inequalities [6].

4.2.1 Purpose and Impact

Social bonds have been pivotal in tackling pressing societal issues, particularly in developing and underdeveloped regions. For example, a social bond may finance the construction of affordable housing for low-income families or fund healthcare services in rural areas. Social bonds also support educational initiatives by providing scholarships or building infrastructure for schools and colleges.

The impact of these bonds extends beyond financial returns, creating measurable social value. Social bonds help to accomplish the sustainable development goals (SDGs) by assigning funds to initiatives that benefit underprivileged populations, such as quality education, reduced inequalities, and good health and well-being.

4.2.2 Digital Platforms in Social Bonds

Digital platforms have democratized access to social bonds, enabling individuals and small investors to participate alongside institutional investors. Online bond marketplaces and crowdfunding platforms facilitate the issuance and trading of social bonds, expanding their reach to a diverse pool of investors.

Additionally, e-platforms enhance accountability by enabling issuers to report on the social impact of funded projects. Metrics such as the number of people housed, students educated, or healthcare facilities upgraded can be tracked and shared with investors in real time, ensuring transparency and reinforcing investor confidence.

4.3 Sustainable Bonds and Loans

Sustainable bonds and loans are hybrid financial instruments designed to address both environmental and social objectives. These tools represent a convergence of green and social finance, catering to projects that deliver comprehensive sustainability outcomes.

4.3.1 Purpose and Scope

Sustainable bonds and loans fund initiatives that generate dual benefits—environmental protection and social development. For instance, a sustainable bond might finance the development of a green affordable housing project, where energy-efficient homes are built for low-income families. Similarly, sustainable loans might be extended to businesses adopting sustainable practices while creating employment opportunities in underserved areas [13].

These instruments appeal to investors and organizations seeking to align their financial goals with comprehensive ESG (environmental, social, and governance) principles. By combining green and social objectives, they offer a holistic approach to sustainable development.

4.3.2 Digital Technology in Sustainable Bonds and Loans

Digital platforms play a crucial role in the efficient management of sustainable bonds and loans. They streamline the process of issuance, tracking, and reporting, reducing the administrative burden on issuers. For example, blockchain technology can be used to track the allocation of funds to ensure they are used for their intended purposes [14].

Furthermore, e-platforms enhance the monitoring and evaluation of projects funded by sustainable bonds and loans. Issuers can use data analytics to measure the environmental and social impact of their projects, providing detailed reports to investors and stakeholders. This level of transparency fosters trust and encourages greater participation in sustainable finance.

4.4 Carbon Credits and Offsetting Mechanisms

Carbon credits are tradable permits that represent the right to emit a specific amount of carbon dioxide or other greenhouse gases. These credits are part of carbon offsetting mechanisms, allowing companies to neutralize their carbon footprint by investing in projects that reduce or capture emissions elsewhere (Kaisto, 2024).

4.4.1 Digitalization of Carbon Credits

The digitalization of carbon credits has revolutionized their issuance, trading, and tracking. Blockchain technology, in particular, has enhanced the transparency and integrity of carbon credit transactions [15]. By recording each credit on an immutable ledger, blockchain ensures that credits are not double-counted or misused, providing a robust system for carbon accounting.

Online platforms have also facilitated the trading of carbon credits, making it easier for companies to purchase offsets. These platforms connect buyers with sellers, offering a marketplace for verified carbon credits from projects such as reforestation, renewable energy installations, and methane capture.

4.4.2 Role in Offsetting Mechanisms

Carbon offsetting mechanisms allow companies to compensate for their emissions by supporting initiatives that reduce greenhouse gases elsewhere. For example, a company might purchase carbon credits from a wind farm project to offset emissions from its manufacturing processes [15]. These mechanisms are integral to achieving net-zero goals, enabling organizations to balance their environmental impact.

The integration of digital technology into carbon offsetting mechanisms has made them more accessible and transparent. Companies can now calculate their carbon footprint, purchase credits, and monitor the impact of offset projects through user-friendly online tools. This not only simplifies the process but also enhances accountability, encouraging more organizations to adopt carbon offsetting as part of their sustainability strategies.

5 Impact and Benefits of Sustainable E-Financial Instruments

Sustainable e-financial instruments are transformative tools in modern finance, designed to align economic activities with the pressing demands of environmental preservation and social well-being. By combining the power of financial innovation with digital technology, these instruments have revolutionized how capital is raised, allocated, and monitored for sustainable projects. This section explores the multifaceted impact and benefits of these instruments across global economies, financial markets, and societal structures [13].

5.1 Environmental Impact

Addressing important environmental issues is one of the most significant advantages of sustainable e-financial instruments. Funding initiatives that lower carbon emissions, preserve natural resources, and encourage the use of clean energy is greatly aided by instruments like carbon credits and green bonds.

1. **Mitigation of Climate Change**

Renewable energy projects like wind farms, solar power plants, and hydroelectric installations are frequently financed with green bonds. By redirecting financial resources toward low-carbon initiatives, these bonds contribute to reducing greenhouse gas emissions, a critical goal in the fight against global warming [16].

2. **Conservation of Natural Resources**

Sustainable financing tools encourage the adoption of eco-friendly practices, such as water conservation, afforestation, and sustainable agriculture. For instance, companies raising capital through sustainable loans often commit to resource-efficient operations, leading to long-term environmental benefits [17].

3. **Support for Circular Economies**

Sustainable financial instruments are integral to fostering circular economies, where resources are reused and recycled rather than wasted. Projects funded through green bonds frequently involve waste management, recycling programs, and the development of sustainable supply chains [18].

4. **Enhancement of Transparency in Environmental Funding**

Digital platforms linked to sustainable instruments ensure that funds are allocated to verified projects, fostering accountability and trust among stakeholders. Blockchain technology, for example, provides an immutable record of transactions, confirming that resources are used as intended [19].

5.2 Social Impact

Sustainable e-financial instruments address significant societal challenges by funding projects that improve living standards, promote inclusivity, and reduce inequalities. Social bonds, in particular, have emerged as powerful tools for financing initiatives with measurable social outcomes.

1. **Reduction of Social Inequalities**

Social bonds are used to fund affordable housing, healthcare access, and educational programs. These initiatives directly uplift marginalized communities, helping bridge the socio-economic divide. For instance, social bonds may finance the construction of schools in rural areas, offering children access to quality education [20].

2. **Enhancement of Public Health**

Health-focused projects, such as vaccination programs, hospital infrastructure development, and access to clean water, are often financed through social bonds. These projects improve community health outcomes, particularly in underprivileged regions [21].

3. **Job Creation and Economic Empowerment**

Sustainable e-financial instruments foster employment opportunities by supporting businesses that adhere to sustainable practices. For example, loans tied to sustainable outcomes often require companies to adopt ethical labor practices, ensuring fair wages and safe working conditions [22].

4. **Promotion of Social Equity**

These instruments also encourage financial inclusion by making investments accessible to a broader audience. Digital platforms simplify the process of purchasing bonds or participating in crowdfunding campaigns, empowering individuals to contribute to societal progress.

5.3 Economic Benefits

Beyond their environmental and social contributions, sustainable e-financial instruments drive economic growth and stability by fostering innovation, attracting new investors, and supporting resilient financial systems.

1. **Attracting Investment Flows**

Investors are increasingly drawn to sustainable finance due to growing awareness of environmental, social, and governance (ESG) factors. Sustainable instruments offer a compelling opportunity to achieve competitive returns while supporting meaningful causes [11].

2. **Enhancing Financial Stability**

By promoting long-term investments in sustainable sectors, these instruments reduce exposure to risks associated with unsustainable practices. Projects that prioritize resource efficiency and social equity are less likely to face regulatory penalties, reputational damage, or market disruptions.

3. **Fostering Innovation**

The rise of sustainable e-financial instruments has spurred innovation in the financial sector. Digital platforms, blockchain technology, and data analytics tools have been developed to enhance the issuance, trading, and tracking of these instruments, paving the way for new financial models [11].

4. **Boosting Economic Resilience**

Sustainable investments strengthen economies by diversifying revenue streams and reducing dependency on volatile industries, such as fossil fuels. Green energy projects, for example, create stable, long-term employment opportunities while reducing environmental risks [11].

5.4 Technological Advancements

The integration of digital technology into sustainable finance has amplified the impact of these instruments by enhancing efficiency, transparency, and accessibility.

1. **Streamlined Issuance and Trading**

Digital platforms have revolutionized the issuance and trading of sustainable instruments, making the process faster, more cost-effective, and accessible to a global audience. Online marketplaces allow investors to participate in green and social projects with ease.

2. **Enhanced Transparency and Accountability**

Blockchain technology ensures that funds raised through sustainable instruments are allocated as promised. For example, the use of blockchain in carbon credit transactions provides a tamper-proof record of each credit, eliminating the risk of double-counting.

3. **Data-Driven Impact Measurement**

Digital tools enable issuers to track the environmental and social impact of funded projects in real time. Metrics such as carbon emissions avoided, jobs created, or people served are shared with investors, fostering confidence and encouraging further investments.

4. **Democratization of Finance**

E-financial instruments lower barriers to entry, allowing individuals and small-scale investors to participate in sustainability initiatives. Crowdfunding platforms, for instance, enable citizens to invest in local renewable energy projects or community development programs.

5.5 Regulatory and Policy Benefits

Sustainable e-financial instruments also align with global policy objectives and regulatory frameworks, making them an essential component of modern governance.

1. **Alignment with Sustainable Development Goals (SDGs)**

Many sustainable instruments are explicitly designed to address the United Nations' SDGs, such as climate action, clean energy, and reduced inequalities. These instruments help governments and organizations meet their sustainability commitments [17].

2. **Compliance with Environmental Regulations**

Governments worldwide are introducing stricter regulations to curb environmental degradation. Organizations that adopt sustainable e-financial instruments are better equipped to comply with these regulations, avoiding penalties and enhancing their reputations.

3. **Incentives for Sustainable Practices**

Policymakers often provide tax benefits, subsidies, or grants to issuers of sustainable instruments. For instance, green bond issuers may receive tax breaks to encourage investments in renewable energy or conservation projects.

4. **Support for Global Climate Agreements**

Instruments like carbon credits contribute to international efforts to combat climate change, such as the Paris Agreement. By enabling companies to offset their emissions, these instruments support national and global carbon reduction targets.

5.6 Broader Benefits

The benefits of sustainable e-financial instruments extend beyond specific sectors, influencing the global economy, society, and the environment.

1. **Changing Investor Behavior**

Sustainable instruments are reshaping investor priorities, encouraging a shift from short-term gains to long-term value creation. This paradigm shift is critical for building a more equitable and sustainable financial system.

2. **Encouraging Corporate Responsibility**

Companies that issue or invest in sustainable instruments often adopt higher standards of corporate social responsibility (CSR). This not only improves their public image but also contributes to lasting positive change.

3. **Global Collaboration and Partnerships**

Sustainable finance fosters collaboration among governments, businesses, and nonprofits. For example, international green bond initiatives bring together diverse stakeholders to address climate change on a global scale.

4. **Empowering Communities**

By funding projects that directly benefit local communities, sustainable instruments empower individuals to take charge of their development. This localized approach ensures that financial resources create tangible, meaningful outcomes.

6 Challenges and Risks in Adopting E-Financial Instruments

The adoption of e-financial instruments has transformed the financial landscape, offering unprecedented accessibility, transparency, and efficiency. However, despite their numerous benefits, these instruments also come with significant challenges and risks. As the financial sector evolves, addressing these hurdles is critical to ensuring their effective and sustainable use. This section explores the key challenges and risks associated with e-financial instruments.

6.1 Technological Challenges

a. **Cybersecurity Risks**

One of the most pressing issues in adopting e-financial instruments is cybersecurity. With the digitization of financial systems, these platforms become vulnerable to hacking, data breaches, and malware attacks. Unauthorized access to sensitive financial data can lead to monetary losses and undermine trust in the system.

b. **System Reliability and Downtime**

E-financial instruments rely on robust digital infrastructure to function efficiently. Downtime caused by system crashes, technical glitches, or server failures can disrupt transactions, delaying critical financial operations. Additionally, reliance on complex technologies such as blockchain may introduce operational inefficiencies if systems are not properly maintained.

c. **Digital Divide**

The adoption of e-financial instruments is heavily dependent on access to digital technology. In developing regions with limited internet connectivity and low digital literacy, the benefits of e-financial instruments are not evenly distributed, exacerbating existing economic inequalities.

6.2 *Regulatory and Compliance Issues*

a. **Inconsistent Global Standards**

The regulatory framework governing e-financial instruments varies across regions. This inconsistency creates challenges for global investors and issuers, particularly in cross-border transactions. Disparities in regulations regarding reporting, transparency, and accountability complicate compliance efforts.

b. **Lack of Comprehensive Policies**

The rapid development of e-financial instruments has outpaced regulatory oversight in many jurisdictions. Insufficient or unclear policies around digital finance can lead to ambiguities, exposing investors and issuers to legal risks.

c. **AML and KYC Challenges**

Anti-money laundering (AML) and know your customer (KYC) regulations are critical in financial markets. However, e-financial instruments, particularly those based on decentralized technologies like blockchain, pose challenges in verifying identities and preventing fraudulent activities.

6.3 *Market and Operational Risks*

a. **Market Volatility**

E-financial instruments, especially those linked to environmental or social goals, can be affected by market fluctuations. For instance, the value of carbon credits may fluctuate due to changes in regulations or market demand, creating uncertainty for investors.

b. **Liquidity Risks**

Many e-financial instruments, such as green bonds and social bonds, operate in niche markets, where liquidity can be limited. Investors may face challenges in selling these instruments quickly without significantly impacting their value.

c. **Operational Complexities**

The issuance and monitoring of e-financial instruments often require specialized knowledge and technology. Companies may struggle to meet the operational requirements, such as impact reporting and compliance with sustainability criteria, leading to inefficiencies and delays.

6.4 Trust and Perception Issues

7 A. Greenwashing Risks

a. **Greenwashing Risks**

A major risk associated with sustainable e-financial instruments is the potential for greenwashing, where issuers falsely claim to align their activities with environmental or social goals. This can undermine investor confidence and discredit the market for such instruments.

b. **Skepticism Toward Technology**

Investors and stakeholders may be skeptical about the reliability of technologies like blockchain or AI-powered platforms, particularly if they are unfamiliar with these systems. This hesitancy can slow the adoption of e-financial instruments.

c. **Transparency Concerns**

Although e-financial instruments are designed to enhance transparency, their complexity may make it difficult for investors to fully understand the underlying mechanisms and risks. This lack of clarity can create distrust and limit participation.

7.1 Environmental and Social Risks

a. **Misalignment with Goals**

E-financial instruments are often tied to specific environmental or social objectives. However, there is a risk that funded projects may fail to achieve the desired outcomes, either due to poor execution or external factors. For instance, a renewable energy project financed through green bonds might not deliver the expected reduction in carbon emissions.

b. **Conflict of Interests**

Balancing financial returns with sustainability goals can be challenging. Issuers may prioritize profits over social or environmental impact, leading to compromises in the quality or integrity of funded projects.

c. **Over-Reliance on Offsetting**

Instruments like carbon credits enable companies to offset their emissions instead of directly reducing them. This over-reliance on offsetting mechanisms can delay the adoption of meaningful sustainability practices, undermining long-term goals.

7.2 *Ethical and Societal Challenges*

a. **Exclusion of Marginalized Communities**

While e-financial instruments aim to promote inclusivity, they may unintentionally exclude vulnerable groups who lack access to digital technology or financial literacy. This could perpetuate existing inequalities rather than address them.

b. **Job Displacement**

The automation and digitization of financial services associated with e-financial instruments may lead to job losses in traditional finance roles. This could create societal backlash and resistance to their adoption.

c. **Ethical Dilemmas in Project Selection**

Selecting projects to fund through sustainable instruments often involves subjective judgments. Issuers may face ethical dilemmas in prioritizing certain projects over others, potentially leading to conflicts with stakeholders.

7.3 *Financial and Investment Risks*

a. **High Initial Costs**

Issuing e-financial instruments often involves significant upfront costs for technological infrastructure, compliance measures, and impact assessments. These expenses may deter small or medium-sized enterprises from participating in the market.

b. **Uncertain Returns**

Investments in sustainable projects may not yield immediate financial returns, which could discourage investors looking for quick profits. Long payback periods or unforeseen project delays add to the risk.

c. **Risk of Default**

Instruments like green and social bonds rely on the issuer's ability to generate sufficient returns from the funded projects. Economic downturns or project failures can increase the risk of default, impacting investor confidence.

7.4 *Technological Dependence and Obsolescence*

a. **Rapid Technological Changes**

The fast pace of technological advancements poses a risk of obsolescence for platforms used in e-financial instruments. Companies must continuously update their systems to remain competitive, which can be costly and time-consuming.

b. **Interoperability Issues**

E-financial instruments often involve multiple technologies, such as blockchain, IoT, and AI. Ensuring that these systems work seamlessly together can be a challenge, leading to inefficiencies or compatibility issues.

c. **Dependency on Third-Party Platforms**

Reliance on third-party digital platforms for trading or monitoring e-financial instruments exposes issuers and investors to risks related to service outages, data breaches, or operational failures by the platform providers.

8 Future Outlook of Social, Green, and Sustainable E-Financial Instruments

As global awareness around sustainability continues to grow, the role of social, green, and sustainable e-financial instruments is poised to expand significantly. These instruments are not only reshaping the financial landscape but are also aligning with global efforts to address environmental and social challenges. The future holds immense potential for these instruments as advancements in technology, policy, and market practices converge to promote sustainable finance. Below is a detailed exploration of the expected trends and developments in this sector.

8.1 *Increasing Market Adoption and Integration*

a. **Mainstreaming of Sustainable Instruments**

Social, green, and sustainable e-financial instruments are expected to transition from niche markets to mainstream financial practices. Institutional investors, including pension funds, mutual funds, and sovereign wealth funds, are likely to increase their allocations to these instruments as part of their environmental, social, and governance (ESG) strategies.

b. **Integration Across Industries**

While traditionally focused on renewable energy, affordable housing, and healthcare, sustainable financial instruments are likely to extend into other sectors, such as agriculture, manufacturing, and technology. This diversification will enable broader adoption and greater impact across the global economy.

8.2 *Technological Innovations Driving Growth*

a. **Blockchain and Smart Contracts**

Blockchain technology is set to play a pivotal role in enhancing transparency, traceability, and efficiency in sustainable finance. Smart contracts will automate the monitoring and reporting of sustainability metrics, ensuring that funds are used as intended.

b. **Artificial Intelligence and Big Data Analytics**

AI-powered tools and data analytics will provide better insights into the environmental and social impact of projects funded through e-financial instruments. This will enable investors and regulators to make more informed decisions and improve accountability.

c. **Digital Platforms for Wider Participation**

The rise of digital platforms will democratize access to social, green, and sustainable financial instruments. Retail investors, who were previously excluded from these markets, will gain access through user-friendly apps and platforms, increasing participation and liquidity.

8.3 *Strengthening of Regulatory Frameworks*

a. **Global Standardization**

As these instruments gain prominence, there will likely be increased efforts to harmonize regulations and standards across countries. International organizations, such as the United Nations and the International Finance Corporation, are expected to play a central role in developing globally accepted frameworks for sustainable finance.

b. **Mandatory Reporting and Disclosures**

Governments and regulatory bodies may introduce stricter requirements for impact reporting and disclosures. These regulations will ensure greater transparency and reduce the risk of greenwashing, fostering trust in the market.

8.4 *Alignment with Global Goals*

a. **Support for Sustainable Development Goals (SDGs)**

Social, green, and sustainable financial instruments will remain closely aligned with the United Nations' SDGs. Future financial products will likely target specific goals, such as climate action (Goal 13), reduced inequalities (Goal 10), and sustainable cities (Goal 11), creating a direct link between investments and measurable outcomes.

b. **Contribution to Net-Zero Goals**

As countries and corporations commit to achieving net-zero carbon emissions, the demand for sustainable financial instruments will surge. Green bonds, carbon credits, and other tools will play a critical role in financing renewable energy projects and carbon reduction initiatives.

8.5 Challenges to Overcome

a. **Balancing Financial and Social Reforms**

A key challenge for the future will be ensuring that financial returns remain competitive while achieving social and environmental goals. Striking this balance will be crucial to attracting and retaining investors.

b. **Overcoming Greenwashing**

The risk of greenwashing—misrepresenting the sustainability of financial instruments—will need to be addressed through stricter monitoring, independent audits, and advanced technological tools for verification.

c. **Addressing Digital Inequality**

To fully leverage the potential of e-financial instruments, efforts must be made to bridge the digital divide. Ensuring equitable access to digital platforms and technologies will be vital for inclusive growth.

8.6 Emerging Trends and Opportunities

a. **Rise of Impact Investing**

Impact investing, which focuses on generating measurable social and environmental benefits alongside financial returns, will become a major driver of sustainable e-financial instruments. Investors will increasingly seek opportunities that align with their values and goals.

b. **Collaboration Between Public and Private Sectors**

Partnerships between governments, financial institutions, and private companies will create new opportunities for innovation in sustainable finance. Public–private collaborations can drive large-scale projects, such as smart cities and green infrastructure.

c. **Expansion into Developing Markets**

Developing economies, with their growing populations and urgent sustainability needs, will become significant markets for these instruments. Tailored solutions will be developed to address local challenges, such as energy access and poverty alleviation.

9 Results and Discussion

1. **Environmental Impact**: Sustainable e-financial instruments significantly contribute to environmental protection by funding renewable energy, conservation, and climate-resilient projects. Green bonds have successfully financed solar and wind energy projects, reducing carbon emissions. Digital verification methods, such as blockchain, have improved transparency in fund allocation, ensuring that investments meet environmental goals.
2. **Social Impact**: Social bonds have played a critical role in financing affordable housing, healthcare, and education initiatives. Case studies show that these bonds have enabled better access to essential services, particularly in developing economies. Digital platforms have democratized investment in social projects, allowing individuals to participate in funding impactful initiatives.
3. **Economic Benefits**: Sustainable e-financial instruments attract investors seeking long-term stability and ethical investments. Green and social bonds provide financial returns while addressing pressing global challenges. However, market volatility remains a concern, requiring better risk management strategies to ensure sustainable growth in the sector.
4. **Technological Advancements**: Digital platforms, blockchain, and artificial intelligence have enhanced the efficiency of sustainable finance. Blockchain technology ensures transparency in green and social bond transactions, reducing the risk of greenwashing. AI-powered tools help assess the environmental and social impact of funded projects, enabling better decision-making.
5. **Challenges and Recommendations**: Despite the growth of sustainable e-financial instruments, challenges such as regulatory inconsistencies, greenwashing, and limited investor awareness hinder their full potential. Standardized global regulations and enhanced monitoring mechanisms are needed to strengthen credibility. Governments and financial institutions must collaborate to create more inclusive policies that promote sustainable investments.

10 Conclusion

The study demonstrates that social, green, and sustainable e-financial instruments are pivotal in transforming the financial sector to align with global sustainability efforts. These instruments, facilitated by digital platforms and technologies like blockchain and AI, enhance transparency, accessibility, and efficiency. They contribute significantly to addressing pressing global challenges, such as climate change, social inequality, and financial exclusion. However, their potential is contingent upon overcoming challenges like greenwashing, regulatory inconsistencies, and digital inequality. The findings underscore the critical role of innovation, collaboration, and global regulatory alignment in fostering the growth and impact of these instruments.

11 Implications of the Study

1. **Practical Implications for Financial Institutions**

Financial institutions can leverage the insights to design and promote sustainable e-financial instruments that align with investor demand and regulatory frameworks.

2. **Policy Implications for Regulators**

Regulators can utilize the findings to establish standardized guidelines, ensuring transparency and reducing risks like greenwashing.

3. **Technological Implications for Developers**

Developers of digital platforms and blockchain-based solutions can address challenges like traceability and scalability, fostering trust and efficiency in sustainable finance.

4. **Academic Implications for Researchers**

The study offers a foundation for further exploration of the intersection between technology and sustainability in the financial domain.

12 Recommendations for Future Study

1. **Exploring Regional Adoption**

Future research can analyze the adoption and impact of sustainable e-financial instruments in different regions, particularly in developing economies.

2. **Focus on Technological Innovations**

Investigate the role of emerging technologies, such as quantum computing and advanced AI, in enhancing the functionality and impact of these instruments.

3. **Longitudinal Studies on Impact**

Conduct long-term studies to measure the sustained impact of green bonds, social bonds, and other instruments on environmental and social outcomes.

4. **Stakeholder Perception and Behavior**

Examine how different stakeholders, including investors, policymakers, and communities, perceive and engage with sustainable e-financial instruments.

5. **Sector-Specific Analysis**

Analyze the role of sustainable financial instruments in specific industries, such as agriculture, healthcare, and energy, to identify tailored solutions and opportunities.

By addressing these areas, future studies can contribute to a deeper understanding of the transformative potential of sustainable e-financial instruments and guide stakeholders in leveraging them effectively.

References

1. Indjikian, R. (2002). E-finance for SMEs: Global trends and national experiences. *Journals of Electronic Commerce, 1*, 1–38.
2. Zhou, W. (2024, February). The transformative impact of FinTech on financial services: A comprehensive analysis. In *2023 5th International conference on economic management and cultural industry (ICEMCI 2023)* (pp. 85–91). Atlantis Press.
3. Malhotra, R., & Malhotra, D. K. (2006). The impact of internet and e-commerce on the evolving business models in the financial services industry. *International Journal of Electronic Business, 4*(1), 56–82.
4. Wanof, M. I. (2023). Digital technology innovation in improving financial access for low-income communities. *Technology and Society Perspectives (TACIT), 1*(1), 26–34.
5. Li, C., & Umair, M. (2023). Does green finance development goals affects renewable energy in China. *Renewable Energy, 203*, 898–905.
6. Rizzello, A., & Kabli, A. (2020). Sustainable financial partnerships for the SDGs: The case of social impact bonds. *Sustainability, 12*(13), 5362.
7. Tao, H., Zhuang, S., Xue, R., Cao, W., Tian, J., & Shan, Y. (2022). Environmental finance: An interdisciplinary review. *Technological Forecasting and Social Change, 179*, Article 121639.
8. Clark, R., Reed, J., & Sunderland, T. (2018). Bridging funding gaps for climate and sustainable development: Pitfalls, progress and potential of private finance. *Land Use Policy, 71*, 335–346.
9. Almeshqab, F., & Ustun, T. S. (2019). Lessons learned from rural electrification initiatives in developing countries: Insights for technical, social, financial and public policy aspects. *Renewable and Sustainable Energy Reviews, 102*, 35–53.
10. Yadav, S., & Kapoor, S. (2024). Technological innovations and their impact on financial inclusion. In *E-financial strategies for advancing sustainable development: Fostering financial inclusion and alleviating poverty* (pp. 281–298). Springer Nature Switzerland.
11. Choudhary, V., & Trivedi, S. (2023). The role of internal audit in risk management with a perspective of sustainability development. In *The impact of climate change and sustainability standards on the insurance market* (pp. 287–315).
12. Singhania, M., & Saini, N. (2023). Institutional framework of ESG disclosures: Comparative analysis of developed and developing countries. *Journal of Sustainable Finance & Investment, 13*(1), 516–559.
13. Sharma, B., Meenakshi, Bala, R., & Awasthi, Y. (2024). E-financial strategies for sustainable development: empirical evidence from India. In *E-financial strategies for advancing sustainable development: Fostering financial inclusion and alleviating poverty* (pp. 299–315). Springer Nature Switzerland.

14. Choudhary, V., Kamboj, N., & Trivedi, S. (2023). Scoping review of literature on the application of blockchain in finance. In *Perspectives on blockchain technology and responsible investing* (pp. 1–28).
15. Kaisto, J. (2024). Emission allowances and carbon credits as property: A Finnish perspective. *European Property Law Journal, 12*(2–3), 148–210.
16. Pathak, H. (2023). Impact, adaptation, and mitigation of climate change in Indian agriculture. *Environmental Monitoring and Assessment, 195*(1), 52.
17. Edunjobi, T. E. (2024). Sustainable supply chain financing models: Integrating banking for enhanced sustainability. *International Journal for Multidisciplinary Research Updates 2024, 7*(02), 001–011.
18. Lei, Z., Cai, S., Cui, L., Wu, L., & Liu, Y. (2023). How do different Industry 4.0 technologies support certain circular economy practices? *Industrial Management & Data Systems, 123*(4), 1220–1251.
19. Honig, D., Lall, R., & Parks, B. C. (2023). When does transparency improve institutional performance? Evidence from 20,000 projects in 183 countries. *American Journal of Political Science, 67*(4), 1096–1116.
20. Vásquez-Vera, H., León-Gómez, B. B., Borrell, C., Jacques-Aviñó, C., López, M. J., Medina-Perucha, L., Pasarin, M., Sánchez-Ledesma, E., Pérez, K. (2023). Inequities in the distribution of COVID-19: An adaptation of WHO's conceptual framework. *Gaceta Sanitaria, 36*, 488–492.
21. Oman, D. (2023). Mindfulness for global public health: Critical analysis and agenda. *Mindfulness,* 1–40.
22. Singh, B., & Kaunert, C. (2024). Futuristic digital banking and e-financial strategies stimulating financial inclusion: Satellite SDG-1 alleviating poverty compliance sustainable development lensing global perspectives. In *E-financial strategies for advancing sustainable development: Fostering financial inclusion and alleviating poverty* (pp. 199–214). Springer Nature Switzerland.

Rising Beyond Traditional Barriers: Challenges and Opportunities for Women in Investment Banking and Capital Markets

Apurva A. Chauhan

Abstract Women invest with a purpose beyond wealth. They want their investments to have a meaningful impact on society, the environment, and future generations - Andy Serwer, Journalist and Former Editor in Chief, Yahoo Finance. The statement is quoted by Andy Serwer, Journalist and Former Editor in Chief, Yahoo Finance. The participation of women in investment decision leads to well-managed and resilient portfolios. And women's influence in investment decisions also drives towards societal change as they invest in sustainable, ethical, and community focused initiatives. Present research attempts to examine gender specific dynamics in investment banking and capital markets. The research questions identified are as follows: RQ1: what are the gender specific challenges in investment banking and capital markets? RQ2: What are the opportunities available for women in investment banking and capital market? The chapter further includes the success stories of few female leaders who have risen beyond the traditional barriers and have significantly impacted the industry. The study provides insights for organizations to include more inclusive practices for making organization a better place for women and help them to balance work–life.

Keywords Gender specific challenges · Opportunities for growth · Success stories

1 Introduction

Financial sector is considered as most dynamic and competitive sector which contributes to economic growth and fasters the innovation. Digitalization has transformed the financial sector and significantly increased the age more focusing on financial technology (Fintech). Investment banking and capital markets are most significant areas of financial sector, where professionals handle complex financial transactions, manage risk of clients, and help them in wealth creation. Traditionally

A. A. Chauhan (✉)
School of Leadership and Management, Manav Rachna International Institute of Research and Studies, Faridabad, India
e-mail: apurvachauhan.slm@mriu.edu.in

K. Khanna et al. (eds.), *From Poverty to Prosperity*, Information Systems Engineering and Management 64, https://doi.org/10.1007/978-3-031-99166-0_17

financial sector was dominated by men; however, in recent years it is observed that females are also contributing to the field very efficiently, and the sector also needs greater gender parity when it comes to investment management and capital market decisions.

Women can bring new perspectives in the financial sector which can bring holistic approach in decision-making. A team with greater gender diversity helps in making decisions better and approaches to risk differently. Women at the higher authorities can influence the organizational policies making them fairer, more supportive, strong documentation, etc. However, women face many challenges at the time of entry into the financial sector, and it continuous at the time of growth period also. These challenges arise due to biases and cultural expectations. According to a study of 134 companies by CFA Institute and CFA Society of India, only 1 women out of 8 people are on payroll, and only 21.7% of employees are women employees in financial companies. According to data provided by (Association of Mutual funds in India (AMFI), the number of women investors in mutual fund industry has increased from 46.66 lakh in Dec 2019 to 74.49 Lakh in Dec 2022.

This chapter identifies various opportunities available for women in the field of investment banking and capital market and attempts to find the challenges faced by women in the field. The study examines women as working professional in financial sector and women as investor in financial sector. With the help of various literatures, the study attempts to identify workplace practices, change in society attitude, and explain where we stand when it comes to gender parity in the field of investment banking and capital market.

1.1 Research Methodology

The study is descriptive in nature explaining various opportunities and challenges faced by women in investment banking and capital market with the help of literature reviews and various articles. The literature reviewed for study is explained in the next section.

2 Literature Reviews

This section explains the literature review referred for opportunities and challenges faced by women in investment management and capital market. Literatures reviewed are explained as:

1. Opportunities and challenges for women working professional in financial sector.
2. Opportunities and challenges for women investor in financial sector.

2.1 *Opportunities and Challenges for Women Working Professional in Financial Sector*

Women who are working Indian banking sector or investment sector must be aware about the term "glass ceiling" which means the unforeseen roadblocks which women faces in their career. The gender bias is present in Indian culture from ancient times, and it is the main root cause of discrepancy at work also. There are challenges for women to reach to higher management, and it becomes difficult to achieve gender equality in financial sector [1]. Banking sector acts as the lifeblood of economic development of country, and working women plays a pivotal role in smooth functioning of banking sector. Less number of women are working in bank irrespective of position such as officer, clerk, or subordinate. There are many factors which are responsible for this such as harassment at workplace, too much target pressure, transfer threats, and unavailability of childcare Centre [2]. Investment and banking sector is safe for women, provides good wages, and equal employment opportunities available. Few skills such as nurturing, listening, collaboration, and partnership are observed in women more and are essential in corporate world. However, lack of strong family support, work–life issues, etc., are few factors which hinders the growth of women in Indian investment sector and capital market [3]. Structure of society, male domination, networking, and male domination are the key themes which affect professional women growth as investment banker. Men in the industry should turn out to be supporter and act as a Fairy Godmother for women professional in investment banking. The success of investment banker depends on how client relationship is made and how much revenue is generated from them, so it is equally important that clients should remain bias free [4].

Underrepresentation of women in senior position in banking industry drives towards study of gender equality in the financial sector. Women's empowerment in banking sector is affected by traditional gender norms and bias, cultural norms, and social constraints [5]. Lack of opportunities and difference in compensation act as barriers for women in India, and these challenges also affect their participation in stock market as a professional also and as an investor also [6].

2.2 *Opportunities and Challenges for Women Investor in Financial Sector*

Lack of strategy and information unavailability is greatest obstacles faced by female investors. To be financial financially independent and gain tax advantage are the primary drivers of investment by females. Females as an investor are considered as fearful and take less risk [7]. The behavioural pattern of Indian women investors is found as risk averse, and they prefer to prioritize safety over attractive returns in the stock market. Gold and bank deposits are generally preferred form of investment by women investor [8]. The preference of women investors is affected by many demographic factors as well such as age of investor, their marital status, occupation, and salary [9]. Although women participate less in investment management, it is observed that non-investors also show interest in stock investment [10].

Despite of increase in financial independence of women and increase in education, few psychological barriers hinders their investment activities. Barriers such as risk aversion limit their decision-making power [11].

The financial independence of women is crucial for progress of society, due to increase in education and empowerment. As compared to men, the investment decisions of women are more conservative [12]. It is important for women investors to become mentor for other women investors and guide them regarding investment pattern. The economic can boost globally by 28 $ trillion if such gender bias can be removed from investment management [13].

3 Contribution of Women in Financial Industry

This section explains contribution of women in Indian financial sector with the help of real examples of few women leaders in the sector:

There are many examples which can be added to above section of chapter which highlights the remarkable presence of female leaders in financial industry. The commitment, achievements, dedication, visionary leadership, and focus on technology are few traits which are presented in Table 1 which also serves as powerful source of inspiration of next generation also.

It is important to know percentage of women holding leadership position for further commenting on women status in financial sector, next section of chapter explains the same!

4 Present Status of Women Professionals in C-suit Roles in Financial Sector

Executive roles performed in a company are referred as C-suit position. Chief executive officer, chief financial officer, chief operating officer, chief marketing officer, etc., are few examples of the same. Figure 1 depicts the % of positions hold by women in C-suit worldwide.

Figure 1 shows % share of women representatives on leadership position worldwide. It can be seen from the data that % of positions held is increasing but growth rate is very less. Reaching from 12.1 to 18.9% in 13 years of time span is a remarkable achievement for women leaders; however, it also makes readers think about why % change is less. The next section of the chapter examines women as an investor in capital market focusing on what motivates them to invest and what barriers they faces at the time of investment.

Table 1 Contribution of women in financial industry

S. No.	Name	Contribution in financial industry
1	Shanti Ekambaram	She worked in Kotak Mahindra Bank for 27 years and became Executive Director and CEO of Kotak Investment Banking. She has introduced digital banking and listed as "Most powerful women in Indian Business" for many years
2	Naina Lal Kidwai	She is a 66 years old chartered accountant and board member of Rothschild7 Co India. She is the first women to graduate from Harvard school and lead the foreign bank in India. Her 9 years of journey with Morgan Stanley impacted the Indian banking sector as she led the creation of JM Morgan Stanley, a partnership with JM Finance
3	Renu Sud Karnad	She is the director of India's biggest mortgage financer, Housing Development Finance corp. for 30 years, and Renu has been playing a fundamental role in HDFCs financial operations. She has been listed as "Top 10 women to watch out for in Asia" by Wall Street Journal
4	Kalpana Morparia	She has worked as CEO of JP Morgan after working for 33 years as a banker at ICICI. She has also played crucial role as Director of Dr. Reddy's Lab, CMC ltd to TATA Consultancy
5	Arundati Bhattacharya	She is highly influential Leader and CDO of sales force, and she has also worked as General Manager at ICICI. Forbes has listed as name in "Top 50 powerful women in international business"
6	Madhabi Puri Buch	She is the first female chair of SEBI, believing in collaboration of technology and finance. She is the only female member of regulatory bodies

Source Authors compilation

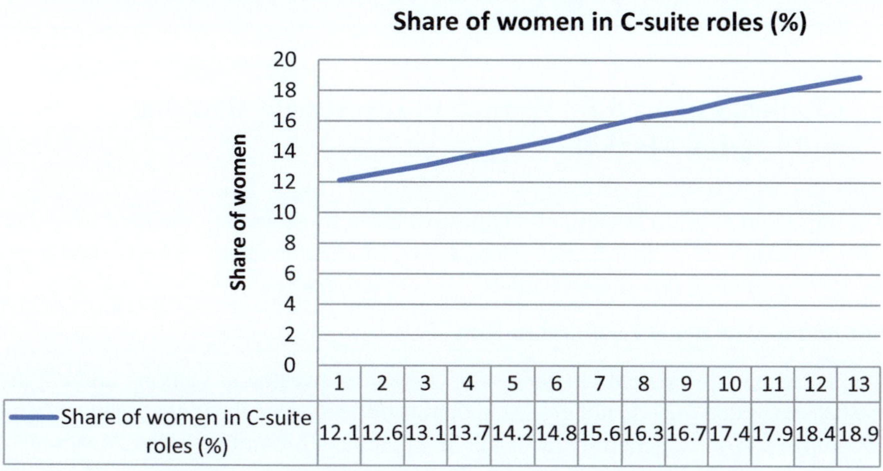

Fig. 1 Share of women in C-suit roles. *Source* Statista Research Department [14]

Table 2 Women as an investor

What motivates women to invest?	Challenges faced by women in investment decision	Strategies to overcome the challenges of investment
• Achieving financial security • Education, retirement planning • Desire of financial independence • Wealth creation • Peer influencers • Social encouragement	• Lack of confidence • Risk aversion attitude • Limited access to resources • Market perception • Gender bias	• Promoting financial literacy • Providing mentorship • Networking

Source Author compilation

5 Women as an Investor in Capital Market

This section of chapter examines women as an investor in the capital market. Table 2 represents 3 areas which need to be addressed while looking at women investors, namely motivators for investment, challenges in investment, and strategies to overcome challenges.

Over the period of years, role of women in financial sector has noticeably increased. It is observed that women brings new perspectives influencing the decision-making process. Investment in stock market by women investors will increase if all above-mentioned motivating factors and strategies to overcome challenges are implemented well. Next section of the chapter explains few challenges faced by women in the field of investment banking and capital market.

6 Challenges Faced by Women in Investment Banking and Capital Market

It is important to know issues and challenges faced by women in the field of investment banking and capital market. This section of chapter shows various challenges faced by women in investment banking and capital market.

Not Being Able to get Leadership Role

This is the most persistent challenge for women in investment banking and capital market sector that they do not get equal opportunities to lead the higher management of company. Women generally faces bias at the time of promotion either consciously not offered the position or unconsciously not considered for promotion. There is a lack of mentors for female entrepreneurs leading to less roles for women's at higher level.

Work–Life Balance

In one of the interviews Pepsi Company CEO Indra Nooyi once stated in her interview that the day she got offer as CEO of such a multinational company, she reached home and being asked about household things only and even mentioned that "You may be CEO of a company but here at home, You are a mother, wife and daughter-so fulfill those responsibilities." It's extremely hard for women to balance work and life together when both the dimensions of life need equal attention and commitment. Success or failure road is not equal for male and female professionals [15].

Bias at Workplace and Gender Pay Gap

Financial sector is mostly dominated by male, and hence, women face a lot of bias at workplace. Such unconscious bias affects the roles given to women, payment offered, less recognition, less opportunities, etc. Many times it is observed that for same roles performed in organization, male and females are paid different amount. Such discrepancies must be addressed by the management team.

Client Bias

Sometimes even the client of investment banking firm asks for male advisor, and such issue can be treated as lack of trust as well as traditional stereotypes people have in their mind regarding certain job roles. Such issues may result in generating less clients and revenue opportunities for firm which may also result in lack of promotional opportunities.

Although various steps have been taken for addressing gender disparities, women in investment banking and capital market continue to struggle for their job roles, equal pay, workplace bias, and work–life battles as mentioned.

7 Strategies for Fostering Gender Diversity in Financial Institutions

Various strategies can be used by financial institutions for fostering the gender diversity in financial institutions. Few strategies for fostering gender diversity are explained in this section of chapter.

Increase the percentage of female board members: By increasing the percentage of female on board members will increase the female leaders in financial sector. Management team should comprise more female leaders for promoting gender diversity. Increase the number of female employees: Financial institutions can increase the number of female employees in management roles, technical roles, or as interns for providing more career opportunities to females in the sector. Enhancing work environment for sustaining female employees: Female employees face lot of issues in managing work–life together; hence, it is very important for a financial institution to provide all necessary support to female employees for their sustained career. Developing and training anti-sexual harassment policies, flexible working hour, support for child care, maternity leave, healthy workplace, etc., are few of the policies which are recommended based on literatures referred for the study [16].

Consider Global Perspective on Gender Diversity in Financial Institutions: Financial institutions should take global perspectives on gender diversity for improving at national level. UAE and USA have introduced the mandatory female representative policy at board positions. Africa has higher female entrepreneurs. Japan and South Korea have introduced innovative policy on less holding fees for women in finance for promoting gender diversity in financial sector [17].

It is important for financial institutions to increase number of female representatives at all levels such as leadership position, management roles, technical roles, and front desk roles for promoting gender diversity in the financial sector.

8 Conclusion and Practical Implication of Study

Women have made significant impact, and their contribution has been remarkable in the field of investment management and capital market. The visibility of contribution has increased over last few years as they have excellent command in analytical skills, strategic thinking, and collaborative work. Successful deals have been made by women professionals, and they have handled the institutions at the time of financial crisis and proved that gender has no impact on work performance. However, as mentioned in the chapter also, percentage share of women leaders worldwide is very less as compared to male leaders which show the road is not same for both the genders.

Various challenges such as facing gender bias at workplace, bias from clients, work–life issues, societal norms, etc., affects the performance of women in any sector. If an organization can address such issues, then it will be a milestone for all the working women [18]. Support from organization such as flexible work arrangement and mentorship commitment from leader towards maintaining inclusive culture will motivate female professionals to not only join the organization but also reach to the leadership position. Women as investors in capital market are found to be clearer with the objectives and take safe decisions based on calculations. They are risk averse at the time of investment preferring safe investment avenues more such as mutual fund and gold.

Implication of study can be seen on individuals, organizations, and policymakers. Various challenges such as gender bias, work–life issues, societal, and cultural bias provide insights to women for developing strategies and overcoming them. The study can also be used by employers for implementing ensuring equal treatment to both genders at the time of policymaking. With the help of issues presented in chapter, companies can also design various training programs which can help women reach the leadership role in capital market. It is important for financial industry to encourage women to participate in the women to participate more in investment management to promote more ethical practices and nurture client-based business. The study shows the requirement of policymakers to make sure good amount of transparency and equal opportunities for men and women for professional growth as well as investment opportunities.

References

1. Rath, S., & Mishra, A. (2022). Study on the glass ceiling faced by women in the banking sector. *International Journal of Research Publication and Reviews,* 3864–3867. https://doi.org/10.55248/gengpi.2022.3.6.45
2. Nisha Ansari, A. (2024). Work culture and status of women's identity in the Indian banking sector. *BSSS Journal of Commerce, 16*(1), 22–29. https://doi.org/10.51767/joc1603
3. Menon, R. U, & Sukumaran, V. P. (2015). Women leadership in indian banking industry—an overview. *International Journal of Research in Management, 1*(5).
4. Tracey Rowe, A. C. (2003). A study of barriers to career advancement for professional women in investment banking. *Journal of Human Resource Management, 1*(2), 21–27.
5. Vandana Bansal, D. S. (2021). Barriers and facilitators for women advancement in banking leadership: An empirical research paper. *International Journal of Creative Research Thoughts (IJCRT), 9*(12).
6. Ramachandran, S., Balva, J., & Siva, M. (2023). *Mind the gender gap: Analysis of women's participation, pay, and other measures in Indian public companies.* https://doi.org/10.56227/23.1.4
7. Umadevi, P., & Joy, N. (2022). Study on factors influencing women investors in capital market. *Comfin Research, 10*, 47–53.
8. Harinakshi, & Goveas, C. (2023). A systematic literature review and research agenda of financial literacy, investment behavior and risk tolerance of Indian women investors. *International Journal of Case Studies in Business, IT, and Education.* https://doi.org/10.47992/ijcsbe.2581.6942.0323
9. Rishi, S., Ajay, K., & Namrata, S. (2022). Relationship between demographics of women and their preferences for self-investment options in financial markets. *International Journal of Engineering Applied Sciences and Technology.* https://doi.org/10.33564/ijeast.2022.v07i01.019
10. Mandeep, K., & Tina, V. (2022). Investment Preferences of Women: A Comparative Study of Women Stock and Non Stock Investors. *Management Dynamics.* https://doi.org/10.57198/2583-4932.1074
11. Prasad, D., Shollapur, M. R., & Patted, S. V. (2014). Indian women investors: Emotional decision makers. *Journal of Small Business and Entrepreneurship Development, 2*(1), 31–48.
12. Alawi, A. I. (2016). Status of Bahraini women in the banking and financial sector: Challenges and opportunities. *Journal of International Women's Studies, 17*(4), 210–228.
13. Nivedita, M., & Alba, C. (2023). The gender bias within investors; an opportunity cost of $4 trillion. *Journal of Student Research, 11*(1).
14. Statista Research Department. (2024). https://www.statista.com/statistics/1322160/share-of-women-financial-services-leadership-worldwide/
15. Geeta, S. (2014). Glass ceiling converted into glass transparency: A study on the factors making women executives successful in the Indian banking sector. *Journal Of Business Management & Social Sciences Research, 3*(4), 46–51.
16. Loibl, C., & Hira, T. K. (2016). Financial issues of women. In *Handbook of consumer finance research* (pp. 195–203).
17. Agyemang-Mintah, P., & Schadewitz, H. (2018). Gender diversity and firm value: Evidence from UK financial institutions. *International Journal of Accounting & Information Management, 26*(3).
18. Ward, A. M., & Forker, J. (2017). Financial management effectiveness and board gender diversity in member-governed. *Community Financial Institutions,* 351–366. https://doi.org/10.1007/s10551-015-2699-9

Understanding Behavioral Intention to Adopt Digital Payments: A TAM Framework Study in the Context of India's Response to the Past Pandemics

Parul Gaur, Ankit Saxena, Rashmi Singh, and Manish Upadhyay

Abstract Due to its convenience, reliability, and contactless nature, mobile payment (M-payment) has been widely embraced in India during the past pandemic. This implementation aims to reduce direct and indirect interactions in transactions, enabling social distancing and promoting stability in the social economy. The study seeks to examine the factors influencing behavioral intentions to use digital payment methods in India, specifically in light of the future pandemic. The study used stratified random sampling to choose 385 respondents as the sample size for the investigation. Data is collected using structured questionnaires and in-person administration, primary data on payment behavior, social variables, demographics, and other elements as mentioned in the technology acceptance model (TAM) framework. The findings underscored the substantial impact of the pandemic notion in accelerating the adoption of digital payments, particularly due to the perceived ease of use as well as perceived utility in shaping customers' intentions. In conclusion, the study, delving into the behavioral intention to adopt digital payments in India, specifically in light of the future pandemic, contributes significant insights into the evolving landscape of financial transactions.

Keywords Behavioral intentions · Digital payment · Pandemics · India · Technology acceptance model (TAM)

P. Gaur · A. Saxena · R. Singh (✉)
GLA University, Mathura, India
e-mail: rashmisinghmathura@gmail.com

P. Gaur
e-mail: gaurparul430@gmail.com

A. Saxena
e-mail: ankit.saxena@gla.ac.in

R. Singh · M. Upadhyay
Rajiv Academy for Technology and Management, Mathura, India
e-mail: manishupadhyay@ratm.in

K. Khanna et al. (eds.), *From Poverty to Prosperity*, Information Systems Engineering and Management 64, https://doi.org/10.1007/978-3-031-99166-0_18

1 Introduction

The all global past pandemic has brought about a significant shift in user experiences by promoting remote work and altering user behavior [1]. During the past pandemic, mobile communication and cutting-edge information technology were crucial. The pandemics have accelerated the widespread use of mobile technology-driven corporate solutions, building upon a preexisting trend [2]. The mobile wallet, often known as an m-wallet, is one of these digital payment initiatives that has evolved into an innovative contactless payment mechanism [3]. The capabilities of m-wallet, which include payments and loyalty cards, peer-to-peer payments, e-receipt keeping, bill payment, and coupon utilization directly from the application, allow users' money to be carried with their mobile phones.

Travel documents, insurance contracts, and e-commerce profiles are among the sensitive data that may be encrypted and stored in m-wallets [4]. Representatives who stand contend that m-wallets facilitate cashless transactions and enable both local and remote payments. They are better than traditional payment systems and relatively easier to use [5]. India had severe payment problems, particularly after demonetization, as a result of a shortage of hard currency and a shift in public sentiment toward digital payments [6]. Customers' positive intentions regarding m-wallets have been further enhanced by the past epidemic [7]. Social distancing and other preventive measures forced users to send and receive money using m-wallets. In India, the use of m-wallets increased by 44% during the national shutdown. The two most widely used m-wallet apps are Google Pay, Paytm, and PayPal [8].

In the current digital age, technology motivates individuals to engage in diverse online activities and transactions due to its perceived time efficiency and ease of accessing necessary information. Technology has seamlessly integrated into the lives of modern individuals, often without their conscious awareness [9]. The utilization of technology encompasses a wide range of applications, ranging from fundamental elements like electricity to more advanced domains such as finance technology (fin-tech). The ease and rapidity of the technology have prompted individuals to incorporate it into their daily lives [10]. The advent of smartphones has enhanced the accessibility of financial technology apps, such as e-payment, e-wallet, and e-money, enabling individuals to utilize them conveniently on their own devices. Based on a survey conducted by business standard, it was projected that the value of m-wallet transactions would increase from ₹5500 crores in the years 2015–16 reaching ₹30,000 crores in 2022. Individuals were motivated to make use of mobile wallets by various incentives; for instance, the Indian government has earmarked ₹1500 crore in the budget for 2021 in order to incentivize the adoption of digital payments., loyalty points, and cashback deals provided by service suppliers [11]. While there is growing enthusiasm among Indian users to embrace digital payments, the utilization of m-wallets in the country is not very promising [6]. Urban customers have shown an increased use of m-wallets; however, semi-urban and rural consumers have been more hesitant to accept and utilize m-wallets due to perceived risks associated with the technology. Indian m-wallet suppliers are keen on discovering methods to enhance m-wallet acceptability and usage intention, which has drawn the attention of researchers and practitioners [12].

In India, research on consumers' behavioral intentions about the use of mobile wallets is still in its infancy. Understanding consumer behavioral intents has become even more crucial given the present pandemic's disruptive shifts, extreme uncertainty, and panic disorder [13].

1.1 Technology Acceptance Model (TAM)

The study on the behavioral inclination to embrace digital payments, specifically in light of India's reaction to the past pandemic, is limited, hence highlighting the significance of this study. The worldwide crisis has led to a surge in the adoption of digital payments, and therefore, it is critical to comprehend the particular elements driving this change in a nation where traditional payment systems are strongly ingrained (Fig. 1).

To better understand the behavioral intention to embrace digital payments using the TAM framework study in the light of India's response to the past pandemic, the following study questions were designed:

i How do social factors (peer influence, societal norms, and recommendations) influence online payments in response to the past pandemic?
ii What is the association between demographic factors (age, income, education, and urban/rural residence) and behavioral intention?
iii Do the key factors (perceived risk, attitude toward use, and trust) influence behavioral intention utilizing the (TAM) framework?

In the context of India's reaction to the past epidemic, the primary intention of the study is to use the TAM framework to explore the variables impacting the behavioral intention to accept digital payments. Through the identification of significant factors and evaluation of the TAM framework's suitability for this particular situation, the research seeks to provide useful information to interested parties, advance scholarly

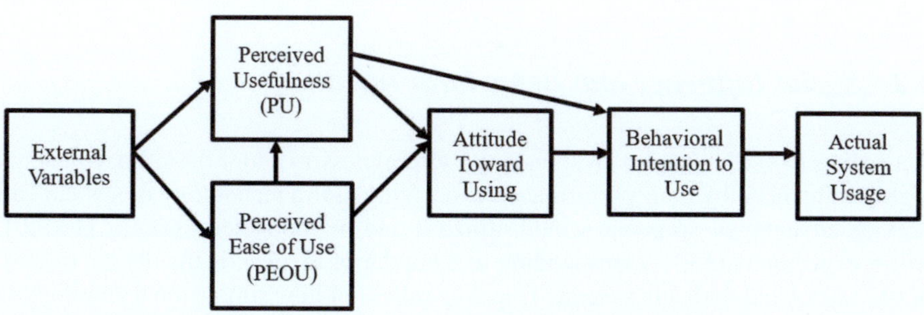

Fig. 1 Framework of the technology acceptance model (TAM). *Source* Self

discourse, and provide guidance to businesses and policymakers' quest to increase the uptake of online payments in India.

2 Review of Literature

2.1 Contactless Payments During Past Pandemic

Digital payment refers to a specific approach whereby payments can be made online. The payer and the payee both employ digital methods for transferring and receiving funds when conducting digital transactions. It is sometimes referred to as electronic payment. According to Wu et al. [14], digital payment refers to an online service that securely maintains payment instrument data for use as payment tools. Additionally, it has the potential to reduce costs, accelerate transactions, and enhance accessibility, usability, efficiency, and efficacy. Sharma et al. [15] claimed that the benefit of digital payment is convenience as it eliminates the need for carrying cash because consumers may conduct payment transactions utilizing smartphones with an Internet connection.

Online payments were a recommended choice during the pandemic due to its user-friendly interface and instantaneous nature, effectively preventing the transfer of viruses. According to Paysafe Insights [16], these payment systems allow users to make payments without the need for physical touch with cash or payment points. The study conducted by Kapoor et al. [17], sought to investigate the inclination to adopt mobile wallets (m-wallets) during the past pandemic by employing a mediated-moderation framework. The study also examined the impact of relative advantage, convenience of use, advantageous infrastructural conditions, security concerns, and touchless transactions on the uptake of mobile wallets. Raman and Aashish [18] developed a theoretical framework that examined the perceived benefits (PB) and perceived risk (PR) of utilizing "mobile payment systems (MPS)" during the pandemic. The study utilized the concepts of (HC) health consciousness and (FOC) fear of Coronavirus to better comprehend individuals' behavior and clarify their inclination to embrace MPS during the pandemic.

2.2 Social Influence and Behavioral Intention

According to earlier studies, respondents' willingness to adopt digital payments is highly influenced by their performance expectations. If a technology or system can provide advantages, someone would utilize it and be motivated to use it [19, 20]. Effort expectancy (EE), corresponding to Chopdar et al. [21], is the degree of ease of use associated with any system. Thus, regardless of how simple the technology is to operate, effort expectation relates to the amount of work required. User intentions to utilize digital payments are influenced by effort expectations, according to the conclusions of earlier studies [14].

According to earlier studies, social influence significantly affected people's intentions to utilize digital payments [22]. When utilizing popular technologies, people would feel more accepted in their social contacts. The findings of Najib and Fahma [23] confirmed that elements such as expected results, expected dedication, facilitating conditions, perceived utility, perceived ease of use, custom, and pricing value have an impression on customer intention to embrace mobile payment services. Moreover, the intention to behave in a certain way has a significant influence on the actual use of mobile payment systems during the pandemic.

Wu et al. [14] defined a facilitating circumstance as an individual's view of the availability of resources to support an activity. Facilitating conditions for diverse technologies include advice, support, and instruction. In a similar vein, Singh et al. [11] research showed that offering customers resources and support in the form of advice, help, and training can boost the implementation of digital payment systems.

Envisioned in the backdrop of the COVID pandemic, Hazarika et al. [24] discovered that organizational commitment and security customization can effectively mitigate consumers' cautious attitudes toward mobile payments. Moreover, offering users the ability to customize their privacy settings can greatly promote the act of revealing personal information, which is essential for verifying transactions and detecting fraudulent activities. The factors influencing students' intentions to utilize mobile banking were studied by Govender and Sihlali [25]. Using the information gathered from 77 samples, multiple regression analysis was performed. The research findings indicated that the desire of IT students to use mobile banking is not influenced by social media. Alalwan et al. [26] conducted a study that similarly demonstrated that customers' behavioral intention to utilize mobile banking is unaffected by social impact.

2.3 Technology Acceptance Model (TAM)

The widely used TAM model was established by Davis [27]. Fishbein and Ajzen [28] created the theory of reasoned action (TRA) and refined it to create the TAM model. Utility and ease of use are the two elements that are used to describe why people utilize technology. The TAM model had been utilized by several academics to investigate contactless payments, either by model extensions or by build extensions to get a more profound comprehension of the acceptance of online payments [29]. By adding concepts like risks, innovativeness, and trust in addition to TAM, Wang and Lin [30] investigated the intention to use contactless cards. In their study, Lule et al. [31] utilized the TAM framework to investigate the determinants that impact the adoption of mobile banking in Kenya. The study specifically concentrated on assessing MKesho, a mobile banking application from Kenya. A survey was conducted to collect data, which was then entered into SPSS. The data was analyzed using confirmatory factor analysis and structural equation modeling.

Different types of studies have explored TAM components concerning the influence of past pandemic on technology system acceptability. The association between Germans' intent to embrace Internet shopping during curfew hours and hedonic, utilitarian, and normative motivations was investigated in a study conducted by Koch et al. [32]. Simultaneously, Sreelakshmi and Prathap [33] discovered that customers' intentions to stick with mobile-based payments are mediated by the perceived utility concerning perceived health danger. Furthermore, comparable findings were found in Aji et al. [34] study that looked at customers' intentions about e-wallets.

The Indonesian government has actively encouraged the digitalization of the payment system as part of its goal to establish itself as the leading digital economy in the Southeast Asian region [35]. The study conducted by Najib and Fahma [36] examined the determinants of digital payment system acceptance in small- and medium-sized restaurants, with a focus on analyzing the elements that influence this adoption. Furthermore, it employed an expanded TAM framework as its theoretical underpinning. The study confirmed that the TAM can be applied effectively by incorporating new variables into the model. This approach was used to assess the acceptance of digital payment systems in restaurants that are small- or medium-sized in developing nations, such as Indonesia.

Furthermore, to the best of our knowledge, very few studies have looked at the intention of users to embrace mobile wallets during past pandemics in India. Prior research mostly neglected the part that various intervening factors play in the adoption process in favor of concentrating on how various internal and external signals impact m-wallet acceptance. According to published research, behavioral intention toward the adoption of new technologies is significantly framed by one's perception of values.

3 Research Methodology

Within the framework of India's reaction to the past epidemic, this study investigates the behavioral intention to adopt digital payments using a cross-sectional methodology and a quantitative research design. The prospective users of digital payment systems in Delhi-NCR are the target audience for this study. The study used stratified random sampling to choose 385 respondents as the sample size for the investigation. Data is collected using structured questionnaires and in-person administration, primary data on payment behavior, social variables, demographics, and other elements as mentioned in the technology acceptance model (TAM) framework. Data analysis uses MS Excel for preliminary processing and SPSS for in-depth statistical studies. Descriptive statistics (mean and standard deviation), regression, and ANOVA are all part of the process (Table 1).

Table 1 Methodical structure of objectives

S. No.	Objective	Hypothesis	Statistical test employed	Description
1	To explore how social factors (peer influence, societal norms, and recommendations) influencedigital payments in response to the past pandemics	H1: There is a significant impact of the social factors (peer influence, societal norms, and recommendations) on digital payments in response to the past pandemics	Regression	Regression is a collection of statistical methods used to determine the level of association between two variables
2	To assess the association between demographic factors (age, income, education, and urban/rural residence) and behavioral intention	H2: There is a significant associationbetween demographicfactors (age, income, education, and urban/ rural residence) and behavioral intention	ANOVA	Analysis of variance (ANOVA) canbe defined as a statistical analysis method, which divides the observed aggregate variability within a dataset into two distinct components: systematic variables and random factors
3	To identify the key factors (perceived risk, attitude toward use, and trust) influencing behavioral intention utilizing the technology acceptance model (TAM) framework	H3: There is a significant impact of the key factors (perceived risk, attitude toward use, and trust) on behavioral intention utilizing the technology acceptance model (TAM) framework	Regression	Regression is a collection of statistical methods used to determine the level of association between two variables

Source Authors' compilation

3.1 Results and Analysis

Demographics Table

Table 2 presents a comprehensive overview of the demographic attributes of respondents, including the distribution of genders, age groups, location, total monthly income, and education. The data reveals a small prevalence of males, accounting for 65.7% of the respondents, while females make up 34.3%. The sample is evenly distributed throughout four age groups: 18–25 years (26.2%), 26–35 years (26.0%), 36–45 years (27%), and above 45 years (20.8%). Similarly, the participants are dispersed among different education levels. Specifically, the bachelor's degree, master's degree, and other each make up 34.0%, 38.7%, and 27.3% of the total responses, respectively. The distribution of total monthly income shows less than

Rs. 20,000, Rs. 20,000–40,000, Rs. 40,000–60,000, and above Rs. 60,000 each make up 24.7%, 26.5%, 27.0%, and 21.8%. The majority of respondents belong to urban areas (54.8%) and 45.2% from local areas.

H1: "There is a significant impact of the social factors (peer influence, societal norms, and recommendations) on digital payments in response to the past pandemics".

Table 3 illustrates the results of the regression analysis conducted to examine the hypothesis on the influence of "social factors (peer influence, societal norms, and recommendations) on digital payments in response to the past pandemics". The variable being studied is digital payments, specifically in relation to the past pandemics outbreak, in order to evaluate the hypothesis. $F = 8.425, p < 0.05$, demonstrating the impact of the social factors (peer influence, societal norms, and recommendations) on digital payments in response to the pandemic ($b = 0.147$, and $p < 0.005$). Moreover, the $R^2 = 0.022$ of the digital payment in respondents to the past pandemics, an alternate hypothesis is accepted.

H2: "There is a significant association between demographic factors (age, income, education, and urban/rural residence) and behavioral intention".

- **Age**

Table 4 illustrates the descriptive statistics for the substantial association between age as well as behavioral intention. The average rating for a significant association

Table 2 Demographics table

S. No.	Demographic characteristics	Category	N	%
1	Gender	Male	253	65.7
		Female	132	34.3
2	Age	18–25 years	101	26.2
		26–35 years	100	26.0
		36–45 years	104	27.0
		Above 45 years	80	20.8
3	Location	Urban	211	54.8
		Rural	174	45.2
4	Education	Bachelors	131	34.0
		Masters	149	38.7
		Others	105	27.3
5	Total monthly income	Less than Rs. 20,000	95	24.7
		Rs. 20,000–40,000	102	26.5
		Rs. 40,000–60,000	104	27.0
		Above Rs. 60,000	84	21.8

Source Author compilation

Table 3 Regression analysis

Hypothesis	RW	Beta coefficient	*F*	*R*	R^2	*p*-value	*t*-value	Result of the hypotheses
H1	Social factors → Digital payments in response to COVID-19	−0.147	8.425	0.147	0.022	0.004	25.764	Supported

Source Self calculation
RW = Regression Weights

between age as well as behavioral intention is 10.5766. The standard deviation is 2.33742, indicating moderate variability in ratings.

The above analysis of variance (ANOVA) Table 5 provides information about the statistical analysis conducted to compare the significant association between age and behavioral intention. The ANOVA table indicates that there is no statistically significant variation between age and behavioral intention. The p-value of 0.044 suggests that the observed variations between groups are most likely to be attributed to random variation and are statistically significant (with a significance level of less than 0.05).

- **Location**

Table 6 provides descriptive statistics of the significant association between location and behavioral intention. The mean rating for significant association between location and behavioral intention is 10.5766. The standard deviation is 2.33742, indicating moderate variability in rating.

The above analysis of variance (ANOVA) Table 7 provides information about the statistical analysis conducted to compare the significant association between location

Table 4 Descriptives statistics for substantial association between age and behavioral intention

Behavioral intention

	N	Mean	Standard deviation	Standard error	95% Confidence interval for mean		Minimum	Maximum
					Lower bound	Upper bound		
18–25 years	103	10.0971	2.24270	0.22098	9.6588	10.5354	4.00	15.00
26–35 years	100	10.4800	2.10569	0.21057	10.0622	10.8978	4.00	15.00
36–45 years	99	10.8586	2.38195	0.23939	10.3835	11.3337	6.00	15.00
Above 45 years	83	10.9518	2.57996	0.28319	10.3885	11.5152	5.00	15.00
Total	385	10.5766	2.33742	0.11913	10.3424	10.8108	4.00	15.00

Source Self-calculation

Table 5 ANOVA table to compare the significant association between age and behavioral intention

Behavioral intention					
	df	Sum of squares	F	Mean square	Sig.
Between groups	3	44.173	2.731	14.724	0.044
Within groups	381	2053.817		5.391	
Total	384	2097.990			

Source Self-calculation

Table 6 Descriptives statistics for significant association between location and behavioral intention

Behavioral intention								
	N	Mean	Standard deviation	Standard error	95% Confidence interval for mean		Minim um	Maxim um
					Lower bound	Upper bound		
Urban	211	10.8057	2.29392	0.15792	10.4944	11.1170	4.00	15.00
Rural	174	10.2989	2.36602	0.17937	9.9448	10.6529	4.00	15.00
Total	385	10.5766	2.33742	0.11913	10.3424	10.8108	4.00	15.00

Source Self-calculation

and behavioral intention. The ANOVA table indicates that there is no statistically significant variation between location and behavioral intention. The p-value of 0.034 suggests that the observed variations between groups are most likely to be attributed to random variation and are statistically significant (with a significance level of less than 0.05).

- **Education**

Table 8 provides descriptive statistics of the significant association between education and behavioral intention. The mean rating for a significant association between education and behavioral intention is 10.5766. The standard deviation is 2.33742, indicating moderate variability in rating.

Table 7 ANOVA table to compare the significant association between location and behavioral intention

Behavioral intention					
	df	Sum of squares	*F*	Mean square	Sig.
Between groups	1	24.497	4.525	24.497	0.034
Within groups	383	2073.493		5.414	
Total	384	2097.990			

Source Self-calculation

Table 8 Descriptives statistics for significant association between education and behavioral intention

Behavioral intention

	N	Mean	Standard deviation	Standard error	95% Confidence interval for mean		Mini mum	Maxi mum
					Lower bound	Upper bound		
Bachelors	128	11.0234	2.38916	0.21117	10.6056	11.4413	4.00	15.00
Masters	151	10.3642	2.24940	0.18305	10.0025	10.7259	4.00	15.00
Others	106	10.3396	2.34170	0.22745	9.8886	10.7906	6.00	15.00
Total	385	10.5766	2.33742	0.11913	10.3424	10.8108	4.00	15.00

Source Self-calculation

The above analysis of variance (ANOVA) Table 9 provides information about the statistical analysis conducted to compare the significant association between education and behavioral intention. The ANOVA table indicates that there is no statistically significant variation between education and behavioral intention. The p-value of 0.030 suggests that the observed variations between groups are most likely to be attributed to random variation and are statistically significant (with a significance level of less than 0.05).

- **Income**

Table 10 provides descriptive statistics of the significant association between income and behavioral intention. The mean rating for a significant association between income and behavioral intention is 10.5766. The standard deviation is 2.33742, indicating moderate variability in rating.

The above analysis of variance (ANOVA) Table 11 provides information about the statistical analysis conducted to compare the significant association between income and behavioral intention. The ANOVA table indicates that there is no statistically significant variation between income and behavioral intention. The *p*-value of 0.035 suggests that the observed variations between groups are most likely to be attributed to random variation and are statistically significant (with a significance level of less than 0.05).

Table 9 ANOVA table to compare the significant association between education and behavioral intention

Behavioral intention

	df	Sum of squares	*F*	Mean square	Sig.
Between groups	2	38.319	3.553	19.160	0.030
Within groups	382	2059.670		5.392	
Total	384	2097.990			

Source Self-calculation

Table 10 Descriptives statistics for significant association between income and behavioral intention

Behavioral intention

	N	Mean	Standard deviation	Standard error	95% Confidence interval for mean		Minimum	Maximum
					Lower bound	Upper bound		
Less than Rs. 20,000	96	10.7917	2.20963	0.22552	10.3440	11.2394	4.00	15.00
Rs. 20,000–40,000	102	10.8824	2.27467	0.22523	10.4356	11.3291	4.00	15.00
Rs.40,000–60,000	101	10.0099	2.28690	0.22755	9.5584	10.4614	4.00	15.00
Above Rs. 60,000	86	10.6395	2.52476	0.27225	10.0982	11.1808	4.00	15.00
Total	385	10.5766	2.33742	0.11913	10.3424	10.8108	4.00	15.00

Source Self-calculation

Table 11 ANOVA table to compare the significant association between income and behavioral intention

Behavioral intention

	df	Sum of squares	*F*	Mean square	Sig.
Between groups	3	46.752	3.553	15.584	0.035
Within groups	381	2051.237		5.384	
Total	384	2097.990			

Source Self-calculation

H3: "There is a significant impact of the key factors (perceived risk, attitude toward use, and trust) on behavioral intention utilizing the technology acceptance model (TAM) framework".

Table 12 illustrates the outcomes of the regression analysis that was carried out to investigate whether the important components have a substantial impact on the hypothesis (perceived risk, attitude toward use, and trust) on behavioral intention employing the technology acceptance model (TAM) framework. The dependent variable is behavioral intention employing the TAM framework. $F = 4.732$, $p < 0.05$, demonstrating the impact of the key factors (perceived risk, attitude toward use, and trust) on behavioral intention employing the TAM framework ($b = 0.110$, and $p < 0.005$). Moreover, the $R^2 = 0.012$ of the behavioral intention utilizing the TAM framework.

Table 12 Impact of the key factors (perceived risk, attitude toward use, and trust) on behavioral intention utilizing the TAM framework

Hypothesis	RW	Beta coefficient	*F*	*R*	R^2	*p*-value	*t*-value	Result of the hypotheses
H1	Key factors → Behavioral intention	−0.147	4.732	0.110	0.012	0.030	18.511	Supported

Source Self-calculation
RW = Regression weights

4 Discussion

The study Understanding Behavioral Intention to Adopt Digital Payments: A TAM Framework Study in the Context of India's Response to the past pandemic seeks to examine the factors that influence people's willingness to use digital payment methods in India, specifically in light of the past pandemics. Multiple research studies have expanded upon the technology acceptability model (TAM) to gain insights into the acceptability of technology in other scenarios, such as healthcare and mobile money services, as demonstrated in the research conducted by Refs. [18, 37], respectively.

The study undertaken by Dhagarra et al. [38] investigated the influence of behavioral characteristics, such as privacy concerns and trust, on the acceptance of technology in the healthcare sector in India. The results of the study revealed that perceived importance, perceived simplicity of use, trust, and privacy concerns directly influence patients' behavior in accepting technology for healthcare services. In India, a study by Suyunchaliyeva et al. [39] revealed that the perceived ease of use plays a crucial role in determining the adoption and utilization of mobile money services, aligning with the TAM.

The study conducted by Prakash et al. [40] examined how the coronavirus pandemic affected the acceptance and spread of mobile payment methods among internal migratory laborers in India. The models of the unified theory of acceptance and use of technology (UTAUT) and protection motivation theory (PMT) were employed to assess the impact of the pandemic on the acceptance of MPS among internal migrant workers in India. In contrast, our study utilized the TAM framework to analyze the effects of key factors (perceived risk, attitude toward use, and trust) on behavioral intention.

Other studies corroborate the findings of the current study, although it does not endorse the specific makeup of the study. The variable of perceived trust had a significant role in the study [18, 41], and this finding was corroborated by the study itself. The perceived simplicity of use and perceived utility played a crucial role in the development of mobile money services. In [42] study, it was shown that perceived usefulness had an impact on consumers' trust in embracing mobile money. However, our current study does not support that finding. Nevertheless, the perception of trust had a detrimental impact on consumers' attitudes, resulting in a persistent negative utilization of mobile money services.

5 Conclusion

In conclusion, the study examines the inclination to embrace digital payments in India during the past pandemics, providing valuable insights into the changing world of financial transactions. The study utilizes the TAM framework to analyze the factors that impact users' inclination to use digital payment solutions. The findings underscored the substantial impact of the pandemic notion in accelerating the adoption of digital payments, particularly due to the perceived ease of use as well as perceived utility in shaping customers' intentions.

The results of the study show that all three hypotheses have been verified, indicating the strong influence of social variable on digital payments in response to the past pandemic pandemic as well as the important influence of three important factors (perceived risk, attitude towards use, and trust) on behavioral intention using the TAM framework. There is a significant impact of the key factors (perceived risk, attitude toward use, and trust) on behavioral intention utilizing the TAM framework.

Further study is warranted to explore the long-term effects of digital payment uptake after the past pandemics, taking into account the changing patterns of user behavior and improvements in technology. Examining the impact of trust, security, and user education on behavioral intentions would deepen our comprehension. Furthermore, longitudinal studies have the potential to provide valuable information about the lasting impacts, assisting policymakers and businesses in adjusting their policies to promote ongoing growth and acceptance of digital payments in the ever-changing ecosystem influenced by the epidemic.

References

1. Shishah, W., & Alhelaly, S. (2021). User experience of utilising contactless payment technology in Saudi Arabia during the past pandemics. *Journal of Decision Systems, 30*(2–3), 282–299.
2. Shankar, A., Gupta, M., Tiwari, A. K., & Behl, A. (2021). How does convenience impact showrooming intention? Omnichannel retail strategies to manage global retail apocalypse. *Journal of Strategic Marketing,* 1–22.
3. Mew, J., & Millan, E. (2021). Mobile wallets: Key drivers and deterrents of consumers' intention to adopt. *The International Review of Retail, Distribution and Consumer Research, 31*(2), 182–210.
4. Shin, D. H. (2009). Towards an understanding of the consumer acceptance of mobile wallet. *Computers in Human Behavior, 25*(6), 1343–1354.
5. Sharma, S. K., Mangla, S. K., Luthra, S., & Al-Salti, Z. (2018). Mobile wallet inhibitors: Developing a comprehensive theory using an integrated model. *Journal of Retailing and Consumer Services, 45*, 52–63.

6. Chawla, D., & Joshi, H. (2019). Consumer attitude and intention to adopt mobile wallet in India—An empirical study. *International Journal of Bank Marketing, 37*(7), 1590–1618.
7. Kosim, K. P., & Legowo, N. (2021). Factors affecting consumer intention on QR payment of mobile banking: A case study in Indonesia. *The Journal of Asian Finance, Economics and Business, 8*(5), 391–401.
8. Undale, S., Kulkarni, A., & Patil, H. (2021). Perceived eWallet security: Impact of past pandemics. *Vilakshan-XIMB Journal of Management, 18*(1), 89–104.
9. Kaur, P., Dhir, A., Bodhi, R., Singh, T., & Almotairi, M. (2020). Why do people use and recommend m-wallets? *Journal of Retailing and Consumer Services, 56*, Article 102091.
10. Legi, D., & Saerang, R. T. (2020). THE analysis of technology acceptance model (TAM) on intention to use of e-money in Manado (Study on: GOPAY, OVO, DANA). *Jurnal EMBA: Jurnal Riset Ekonomi, Manajemen, Bisnis dan Akuntansi, 8*(4).
11. Singh, N., Sinha, N., & Liébana-Cabanillas, F. J. (2020). Determining factors in the adoption and recommendation of mobile wallet services in India: Analysis of the effect of innovativeness, stress to use and social influence. *International Journal of Information Management, 50*, 191–205.
12. Shankar, A., & Behl, A. (2023). How to enhance consumer experience over mobile wallet: A data- driven approach. *Journal of Strategic Marketing, 31*(4), 838–855.
13. Hossain, M. A., Jahan, N., & Kim, M. (2023). A multidimensional and hierarchical model of banking services and behavioral intentions of customers. *International Journal of Emerging Markets, 18*(4), 845–867.
14. Wu, L. H., Wu, L. C., & Chang, S. C. (2016). Exploring consumers' intention to accept smartwatch. *Computers in Human Behavior, 64*, 383–392.
15. Sharma, S., Singh, G., Pratt, S., & Narayan, J. (2021). Exploring consumer behavior to purchase travel online in Fiji and Solomon Islands? An extension of the UTAUT framework. *International Journal of Culture, Tourism and Hospitality Research, 15*(2), 227–247.
16. Paysafe Insights (2020) *Combating past pandemics: Contactless payments on the doorstep.* https://www.paysafe.com/blog/combating-pastpandemics-contactless-payments-on-the-doorstep/
17. Kapoor, A., Sindwani, R., & Goel, M. (2022). Assessing mobile banking service quality dimensions using multi-criteria decision making. In *Future of work and business in past pandemics era: Proceedings of IMC-2021* (pp. 131–147). Springer Nature Singapore.
18. Raman, P., & Aashish, K. (2023). Influence of past pandemics on the intention to adopt mobile payment systems in India. *Qualitative Market Research: An International Journal, 26*(4), 368–394.
19. Chayomchai, A., Phonsiri, W., Junjit, A., Boongapim, R., & Suwannapusit, U. (2020). Factors affecting acceptance and use of online technology in Thai people during past pandemics quarantine time. *Management Science Letters, 10*(13), 3009–3016.
20. Rosnidah, I., Muna, A., Musyaffi, A. M., & Siregar, N. F. (2019, March). Critical factor of mobile payment acceptance in millenial generation: Study on the UTAUT model. In *International symposium on social sciences, education, and humanities (ISSEH 2018)* (pp. 123–127). Atlantis Press.
21. Chopdar, P. K., Korfiatis, N., Sivakumar, V. J., & Lytras, M. D. (2018). Mobile shopping apps adoption and perceived risks: A cross-country perspective utilizing the unified theory of acceptance and use of technology. *Computers in Human Behavior, 86*, 109–128.
22. Sivathanu, B. (2019). Adoption of digital payment systems in the era of demonetization in India: An empirical study. *Journal of Science and Technology Policy Management, 10*(1), 143–171.
23. Nandru, P., Chendragiri, M., & Senthilkumar, S. A. (2023). Factors affecting the adoption of mobile payment services during the past pandemics: An application of extended UTAUT2 model. *Journal of Science and Technology Policy Management.*
24. Hazarika, B., Shrivastava, U., Singh, V. K., & Rea, A. (2023). Motivating mobile payment adoption during global pandemic: Insights from protection motivation and theory of planned behavior. *Global Knowledge, Memory and Communication.*

25. Govender, I., & Sihlali, W. (2014). A study of mobile banking adoption among university students using an extended TAM. *Mediterranean Journal of Social Sciences, 5*(7), 451.
26. Alalwan, A. A., Dwivedi, Y. K., & Rana, N. P. (2017). Factors influencing adoption of mobile banking by Jordanian bank customers: Extending UTAUT2 with trust. *International Journal of Information Management, 37*(3), 99–110. From: https://www.securetechalliance.org/resources/pdf/Contactless_Payments_FAQ.pdf
27. Davis, F. D. (1989). Perceived usefulness, perceived ease of use, and user acceptance of information technology. *MIS Quarterly,* 319–340.
28. Fishbein, M., & Ajzen, I. (1977). *Belief, attitude, intention, and behavior: An introduction to theory and research.*
29. Cocosila, M., & Trabelsi, H. (2016). An integrated value-risk investigation of contactless mobile payments adoption. *Electronic Commerce Research and Applications, 20*, 159–170.
30. Wang, Y. M., & Lin, W. C. (2019). Understanding consumer intention to pay by contactless credit cards in Taiwan. *International Journal of Mobile Communications, 17*(1), 1–23.
31. Lule, I., Omwansa, T. K., & Waema, T. M. (2012). Application of technology acceptance model (TAM) in m-banking adoption in Kenya. *International Journal of Computing & ICT Research, 6*(1).
32. Koch, J., Frommeyer, B., & Schewe, G. (2022). Managing the transition to eco-friendly packaging–An investigation of consumers' motives in online retail. *Journal of Cleaner Production, 351*, Article 131504.
33. Sreelakshmi, C. C., & Prathap, S. K. (2020). Continuance adoption of mobile-based payments in past pandemics context: An integrated framework of health belief model and expectation confirmation model. *International Journal of Pervasive Computing and Communications, 16*(4), 351–369.
34. Aji, H. M., Berakon, I., & Riza, A. F. (2020). The effects of subjective norm and knowledge about riba on intention to use e-money in Indonesia. *Journal of Islamic Marketing, 12*(6), 1180–1196.
35. Uche, D. B., Osuagwu, O. B., Nwosu, S. N., & Otika, U. S. (2021). Integrating trust into technology acceptance model (TAM), the conceptual framework for e-payment platform acceptance. *British Journal of Management and Marketing Studies, 4*(4), 34–56.
36. Najib, M., & Fahma, F. (2020). Investigating the adoption of digital payment system through an extended technology acceptance model: An insight from the Indonesian small and medium enterprises. *International Journal on Advanced Science, Engineering and Information Technology, 10*(4), 1702–1708.
37. Kelly, A. E., & Palaniappan, S. (2023). Using a technology acceptance model to determine factors influencing continued usage of mobile money service transactions in Ghana. *Journal of Innovation and Entrepreneurship, 12*(1), 34.
38. Dhagarra, D., Goswami, M., & Kumar, G. (2020). Impact of trust and privacy concerns on technology acceptance in healthcare: An Indian perspective. *International Journal of Medical Informatics, 141*, Article 104164.
39. Suyunchaliyeva, M. M., Nautiyal, R., Shaikh, A. A., & Sharma, R. (2021). The use of mobile payment systems in post-past pandemics economic recovery: Primary research on an emerging market for experience goods. *Sustainability, 13*(24), 13511.
40. Prakash, A., Mahajan, Y., & Gadekar, A. (2022). Adoption of mobile money among internal migrant workers during the corona pandemic in India: A study focused on moderation by mode of payments. *International Journal of Electronic Banking, 3*(2), 144–162.
41. Mariki, A. J., & Mbamba, U. O. *Adoption of mobile financial services by the banking customers in Tanzania.*
42. Maradung, P. (2013). *Factors affecting the adoption of mobile money services in the banking and financial industries of Botswana* (Doctoral dissertation).

Green Bonds and Their Role in Sustainable Development

Shreya Arora and B. Veeralakshmi

Abstract Green bonds are essential financial instruments for promoting sustainable development goals (SDGs) and financing projects that benefit the environment. To investigate their function and influence, this chapter integrates case studies and bibliometric data. In order to uncover knowledge gaps and new areas of concentration, the bibliometric analysis looks into important contributions, major themes, and worldwide research trends in the green bond literature. The study emphasizes the development of green finance research and its relationship to sustainability through the use of bibliometric tools. To provide practical insights, case studies across sectors such as renewable energy and sustainable infrastructure are presented. These illustrate their implementation, challenges, and measurable environmental and social benefits. By integrating quantitative bibliometric insights with qualitative case studies, this chapter offers a comprehensive understanding of green bonds' contributions to sustainable development.

Keywords SDG · Green bonds · Sustainable finance · Bibliometric

1 Introduction

Green bonds have become revolutionary financial tools in the worldwide pursuit of the sustainable development goals (SDGs). Intended to fund initiatives that have a favorable environmental impact, green bonds represent a critical nexus between financial markets and sustainable development, enabling the allocation of capital

S. Arora (✉)
Assistant Professor, Mangalayatan University, Jabalpur, MP, India
e-mail: Sherya.arora14dec@gmail.com; shreya.arora@mangalayatan.ac.in

B. Veeralakshmi
College of Business Studies, COER University, Roorkee, India

K. Khanna et al. (eds.), *From Poverty to Prosperity*, Information Systems Engineering and Management 64, https://doi.org/10.1007/978-3-031-99166-0_19

toward projects that align with climate and ecological priorities [1]. Their contribution to tackling urgent global issues including urbanization and climate change underscores their increasing relevance in academic research and practical implementation [2].

Green bonds were first proposed in the early 2000s, with their formal introduction by the European Investment Bank in 2007, marking a milestone in climate finance [3]. The market has grown significantly since then, and by 2022, its total issuance will have surpassed $1 trillion, with significant contributions from regions like Europe, North America, and Asia [3]. This expansion is indicative of a larger movement in financial decision-making to incorporate environmental, social, and governance (ESG) considerations. This chapter investigates the dual facets of green bonds: their theoretical foundations and practical applications. By employing bibliometric analysis, it explores the intellectual landscape of green bond literature, uncovering key themes, research trends, and areas requiring further exploration. Bibliometric tools, such as co-citation analysis and keyword mapping, provide insights into the evolution of green finance research, highlighting its intersections with sustainability science [4]. These tools facilitate the identification of critical knowledge gaps, thereby laying the groundwork for future academic inquiries and policy development [5].

Complementing this quantitative approach, the chapter presents qualitative case studies that illustrate the practical implications of green bonds across two pivotal sectors: renewable energy and sustainable infrastructure. Green bond-financed renewable energy projects, like Norway's hydropower projects and India's solar parks, show their ability to lower greenhouse gas emissions and advance clean energy transitions. [6, 7]. Similarly, green bond-backed urban infrastructure projects, such as France's Metro Expansion and Indonesia's Green Sukuk for water management, showcase their role in enhancing urban resilience and inclusivity [8, 9]. Despite their successes, green bonds face challenges, including regulatory inconsistencies, greenwashing risks, and high upfront costs, which hinder their broader adoption and effectiveness [10]. These issues underscore the need for robust frameworks, enhanced transparency, and comprehensive impact assessments to ensure their continued alignment with sustainability goals [11].

By integrating bibliometric analysis with sector-specific case studies, this chapter provides a holistic perspective on green bonds' contribution to the promotion of sustainable development. It offers a sophisticated comprehension of their contributions, possibilities as well as problems, contributing to the broader discourse on sustainable finance. Policymakers, investors, and other stakeholders involved in promoting sustainability using cutting-edge financial instruments can benefit from this integration's practical advice in addition to its theoretical insights [12]. By channeling investments into renewable energy, energy efficiency, and other climate-aligned initiatives, green bonds have strengthened the role of finance in addressing global environmental challenges while fostering economic growth [13]. The adoption of green bonds aligns closely with the principles of sustainable finance, which emphasize accountability, transparency, and measurable environmental outcomes. Initiatives like the International Capital Market Association (ICMA) and the Climate Bonds Initiative (CBI) provide rigorous standards, ensuring that bond proceeds directly support environmental objectives [12]. These frameworks have built investor confidence and encouraged the diversification of green bond applications, spanning projects like sustainable infrastructure and climate adaptation measures [14].

2 Literature Review

A key factor in bringing financial markets into line with sustainability objectives is the emergence of green bonds as revolutionary financial products. Their purpose is to provide funding for initiatives that have a beneficial impact on the environment or climate, which will help the United Nations achieve its sustainable development goals (SDGs) [2]. Green bonds have become popular as the need for sustainable finance increases, bridging the gap between economic success and environmental stewardship [1]. The European Investment Bank issued the first green bond in 2007, demonstrating the historical development of green bonds, which have their origins in the early 2000s. This marked a shift in global financial paradigms, emphasizing environmental and social governance (ESG) priorities [2]. The subsequent proliferation of green bonds underscores their adaptability across diverse industries, from renewable energy to urban infrastructure [15]. Several studies have highlighted the financial performance of green bonds relative to conventional bonds. For instance, Ehlers and Packer [16] noted that green bonds often trade at a "greenium," reflecting investor preference for sustainable investments. Similarly, Baker et al. [17] found that green bond issuers often benefit from reduced borrowing costs, incentivizing environmentally aligned financial practices. Green bonds' role in renewable energy development has been particularly significant. According to Banga [18], these bonds have catalyzed investments in solar, wind, and hydropower projects, addressing global energy security challenges while reducing greenhouse gas emissions. For example, India's renewable energy expansion, funded through green bonds, showcases their potential to drive transformative change in developing economies [19]. In the infrastructure domain, green bonds have been instrumental in funding sustainable urban development. France's issuance of green bonds to finance its metro expansion is a prominent example, illustrating how such instruments can enhance urban mobility while reducing carbon footprints [8]. Similarly, Indonesia's Green Sukuk initiative highlights the integration of Islamic finance principles with sustainability goals, expanding green finance's cultural and regional relevance [8]. Despite these successes, challenges persist. Regulatory inconsistencies and the lack of standardized green bond frameworks have been frequently cited as barriers to market growth [11]. Greenwashing, misleading claims about the environmental benefits of bonds, remains a critical concern, emphasizing the need for stringent verification mechanisms [20]. Studies by Refs. [19, 20] emphasize the need for robust impact assessment methodologies to ensure transparency and accountability in the green bond market. Additionally, bibliometric analyses of green finance literature reveal an upward trajectory in global research output, with significant contributions from China, the European Union, and the United States [4, 5, 21]. Case studies across sectors underscore the practical applications and challenges of green bonds. For instance, South Africa's wind energy projects reveal both the financial feasibility and regulatory hurdles associated with green bond financing [21]. Similarly, Norway's hydropower initiatives funded through green bonds highlight the dual benefits of energy generation and ecosystem preservation [22]. Another critical area of study is the social dimension of green bonds. Research by Jagriti [23] highlights how these instruments can address social inequities by funding projects that improve water access and urban resilience. For instance, green bonds in Indonesia have been used to enhance water and sanitation services, illustrating their potential to tackle pressing social challenges [24]. The integration of green bonds into sustainable development strategies has spurred innovations in financial instruments. Transition bonds, blue bonds, and social bonds have emerged as complementary tools, broadening the scope of impact finance [4]. These innovations highlight the dynamic nature of green finance and its capacity to evolve in response to emerging challenges. Quantitative analyses further reinforce the systemic importance of green bonds in the global financial ecosystem. Studies by Nguyễn [25] indicate that green bonds contribute to market stability by attracting long-term investors committed to sustainable outcomes. This aligns with findings by Karpf and Mandel [26], who argue that green bonds are critical for aligning private capital with public policy objectives.

The literature on green bonds has significantly grown, aligning with the rising urgency to address climate change and achieve the sustainable development goals (SDGs). Green bonds emerged as a critical financial tool to mobilize capital for environmentally sustainable projects. Their inception traces back to 2007 when the European Investment Bank and the World Bank issued the first bonds to fund climate-resilient projects [27]. Since then, their issuance has accelerated globally, supported by strong policy frameworks like the [28], which emphasized a transition toward low-carbon economies [28]. Studies have highlighted green bonds' role in reducing carbon emissions and promoting renewable energy production. For instance, empirical evidence shows a negative correlation between green bonds and carbon emissions while positively influencing renewable energy adoption [27]. This underscores the potential of green bonds in fostering global sustainability through financial innovation. The issuance of green bonds is also a measure of a country's commitment to sustainable development. Research indicates that nations with higher green bond issuance demonstrate notable progress toward sustainability, especially in reducing greenhouse gases and increasing renewable energy capacity [28]. However, barriers such as the lack of standardization in green bond certification and limited investor awareness persist as challenges to their widespread adoption [29]. Bibliometric analyses reveal that academic interest in green bonds has proliferated, with significant contributions from developed nations like the United States and Europe. Notably, themes such as green finance, climate risk mitigation, and sustainable infrastructure dominate the scholarly discourse [28]. Studies also examine the economic implications of green bonds, asserting that they lower the cost of capital for environmentally sustainable projects while fostering investor confidence in green innovations [27]. Furthermore, the performance of green bonds in financial markets is an emerging area of interest. Research suggests that green bonds often yield similar or marginally higher returns compared to traditional bonds, making them an attractive option for socially responsible investors [27]. This is crucial in driving private sector participation in financing sustainability. Green bonds also intersect with corporate social responsibility (CSR). Companies utilizing green bonds report enhanced reputational benefits and stakeholder trust, further incentivizing sustainable business practices [28]. However, concerns about greenwashing, where bonds are labeled green without substantial environmental impact, require stricter regulatory oversight and transparent reporting mechanisms. Finally, regional variations in green bond adoption reveal diverse trajectories. While Europe leads in total issuance, emerging markets like China and India show rapid growth, propelled by government-backed initiatives and international collaborations [30]. These regions highlight the potential for green bonds to bridge financing gaps in addressing global environmental challenges.

3 Research Methodology

Using a mixed-methods approach, this chapter offers a thorough examination of green bonds and how they contribute to the sustainable development goals (SDGs). The methodology is structured around two key components: bibliometric analysis and qualitative case studies. This dual approach ensures both a macro-level understanding of global research trends and a micro-level exploration of sector-specific applications.

3.1 Bibliometric Analysis

The bibliometric analysis employs quantitative methods to evaluate the academic literature on green bonds. VOSviewer, a crucial bibliometric tool, is used to examine citation patterns, co-authorship networks, publishing trends, and keyword co-occurrence. Data for the analysis is sourced from prominent database, including Scopus, ensuring comprehensive coverage of high-quality peer-reviewed publications.

The bibliometric analysis follows a systematic process:

1. **Data Collection**: The keywords such as "green bonds," "sustainable finance," and "climate finance" were used to retrieve relevant articles published between 2007 and 2023.
2. **Data Cleaning**: Duplicate entries and irrelevant publications were excluded to refine the dataset.
3. **Visualization**: Networks of keywords, authors, and countries were mapped to identify major contributors, thematic trends, and emerging areas in green bond research.

4. **Trend Analysis**: The evolution of green bond research was assessed to highlight shifts in focus areas, such as the increasing emphasis on renewable energy and urban infrastructure.

3.2 Case Study Analysis

To complement the quantitative insights, the chapter incorporates qualitative case studies from two critical sectors: renewable energy and sustainable infrastructure. This method is chosen to provide practical examples of green bond applications, focusing on their implementation, challenges, and measurable environmental and social benefits.

The selection of case studies was predicated on their pertinence, diversity, and availability of reliable data. Key criteria for selection include:

1. **Sectoral Relevance**: Projects funded through green bonds in renewable energy (e.g., solar and wind) and sustainable infrastructure (e.g., urban transit and water management).
2. **Geographic Diversity**: Inclusion of cases from developed and developing countries to capture a broad spectrum of contexts.
3. **Documented Impact**: Availability of reports, evaluations, or studies highlighting measurable outcomes.

For each case study, the analysis focuses on:

- **Implementation Process**: The use of green bonds in project financing, including stakeholder involvement and financial structuring.
- **Challenges**: Regulatory barriers, greenwashing risks, and other obstacles encountered during project execution.
- **Benefits**: Environmental outcomes (e.g., CO_2 reduction), social impacts (e.g., job creation), and alignment with SDGs.

4 Data Collection and Data Analysis

The publications have been collected from the Scopus database, and the criteria of selection of publications have been shown with the help of a flowchart. Analysis has been done using the VOSviewer software, and the results have been discussed below (Fig. 1).

The publications on Green bond and sustainability have been searching on the Scopus database using the keywords "Green Bond," resulting in 1474 Documents. Next, the keywords "Green Bond" AND "sustainable development goals" resulting in 105 documents. One hundred and twelve documents become the final dataset for the purpose of analysis for this chapter resulting from the keywords "Green Bond" AND ("sdg" OR "Sustainable development goals").

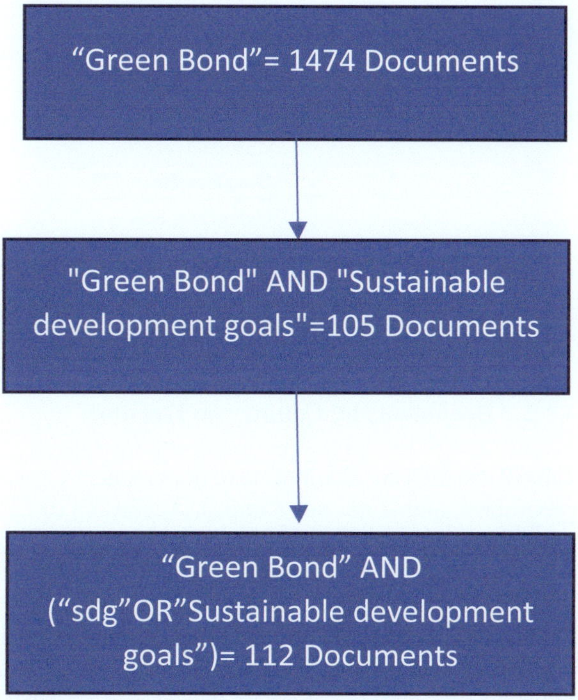

Fig. 1 Criteria for selection of papers. *Source* Author's creation

4.1 Results of Scopus

The number of documents has increased significantly, peaking in 2023 with 32 documents and slightly declining to 30 in 2024. This suggests a sustained period of high productivity. There is a gradual rise in document creation, from 3 documents in 2019 to 23 in 2022, indicating a period of steady development. Document creation was minimal, with only 1 document in 2016, none in 2017, and 3 in 2018. This suggests that the process was either newly initiated or less emphasized during this time. There is a clear upward trajectory in document creation over the years, reflecting increased activity or emphasis on document generation in recent years.

4.1.1 Author-Wise Documentation

Table 1 presents a list of authors alongside the number of documents they have contributed to. It highlights the productivity of each author based on the number of publications attributed to them. Publications in the related field of Green Bonds and Sustainability, where Taghizadeh-Hesary, F. have published 3 impactful papers in Scopus as against other authors. Minimum number of publications per author is two.

Table 1 Author-wise documentation

Author	Documents
Taghizadeh-Hesary, F.	3
Bisultanova, A.	2
Chadha, G.	2
González-Ruiz, J.D.	2
Gyamerah, S.A.	2
Sarker, T.	2
Sholoiko, A.	2
Singhania, M.	2

4.1.2 Documents by Country or Territory

Table 2 displays the distribution of documents by country or territory, illustrating the geographic spread of contributions. According to the analysis, India leads with 19 documents, followed by China with 15. These two countries are the most significant contributors. Countries like the Russian Federation and the United Kingdom have 9 documents each, while Australia and Vietnam each contributed 7 documents. These regions also demonstrate notable involvement in document indexing in the Scopus database. Several countries, such as Pakistan, Spain, and the United States, have 6 documents each, while others like France, Indonesia, and Ukraine contributed 5 documents. This indicates a relatively balanced spread of contributions from a variety of regions. A number of countries, including Brazil, Czech Republic, Kazakhstan, and Poland, have 2 documents each. These contributions, though smaller, still represent global participation. Table 2 reflects contributions from diverse regions, including Asia, Europe, the Americas, and Africa.

4.2 Results of VOSviewer

4.2.1 Co-authorship-Authors Network

Minimum number of documents of an author is chosen as "1". Also minimum number of citations of an author is also set as "1". Eighty five authors have met this threshold (Fig. 2).

Table 3 presents data related to authorship, documents, citations, and total link strength. Total link strength represents a measure of the connections or influence of the authors within the network. Here, it is consistently 0, indicating the absence of co-authorship network data. The total link strength is 0 for all entries, suggesting no recorded collaborative relationships or interactions in this dataset. In co-authorship networks, total link strength typically measures the intensity of connections between

Table 2 Country-wise documentation

Country/Territory	Documents
India	19
China	15
Russian Federation	9
United Kingdom	9
Australia	7
Viet Nam	7
Pakistan	6
Spain	6
United States	6
France	5
Indonesia	5
Ukraine	5
Japan	4
Malaysia	4
Nigeria	4
Canada	3
Colombia	3
Germany	3
Ghana	3
Hong Kong	3
Turkey	3
Brazil	2
Brunei Darussalam	2
Czech Republic	2
Kazakhstan	2
Macao	2
Peru	2
Poland	2
Saudi Arabia	2
Slovakia	2
South Korea	2

authors. Collaboration appears to be a hallmark of impactful publications, especially for authors with high citation counts.

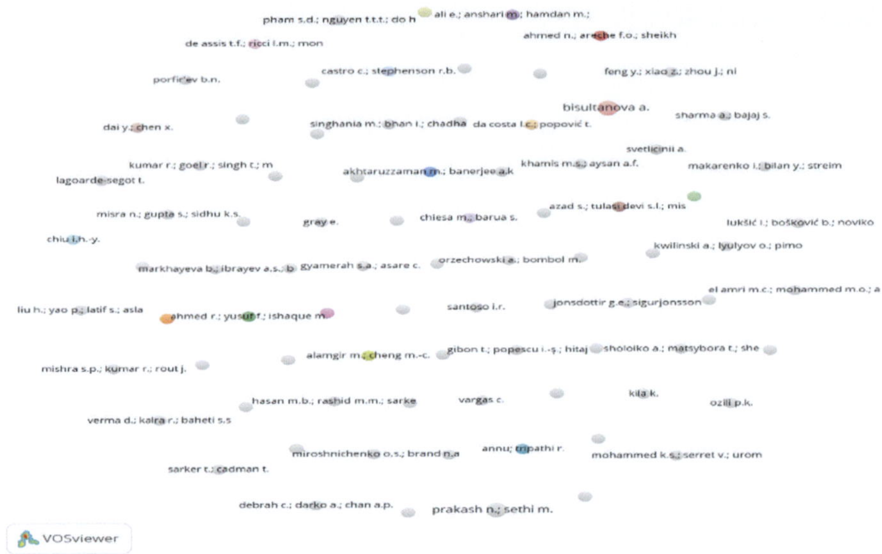

Fig. 2 Co-authorship-authors network

4.2.2 Analysis of Co-authorship with Organization

Two hundred and eighteen of the 259 organizations have at least one document with at least one citation. The overall strength of co-authorship relationships with other organizations is determined for each of the 218 organizations. The organizations with the highest link and length total strengths have been chosen. Only 7 items have been connected among themselves which has been shown in the Fig. 3.

4.2.3 Co-authorship Within the Countries

Fifty four of the 56 nations have at least one document from their country that also has at least one citation. The overall strength of the co-authorship linkages with the other 25 nations is determined for each country, and 54 countries are chosen as having the highest total link strength (Fig. 4).

4.2.4 Co-authorship-All Keywords

Minimum number of occurrences of a keyword is "1" whereby 634 keywords meet the threshold.

Figure 3. Keywords co-occurrence map. *Source* VOSviewer. *Note* The size of the nodes represents the relevance of the terms in the papers. The thickness of the lines

Table 3 Year-wise documentation

ID	Author	Documents	Citations	Total link strength
2	"ahmed n.; areche f.o.; sheikh a.a.; lahiani a."	1	34	0
3	"ahmed r.; yusuf f.; ishaque m."	1	10	0
5	"akhtaruzzaman m.; banerjee a.k.; ghardallou w.; umar z."	1	35	0
6	"alamgir m.; cheng m.-c."	1	22	0
7	"ali e.; anshari m.; hamdan m.; ahmad n.; surieshtino y."	1	1	0
8	"annu; tripathi r."	1	6	0
9	"argandoña l.c.b.; rambaud s.c.; pascual j.l."	1	9	0
10	"azad s.; tulasi devi s.l.; mishra a.k."	1	2	0
11	"basak s.; sana a.k."	1	1	0
14	bisultanova a	2	3	0
15	"bosmans p.; de mariz f."	1	9	0
16	"castro c.; stephenson r.b."	1	2	0
17	"chang l.; taghizadeh-hesary f.; chen h.; mohsin m."	1	98	0
18	"chiesa m.; barua s."	1	66	0
19	chiu i.h.-y	1	7	0
20	"da costa l.c.; popović t."	1	9	0
21	"dai y.; chen x."	1	7	0
22	"de assis t.f.; ricci l.m.; monteiro t.g.m.; de abreu v.h.s.; d'agosto m.a.; santos a.s."	1	2	0
23	"debrah c.; darko a.; chan a.p.c."	1	41	0
24	"el amri m.c.; mohammed m.o.; abdi m.h."	1	5	0
25	"feng y.; xiao z.; zhou j.; ni g."	1	29	0
26	"gabr d.h.; elbannan m.a."	1	2	0
27	"gibon t.; popescu i.-ş.; hitaj c.; petucco c.; benetto e."	1	47	0
28	"gonzález-ruiz j.d.; mejía-escobar j.c.; rojo-suárez j.; alonso-conde a.-b."	1	8	0
29	gray e	1	2	0
31	gutiérrez-torrenova m.g	1	3	0
33	"gyamerah s.a.; asare c."	1	1	0
35	"hasan m.b.; rashid m.m.; sarker t.; shafiullah m."	1	1	0

(continued)

Table 3 (continued)

ID	Author	Documents	Citations	Total link strength
36	"jonsdottir g.e.; sigurjonsson t.o.; alavi a.r.; mitchell j."	1	23	0
37	"joshipura m.; mathur s.; kedia n."	1	6	0
38	"khamis m.s.; aysan a.f."	1	4	0
40	kila k	1	3	0
42	kocaarslan b	1	30	0
43	"kumar r.; goel r.; singh t.; mohanty s.m.; gupta d.; alkhayyat a.; khanna r."	1	2	0
44	"kwilinski a.; lyulyov o.; pimonenko t."	1	36	0
45	lagoarde-segot t	1	49	0
46	"liu h.; yao p.; latif s.; aslam s.; iqbal n."	1	153	0
47	"lukšić i.; bošković b.; novikova a.; vrbensky r."	1	16	0
48	"makarenko i.; bilan y.; streimikiene d.; rybina l."	1	11	0
49	"markhayeva b.; ibrayev a.s.; beisenova m.; serikbayeva g.; arrieta-lópez m."	1	10	0
50	"marín-rodríguez n.j.; gonzález-ruiz j.d.; valencia-arias a."	1	10	0
52	"miroshnichenko o.s.; brand n.a."	1	12	0
53	"mishra s.p.; kumar r.; rout j."	1	1	0
54	"misra n.; gupta s.; sidhu k.s.; kumar a.; anandaram h.; rao t.j.; singh r.; gehlot a.; akram s.v.; joshi k."	1	3	0
55	"mohammed k.s.; serret v.; urom c."	1	13	0
56	"negra c.; remans r.; attwood s.; jones s.; werneck f.; smith a."	1	38	0
57	"ng a.w.; nathwani j.; fu j.; zhou h."	1	35	0
58	"nguyen a.h.; hoang t.g.; nguyen d.t.; nguyen l.q.t.; doan d.t."	1	17	0
59	"nguyen n.m.; luu n.h.; hoang a.; nguyen m.t.n."	1	2	0
60	"nurgaliyeva a.m.; kazbekova z.s.; bokenchina l.k.; bekniyazova d.; bokenchin k.k."	1	15	0
62	"oguntuase o.j.; windapo a."	1	16	0
63	"orzechowski a.; bombol m."	1	7	0

(continued)

Table 3 (continued)

ID	Author	Documents	Citations	Total link strength
64	ozili p.k	1	71	0
66	"pham s.d.; nguyen t.t.t.; do h.x."	1	2	0
67	phan t.c	1	5	0
68	porfir'ev b.n	1	34	0
69	"prakash n.; sethi m."	2	36	0
70	"quang p.t.; thao d.p."	1	26	0
71	"raimi l.; olowo r.; shokunbi m."	1	10	0
72	"rasoulinezhad e.; taghizadeh-hesary f."	1	366	0
73	"sadiq m.; le-dinh t.; tran t.k.; chien f.; phan t.t.h.; quang huy p."	1	15	0
75	santoso i.r	1	9	0
76	"sarker t.; cadman t."	1	1	0
77	"sharma a.; bajaj s."	1	1	0
78	"sharma g.d.; shahbaz m.; singh s.; chopra r.; cifuentes-faura j."	1	29	0
79	"sholoiko a.; matsybora t.; shevchenko l."	1	1	0
81	"singhania m.; bhan i.; chadha g."	1	3	0
82	"singhania m.; chadha g.; prasad r."	1	8	0
83	"sinha a.; mishra s.; sharif a.; yarovaya l."	1	212	0
86	"sreelekshmi g.; biju a.v."	1	2	0
87	svetlicinii a	1	2	0
88	"taghizadeh-hesary f.; yoshino n.; otsuka m."	1	3	0
90	thompson b.s	1	26	0
92	"tolliver c.; keeley a.r.; managi s."	1	132	0
93	"tran n.t.; nguyen h.-t.; tran t.k.; diep g.l.; nguyen t.h.n.; tran thai ha n."	1	1	0
94	"ul haq i.; maneengam a.; chupradit s.; huo c."	1	19	0
95	"ulfah i.f.; sukmana r.; laila n.; sulaeman s."	1	4	0
96	vargas c	1	2	0
97	"verma d.; kalra r.; baheti s.s."	1	1	0
98	"versal n.; sholoiko a."	1	32	0
101	wayan budiasa i	1	8	0
104	"yang q.; du q.; razzaq a.; shang y."	1	124	0

(continued)

Table 3 (continued)

ID	Author	Documents	Citations	Total link strength
105	"zhang d.; chen x.h.; lau c.k.m.; cai y."	1	72	0
107	"zhang x.; aranguiz m.; xu d.; zhang x.; xu x."	1	28	0
109	"zhukovska a.; zheliuk t.; zhuk i.; borshch v.; makarenko m.; vozniakovska k."	1	1	0

Fig. 3 Co-authorship with organization

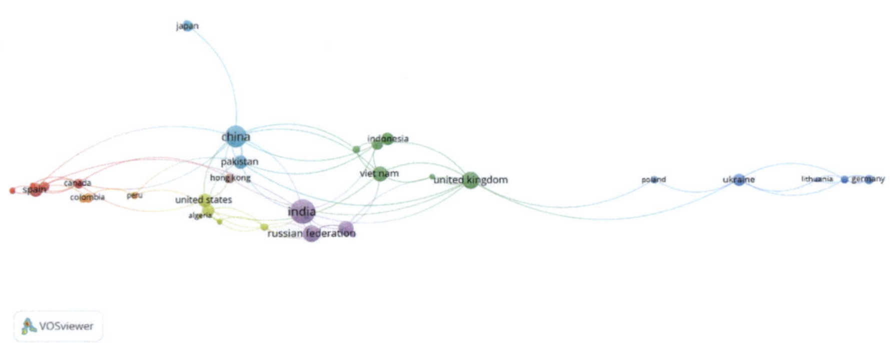

Fig. 4 Co-authorship within the countries

means the bonding force between them. Finally, the colors indicate the number of groups.

5 Findings and Discussion

The bibliometric analysis reveals an increasing trend in publications, with renewable energy and infrastructure finance being the most frequently studied areas. Research and issuance of green bonds are concentrated in developed economies, though emerging markets are showing gradual growth. Case studies confirm significant environmental outcomes, such as reduced carbon emissions and enhanced renewable energy capacity, driven by green bond investments. Infrastructure projects financed through green bonds often deliver co-benefits, including job creation and improved public services, though these impacts are less studied compared to environmental outcomes. Common challenges include regulatory constraints, the risk of greenwashing, and a lack of standardized reporting mechanisms, which could undermine investor confidence. The findings suggest that while green bonds are pivotal in bridging the financing gap for sustainability projects, their effectiveness could be enhanced through better alignment with policy frameworks and international standards. Future research should focus on evaluating the long-term impacts of green bond investments, particularly in low-income regions, where funding gaps persist (Table 4).

Case studies in renewable energy and sustainable infrastructure sectors demonstrate the tangible benefits of green bonds in real-world contexts. Renewable energy projects funded by green bonds have consistently achieved measurable outcomes, such as increased clean energy capacity and reduced greenhouse gas emissions. Sustainable infrastructure projects, such as urban transit systems and water management frameworks, illustrate the ability of green bonds to drive long-term societal benefits, including enhanced accessibility and job creation.

6 Conclusion

This chapter highlights the multifaceted contributions of green bonds to sustainable development, emphasizing their role as a critical tool for financing environmentally beneficial projects. The integration of bibliometric and qualitative case study analyses provides a comprehensive understanding of green bonds' current landscape, challenges, and potential.

The bibliometric analysis demonstrates the evolution of green finance research, pointing to significant contributions in renewable energy and sustainable infrastructure. Simultaneously, the case studies offer practical insights into the on-ground implementation of green bonds, illustrating both their transformative potential and the challenges they face.

Despite their success, green bonds require enhanced governance, better impact assessment frameworks, and wider geographical adoption to realize their full potential. By addressing these gaps, green bonds can significantly contribute to achieving SDGs, fostering innovation, and promoting global sustainability.

Challenges identified in the case studies include regulatory barriers, greenwashing concerns, and the need for rigorous impact measurement frameworks. These challenges underscore the importance of robust governance and standardized criteria for green bond certification to ensure credibility and alignment with sustainability objectives.

Table 4 Summary of themes identified in case studies

Theme	Subtheme	Description	Examples from case studies	Measurable benefits
Implementation	Framework adherence	Use of established frameworks like the Climate Bond Initiative or Green Bond Principles for project validation and issuance	India's solar parks financed through SECI	Ensures credibility and investor trust
			France's Metro Expansion aligning with EU green standards	Provides a benchmark for environmental impact reporting
	Multi-stakeholder collaboration	Partnerships between governments, private investors, and international organizations to fund projects	Norway's hydropower projects	Increased access to capital
			Indonesia's Green Sukuk initiative	Enhanced project scale through pooled resources
Challenges	Regulatory delays	Long approval processes and inconsistent policies delaying project execution	South Africa's wind energy projects	Loss of investor confidence
				Increased project costs due to time delays
	Transparency and accountability	Difficulty in ensuring the accurate reporting of environmental and financial performance	China's green bond sector facing concerns over greenwashing	Calls for stricter auditing mechanisms
				Reduces overall market trust
	High initial costs	Substantial upfront capital requirements limiting the number of eligible projects	Sustainable urban mobility projects in developing countries	Requires subsidies or incentives
				Discourages smaller developers
Environmental benefits	Carbon reduction	Reduction in greenhouse gas emissions through renewable energy or sustainable practices	India's solar parks reducing CO_2 emissions by 3 million tons annually	Significant contributions to national and global climate targets

(continued)

Table 4 (continued)

Theme	Subtheme	Description	Examples from case studies	Measurable benefits
	Renewable energy adoption	Shift from fossil fuels to renewable energy sources like solar, wind, and hydropower	Norway's hydropower projects	Renewable energy accounts for a greater percentage of the energy mix
			India's solar energy expansion	
Social benefits	Job creation	Employment opportunities generated during project construction and operation	India's solar projects creating over 10,000 jobs	Reduces unemployment in project areas
				Contributes to local economic development
	Urban resilience and mobility	Enhancements in transportation and urban infrastructure to make cities more livable and reduce environmental impacts	France's Metro Expansion	Improved public transportation access
			Indonesia's urban green bond-funded projects	Reduced vehicular emissions
	Accessibility and inclusivity	Improved access to essential resources like water, energy, and transport for marginalized communities	Indonesia's Green Sukuk enhancing water and sanitation services	Improved quality of life for underserved populations
				Greater economic inclusion
Financial impacts	Risk mitigation	Diversifying green bond portfolios to minimize risk associated with market volatility	Adoption of Green Sukuk to spread financial risk	Stable returns for investors
				Ensures long-term project viability
	Cost efficiency	Leveraging economies of scale and green bonds to lower financing costs	Aggregated funding for France's Metro Expansion	Reduced interest rates for borrowers
				Larger-scale projects become feasible

(continued)

Table 4 (continued)

Theme	Subtheme	Description	Examples from case studies	Measurable benefits
Sector-Specific Insights	Renewable energy	Emphasis on solar, wind, and hydropower for energy generation	India's solar parks	Reduces dependency on fossil fuels
			South Africa's wind projects	Promotes energy security
			Norway's hydropower	
	Sustainable Infrastructure	Focus on urban mobility, water access, and urban resilience	France's Metro Expansion	Safer, more sustainable urban environments
			Indonesia's urban planning projects	Lowered urban carbon footprint

Source Self

References

1. Flammer, C. (2021). Corporate green bonds. *Journal of Financial Economics, 142*(2), 499–516. https://doi.org/10.1016/j.jfineco.2021.01.010
2. Tang, D. Y., & Zhang, Y. (2020). Do shareholders benefit from green bonds? *Journal of Corporate Finance, 61*, Article 101427. https://doi.org/10.1016/j.jcorpfin.2018.12.001
3. Climate Bonds Initiative. (2022). *Global state of the market report 2022.* https://www.climatebonds.net/resources/reports/global-state-market-report-2022
4. Liu, J., Zhang, H., & Huang, B. (2021). Exploring the innovation of sustainable bonds: Transition and blue bonds. *Journal of Cleaner Production, 297*, Article 126698. https://doi.org/10.1016/j.jclepro.2021.126698
5. Luo, W., Tian, Z., Zhong, S., Lyu, Q., & Deng, M. (2022). Global evolution of research on sustainable finance from 2000 to 2021: A bibliometric analysis on WoS database. *Sustainability, 14*(15), 9435.
6. African Development Bank. (2021). *Sustainable bond program.* https://www.afdb.org/en/financial-information/investor-resources/capital-markets/sustainable-bond-program
7. Apple Inc. (2020). *Apple environmental progress report 2020.* https://www.apple.com/euro/environment/pdf/a/generic/Apple_Environmental_Progress_Report_2020.pdf
8. Tolliver, C., Keeley, A. R., & Managi, S. (2019). Urban infrastructure and green bond utilization. *Environmental Research Letters, 14*(10), Article 104013. https://doi.org/10.1088/1748-9326/ab4c16
9. Beschloss, R. (2019). Green bonds offer lessons for sustainable finance. *Finance & Development, 56*(4). International Monetary Fund. https://www.imf.org/en/Publications/fandd/issues/2019/12/green-bonds-offer-lessons-for-sustainable-finance-beschloss
10. Debrah, C., Darko, A., & Chan, A. P. C. (2023). A bibliometric-qualitative literature review of green finance gap and future research directions. *Climate and Development, 15*(5), 432–455. https://doi.org/10.1080/17565529.2022.2095331
11. Shishlov, I., Morel, R., & Cochran, I. (2016). Beyond transparency: Unlocking the full potential of green bonds. *Institute for Climate Economics, 2*(32), 1–28.

12. Bhatnagar, S., & Sharma, D. (2022). Evolution of green finance and its enablers: A bibliometric analysis. *Renewable and Sustainable Energy Reviews, 162*, 112405. https://doi.org/10.1016/j.rser.2022.112405
13. Government of Poland. (2021). *Green bond report on the use of proceeds*. https://www.gov.pl/documents/1079560/1080340/Green_Bond_Report_on_the_Use_of_Proceeds.pdf
14. International Hydropower Association. (2017). *Green bonds for hydropower financing*. Hydropower.org. Retrieved December 6, 2024, from https://www.hydropower.org/blog/green-bonds-for-hydropower-financing
15. Climate Bonds Initiative. (2022). *Q3 2022 market summary: Green bonds and sustainable finance*. Climate Bonds Initiative. https://www.climatebonds.net/resources/reports/q3-2022-market-summary
16. Ehlers, T., & Packer, F. (2017). Green bond finance and certification. *BIS Quarterly Review September*. https://ssrn.com/abstract=3042378
17. Baker, M., Bergstresser, D., Serafeim, G., & Wurgler, J. (2018). *Financing the response to climate change: The pricing and ownership of US green bonds*. National Bureau of Economic Research Working Paper Series. https://doi.org/10.3386/w25194
18. Banga, J. (2019). The green bond market: A potential source of climate finance for developing countries. *Journal of Sustainable Finance & Investment, 9*(1), 17–32. https://doi.org/10.1080/20430795.2018.1498617
19. Zerbib, O. D. (2019). The effect of pro-environmental preferences on bond prices: Evidence from green bonds. *Journal of Banking & Finance, 98*, 39–60. https://doi.org/10.1016/j.jbankfin.2018.10.012
20. Maltais, A., & Nykvist, B. (2020). Understanding the role of green bonds in advancing sustainability. *Journal of Sustainable Finance & Investment,* 1–20. https://doi.org/10.1080/20430795.2020.1724864
21. Bhattacharya, A. (2019, June 13). *Aligning G20 infrastructure investment with climate goals: The 2030 agenda*. Boston University Global Development Policy Center. https://www.bu.edu/gdp/2019/06/13/aligning-g20-infrastructure-investment-with-climate-goals-the-2030-agenda/
22. Araminta, D. V., & Timur, Y. P. (2022). The role of Green Sukuk in realizing the sustainable development goals 2030 agenda. *Jurnal Ekonomi & Bisnis Islam, 8*(2). https://doi.org/10.20473/jebis.v8i2.37531
23. Jagriti, M. (2023). Bibliometric analysis of green bonds. *Envision Journal of Commerce Department of ACFA, 17*.
24. Tukhtabayev, A., Dadajonova, M., Sativaldiyeva, G., Kholiyorov, U., & Hamidova, S. (2024). Harnessing green bonds for sustainable development: An empirical impact and insight. In *E3S web of conferences* (Vol. 574, p. 03006). EDP Sciences. https://doi.org/10.1051/e3sconf/202457403006
25. Nguyễn, Q. T. H. (2021). *Essays on climate finance transition risk* (Doctoral dissertation, University of Otago).
26. Karpf, A., & Mandel, A. (2018). The changing value of the 'green' label on the US municipal bond market. *Nature Climate Change, 8*(2), 161–165. https://doi.org/10.1038/s41558-017-0062-0
27. Alamgir, M., & Cheng, M. C. (2023). Do green bonds play a role in achieving sustainability? *Sustainability, 15*(13), 10177. https://doi.org/10.3390/su151310177
28. Alsmadi, A. A., Al-Okaily, M., Alrawashdeh, N., Al-Gasaymeh, A., Moh'd Al-hazimeh, A., & Zakari, A. (2023). A bibliometric analysis of green bonds and sustainable green energy: Evidence from the last fifteen years (2007–2022). *Sustainability, 15*(7), 5778. https://doi.org/10.3390/su15075778
29. Sobkowiak, M., Cuckston, T., & Thomson, I. (2020). Framing sustainable development challenges: Accounting for SDG-15 in the UK. *Accounting, Auditing & Accountability Journal, 33*(7), 1671–1703. https://doi.org/10.1108/AAAJ-01-2019-3810
30. Taghizadeh-Hesary, F., Phoumin, H., & Rasoulinezhad, E. (2023). Assessment of role of green bond in renewable energy resource development in Japan. *Resources Policy, 80*, Article 103272. https://doi.org/10.1016/j.resourpol.2022.103272

MIX
Papier aus verantwortungsvollen Quellen
Paper from responsible sources
FSC® C105338

If you have any concerns about our products,
you can contact us on
ProductSafety@springernature.com

In case Publisher is established outside the EU,
the EU authorized representative is:
Springer Nature Customer Service Center GmbH
Europaplatz 3, 69115 Heidelberg, Germany

Printed by Libri Plureos GmbH
in Hamburg, Germany